Lake Victoria Monitored from Space

Joseph Awange

Lake Victoria Monitored from Space

Springer

Joseph Awange
School of Earth and Planetary Sciences
(Spatial Sciences Discipline)
Curtin University
Perth, WA, Australia

ISBN 978-3-030-60550-6 ISBN 978-3-030-60551-3 (eBook)
https://doi.org/10.1007/978-3-030-60551-3

This Springer imprint is published by the registered company Springer Nature Switzerland AG
The registered company address is: Gewerbestrasse 11, 6330 Cham, Switzerland

To the memories of my late parents (James Odhiambo Awange and Margaret Atieno Awange) and my late grandmothers (Amelea Gwada and Rosalina Omulo Odera). The love you showered on me is still fresh like the waters of Lake Victoria. You are greatly missed.

Joseph L. Awange
Perth Australia, September 2020

Foreword

Joseph L. Awange is truly an interdisciplinary scientist who was trained in geodesy, geomathematics, Geographic Information Science (GIS), and remote sensing, and he excels at addressing contemporary scientific questions and applications including climate change, hydrology, flooding, drought, and water resources management. Throughout his professional career, he has used novel mathematical formalism, geodetic and passive remote sensing observations, and four-dimensional assimilative hydrologic modeling regimes to quantify both surface and groundwater processes, toward solving timely scientific problems and applications to benefit people in need. Like his many excellent peer-reviewed publications and books, this book on *Lake Victoria Monitored from Space* is no exception. Using Lake Victoria or Victoria Nyanza, one of Africa's Great Lakes, as a poster child of water bodies under climate stress in northeastern Africa, he elaborated on the critical societal importance of global abundant and clear water resources, and focused on the need of accurate and timely observations to study, understand the anthropogenically impacted Lake Victoria's water and hazard managements, and the needed sound government policy for sustainability, and revision of obsolete laws to lessen or mitigate regional and international water conflicts. He is correct to postulate the need of spaceborne observations, for whom he is an expert on the exploitation of contemporary multiple passive optical and multi-spectral remote sensing, and innovative geodetic remote sensing. These geodetic sensors include the use of ground-based and potential spaceborne Global Navigation Satellite System (GNSS) Signal of Opportunity in L-Band bistatic radar reflectometry for measuring water level, soil moisture, and wind speed; satellite radar and laser altimeters for measuring water level and wave/wind speed; and the use of satellite gravimetry data collected by the Gravity Recovery And Climate Experiment (GRACE) and its successor, GRACE-Followon (GRACE-FO) twin-satellite missions to invert for temporal gravity fields at monthly sampling and a spatial scale longer than 333 km, half-wavelength. GRACE and GRACE-FO data would be the first satellite sensor to sense groundwater storage changes in aquifers, provided that the surface hydrology is known and remove from the retrieved gravity signal.

In a unique cross-disciplinary approach, the book articulated the various climatic impacts and explanations from natural and anthropogenic origins, which affected Lake Victoria and its vicinity, including the drastic increase and depletion of water level in the lake and dams, floods and droughts, water quality/security, crop health, food security, and economic implications. With no exception as in his many publications, Joseph L. Awange used data analysis methodologies including filtering, adjustment theory, and robust statistics, to quantify the hydrologic and other parameters, and their estimated uncertainties. This book is recommended for readers from a diverse disciplines, including physical and social sciences, policy, law, engineering, and disaster management.

October 2020

C. K. Shum
Ohio State University
Ohio, USA

Preface

This book employs a suite of remotely sensed products and advanced technologies to provide the first comprehensive space-based sensing of Lake Victoria, the world's second largest freshwater lake that supports the livelihood of more than 42 million people, modulates regional climate, but faces myriads of challenges. Proper understanding of the lake and changes in its physical dynamics (e.g., water level, shorelines and areal dynamics) resulting from the impacts of climate variation and climate change as well as anthropogenic (e.g., hydropower and irrigation) is important for its management as well as for strategic development before, during, and after climate extremes (e.g., floods and droughts) in order to inform policy formulations, planning and mitigation measures. Owing to its sheer size and lack of research resources commitment by regional governments that hamper its observation, however, it is a daunting task to undertake studies on Lake Victoria relying solely on in situ "boots on the ground" measurements, which are sparse, missing in most cases, inconsistent or restricted by governmental red tapes. Because of this, changes in its physical dynamics that have occurred due to climatic variation/ change and anthropogenic impacts have not been thoroughly studied. For example, articles written on Lake Victoria referenced various figures for its dimensions (e.g., 66,400–69,485 km^2 for its area; 300–412 km for its maximum length; 240–355 km for its maximum width; and 3300–4828 km for its shorelines). These discrepancies are largely due to the difficulties of obtaining accurate data because of both the size of the lake and the lack of resources that have been committed for exploratory research by regional governments.

In this book, which provides a pioneering compilation of satellite applications to Lake Victoria, a suite of high spatio-temporal remotely sensed data, reanalysis products, as well as those of hydrological models are all employed to sense the lake's precious resource, water. With the advances in satellite technology, i.e., the maturity of the GRACE (Gravity Recovery and Climate Experiment) satellite mission among others, the lake's waters can now be accurately weighed from space. Remote sensing of this precious lake from space, therefore, shows its current state and that of its basin, challenges and potential. The book will be useful to those in water resources management and policy formulations, hydrologists, environmentalists, engineers and researchers.

Recife, Brazil; Karlsruhe, Germany; Kisumu, Kenya; Perth, Australia

Joseph Awange

Contents

Part I
Global and Lake Victoria's Water Resources

Chapter 1
Global Freshwater Resources

Several authors, politicians, leaders of international organizations and journalists have cautioned the world community that the increasing scarcity of freshwater resources might lead to national and international conflicts. When relating this to climate change forecasts – most of which indicate that climate change will have a significant impact on the availability of freshwater resources, on water quality, and on the demand for water – this is alarming news for humankind as it threatens human security.

– Molen and Hildering [1]

Accounting for water for environmental requirements shows that abstraction of water for domestic, food and industrial uses already have a major impact on ecosystems in many parts of the world, even those not considered "water scarce" [2]. In the absence of coordinated planning and international cooperation at an unprecedented scale, therefore, the next half century will be plagued by a host of severe water-related problems, threatening the well being of many terrestrial ecosystems and drastically impairing human health, particularly in the ***poorest regions*** *of the world [3].*

1.1 Diminishing Freshwater Resources

1.1.1 Status

Freshwater, influenced globally by climate variability/change [4–9], human use [10–12] and knowledge deficiency resulting from inadequate hydrometeorological observation stations [13–16], is one of the basic necessities without which human beings cannot survive since it is key to the sustainability of all kinds of life forms. Water

J. Awange, *Lake Victoria Monitored from Space*,
https://doi.org/10.1007/978-3-030-60551-3_1

has multiple uses namely; nutritional, domestic, recreational, navigational, waste disposal and ecological as it is a habitat for living and non-living organisms (biodiversity), etc. And, because it is indispensable to different sectors including manufacturing, agriculture, fisheries, wildlife survival, tourism and hydroelectric power generation, just to list but a few, it is a vital factor of economic production. For many countries, most freshwater endowments encompass surface waters, groundwater, wetlands and glaciers. Surface water bodies include lakes, rivers, swamps, springs, dams and water pans dispersed within different basins. In general, people living in the vast arid and semi-arid parts of the world rely heavily on groundwater resources. Gleeson et al. [17] puts the number of people living in regions where groundwater is threatened to be 1.7 billion. Furthermore, groundwater is also an important supplementary source of water for many urban households in most developing countries.

At a global scale, although much of the Earth is covered by water, most of it is unsuitable for human consumption, since 96% of it is found in the saline oceans. According to the U.N., only 2.5% of the roughly 1.4 billion cubic kilometers of water on Earth is freshwater, and approximately 68.9% of the freshwater is trapped in glacial ice or permanent snow in mountainous regions—the Arctic and Antarctica. Roughly 30.8% is groundwater, much of which is inaccessible to humans, and the remainder 0.3% comprise surface waters in lakes and rivers [18]. Of the 0.3% available for human and animal consumption, much is inaccessible due to unreachable underground locations and depths [19]. Jury and Vaux Jr. [3] caution that focusing on the global freshwater storage resource alone is misleading because much of the water is inaccessible. They suggest that humanity's freshwater resource consisting of rainfall used to grow crops, accessible groundwater, and surface water be considered.

1.1.2 Water Scarcity

Although no common definition of water scarcity exists, Rijsberman [2] defines a water insecure person as an individual who does not have access to safe and affordable water to satisfy his or her needs for drinking, washing, or livelihood. A water scarce area is then said to be an area where a large number of people are water insecure for a significant period of time, e.g., Rijsberman [2]. Rijsberman [2] points out that an area qualifies as a water-scarce area depending on, e.g., (i) the definition of people's needs and whether the definition takes into account the needs of the environment, the water for nature, (ii) the fraction of the resource that is made available (or could be made available) to satisfy these needs, and (iii), the temporal and spatial scales used to define scarcity.

Even though Rijsberman [2] argues that at a global scale, and from a supply and demand perspective, it is still debatable whether water scarcity is fact or fiction, it is incontestable that fresh water is increasingly becoming a scarce resource and shortages could drive conflict as well as negatively hit food and energy production [3, 20, 21] (see Fig. 1.1). That water shortage is emerging as one of the leading challenges of the 21st century has been documented, e.g., in [2, 22–24]. To underscore the

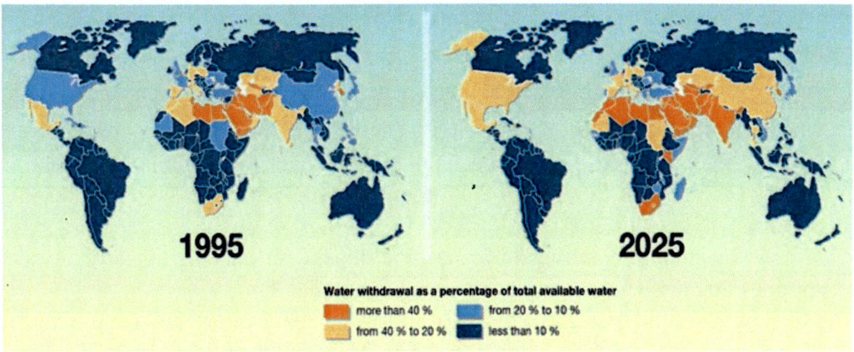

Fig. 1.1 Increased global water stress. *Source* [37]

seriousness of water shortage, the World Bank in its 1992 World Development Report pointed out that 22 countries faced severe water shortage while further 18 were in danger of facing shortages if fluctuation in rainfall persisted [25]. Recent studies, e.g., [26–29, 80, 81] point to fluctuation in rainfall in parts of Greater Horn of Africa where most low income earners depend on rain fed agriculture for food production, thus suggesting that the water shortage problem is not fading away any time soon. A more gloomy picture is the estimate that by 2050, about two billion people will be short of water, a potential cause of conflict [30]. That a large population of the world will face water scarcity is supported by several studies, e.g., [5, 31–34], with the most likely to be affected dwelling in Africa, Asia and the Middle East, see, e.g., [24, 35, 36, 82]. Already, model projections, see e.g., [7] suggest a 40% of the population languishing under water scarcity, with 57% likely to be water stressed by 2015 and 69% by 2075 [23, 31].

By reviewing several publications on water in relation to poverty and environmental degradation nexus, Duraiappah [25] presents activities that lead to water shortages. ***First***, there is the issue of the commercial interest of the rich and wealthy driven primarily by power, greed and wealth that benefits from market and institutional failures (e.g., absence or misuse of water property rights). ***Second***, Duraiappah [25] attribute over usage of water supply by the small holders (the poor) to water subsidies that provide incentives. Of the two groups, the poor are more likely to be affected by water shortages as compared to the rich, a situation that could contribution to environmental degradation on the one hand, which would lead to poverty on the other hand, i.e., endogenous poverty causing environmental degradation, see, e.g., Duraiappah [25].

This picture leads Jury and Vaux Jr. [3] to warn that "without immediate action and global cooperation, water supply and water pollution crisis of unimaginable dimensions will confront humanity, limiting food production, drinking water access, and the survival of innumerable species on the planet". They list the following four factors to support their hypothesis that the world is headed towards a future where billions of people will be forced to live in places where their food and water requirements will not be met [3]:

1. Unlike estimates of the global supply of scarce minerals or underground fuels, which are surrounded by uncertainty, planetary supplies of water are relatively well characterized. No large deposits of groundwater await human detection in readily accessible locations, so that any new resources discovered will be very expensive to develop.
2. Many vital human activities have become dependent on utilizing groundwater supplies that are being exhausted or contaminated.
3. Much of the population growth projected for the next century will occur in areas of greatest water shortages, e.g., Africa and Asia, and there is no plan for accommodating the increases.
4. Global economic forces are luring water and land from food production into more lucrative activities such as biofuel, see, e.g., [38, 39, 79], while at the same time encouraging pollution that impairs drinking water quality for a large and ever-growing segment of the population.

Evidently, the management of water resources' conflicts, focusing on negotiation, mediation and decision-making processes, in order to prevent, manage and resolve water conflicts is emerging as a contemporary and topical research issue.

Physical water scarcity is evident in densely populated arid areas in many parts of Central and West Asia, and North Africa with projected availabilities of less than 1000 m^3/capita/year [2]. This has a wide range of negative impacts and ramifications. For instance, it results in higher incidences of waterborne, water-related or sanitation-related diseases such as malaria, diarrhoea and skin infections. In addition, as has been pointed out before, increasing cases of water conflicts, especially between pastoralist and farming communities along lower and upper river basins, as witnessed in water stressed countries like Kenya [41] (with a renewable freshwater per capita endowment estimated at about 548 m^3/capita/year [40]), are also likely to heighten food insecurity. Alcamo [31] point to the fact that 12% of the global basins are already vulnerable to water withdrawals. On the gender scale, since women are the primary collectors, users and managers of water for domestic use in most developing countries, water scarcity disproportionately affects them because it is they who have to trek long distances, often all day, in search of water. Against the above background, is the growing realization today that availability of sufficient, accessible and quality potable water is not only a matter of great socio-economic and political importance, but also one of fundamental human rights, see e.g., [42–45].

1.1.3 Impacts of Climate Variability/Change on Freshwater

Intergovernmental panel of climate change [83] projects a temperature increase of 1.5 °C by 2030, an increase which will definitely have an enormous impact on the world's freshwater. That temperature has been on the rise due to human's carbon footprint had been noted in the previous IPCC (2007 and 2013) reports [84]. Infact, [85] points to the fact that the global and ocean temperatures rose by 0.85 °C [0.65–1.06 °C] over the period 1880–2012, 0.89 °C [0.69–1.08 °C] from 1901–2012, and 0.72 °C [0.49–0.89 °C] between 1951 and 2012. Change in climate resulting in the temperature rise above as well as climate variability exemplified through indicators such as global teleconnections such as ENSO [86, 87] impact on the availability of freshwater. For instance, climate variability influence extreme whether/climate conditions resulting, e.g., in increased frequency and severity of droughts [81, 88, 89]. Droughts majorly affect the surface water although a long spell could also impact on groundwater [90, 91]. Moreover, the influence of climate variability/change on the hydrological cycle impacts on precipitation, which is the main recharge of freshwater.

1.1.4 Water-Poverty-Environment Nexus

Water plays a significant role both in poverty alleviation and environmental related issues. With regard to poverty, food policy contributes toward the overall goal of eradicating extreme poverty and hunger, and is dependent on water as one of the drivers for realizing such a goal, see e.g., Hanjra and Qureshi [46]. It provides a key component of agricultural requirement for food production as well as for domestic use. If people are able to provide food, then absolute poverty measure of a society could arguably be low. Indeed, that water and food security issues are closely linked has been pointed out, e.g., by [10, 46]. Rijsberman [2] is even more direct by stating that water will be the major constraint for agriculture in the coming decades, more so in the continents that experience high percentage of poverty such as Asia and Africa.

The problem, however, is that more focus is placed on the water crisis, which is likely to be fuelled by increased population leading to about 1.2 billion lacking water. To the contrary, less documentation exist for the large part of the population that live in rural areas below the poverty line. Duraiappah [25], through literature review, identified

- water shortage, and
- water pollution

as the two major issues within the water sector that plays an important role in the poverty-environmental degradation nexus. Understanding the link between poverty and environment in relation to water, therefore, calls for a more closer look at the whole issue of water scarcity in relation to the poor rather than a holistic approach where such scarcity is viewed in terms of the total population. This in essence calls for poverty eradication measures to incorporate issues that will address water scarcity and insecurity. This is partly because if the poor are unable to access clean and safe water, food productivity will be hampered leading to hunger and malnutrition hence less productivity. Also, lack of access to safe drinking water by the poor will only aggravate the risk of water borne diseases such as cholera. When the low income group are faced by such water-borne diseases, their productivity deteriorates and they risk losing their jobs, and hence sources of income. The expected outcome of loss in income is that the low income earners will experience economic and social hardship, which over time results in poverty, thus exemplifying how environmental degradation causes poverty, e.g., [22, 25].

1.2 Water Resource Monitoring

1.2.1 Need for Monitoring

The importance of water as a resource, therefore, calls for sound environmental conservation measures that enhance its protection and management. It is in relation to this that the World Bank, as an emerging priority of its lending framework, decided to broaden the development focus in its 1993 "Water resource management policy paper" to include the *protection* and *management* of water resources in an environmentally sustainable, socially acceptable, and economically efficient manner [40]. The protection and management of water resources calls for an elaborate and well established management and monitoring program, e.g., [47, 48].

Information about water resources and the environment is inherently geographic. Maps, whether on paper or in digital Geographical Information System (GIS, see e.g., [47, 48]) formats, continue to be the medium for the expression of engineering plans and designs. This is because we are basically concerned about the spatial distribution and character of the land and its waters. Johnson [49] argues that weather patterns, rainfall and other precipitation, and resultant water runoff are primary driving forces for land development, water supplies, and environmental impacts and pollution. Our water resources systems comprise dams and reservoirs, irrigated lands and canals, water supply collection and distribution systems, sewers and storm water systems, and flood plains. These systems are designed in response to a complex mix of topography and drainage patterns, population and land use, sources of water, and related environmental factors [49, 92].

In general, the planning and engineering design processes used in the development and management of water resources involve different levels of data abstraction. Data are collected and used to characterize the environment at some level of detail, or scale. In seeking to make decisions about plans and designs, data must be collected to describe the resource, and procedures or models must be developed to predict the resultant changes. These data and models help us understand the real world, and this understanding guides our decision making [49].

According to Taylor and Alley [50], essential components of a water level monitoring program include; *selection of observation wells*, *determination of the frequency of water level measurements*, *implementation of quality assurance*, and *establishment of effective practices for data reporting*. In selecting the observation wells, the authors state that the decisions made about the number and locations of observation wells are crucial to any water-level data collection program [50], see also [91]. In regard to locations, Global Navigation Satellite Systems (GNSS) satellites [51, 52] could contribute in generating a fast and accurate survey of well location-based data.

These data could then be integrated with other information such as water level in a GIS system to enhance the accessibility of water level data, where the GIS plays the role of depicting the locations of the observed wells relative to pertinent geographic, geologic, or hydrologic features, e.g., [50].

Taylor and Alley [50] present areas where the monitored ground water levels could be used. Some of these include: determination of the hydraulic properties of aquifers (aquifer tests); mapping of the altitude of the water table or potentiometric surface; monitoring of the changes in groundwater recharge and storage, e.g., [91]; monitoring of the effects of climatic variability; monitoring of the regional effects of groundwater development; statistical analysis of the water level trends; monitoring of the changes in groundwater flow directions; monitoring of the groundwater and surface water interaction; and numerical (computer) modeling of groundwater flow or contaminant transport.

Information on the spatial and temporal behaviour of terrestrial water storage, therefore, is crucial for the management of local, regional and global water resources. This information will [53]:

- Enhance sustainable utilization of water resources by, e.g., farmers, urban consumers, miners, etc.
- Guide water resource managers and policy makers in the formulation of policies governing its sustainable use, conservation and management. In particular, State water managers are more informed in regulating the utilization of water, e.g., for industrial and irrigation purposes.
- Benefit local environmental monitoring, management policies and practices that ensures a balance between sustainable utilization and environmental conservation and protection. Changes in water availability impacts upon the environment in several ways, e.g., any significant imbalance in its level affects the ecological system by influencing salinity, land subsidence, and the vulnerability of wetlands ecosystem among others.
- Benefit various government agencies at various levels (national, provincial, and local) by providing data that enhance and compliment their works. Such agencies include departments of *water*, *agriculture*, *weather forecasting* and *climate* studies, and so forth.

The conservation and management of water is of paramount importance in areas with arid or semi-arid climates, which include many parts of Australia, especially in times of severe drought, as experienced in Murray Darling Basin [53]. In 2006, Australia faced its worst drought in a century as was seen from daily reports that were emerging in both the local and international media. A more grim picture of the future of the water situation for Australia was to follow from the IPCC [54] report, which stated that Australia's water crisis will worsen in the coming years due to drought! There clearly exists an urgent need to have efficient monitoring technique(s) that will enhance the analysis of water scarcity at river basin or more localized scales. Indeed, this argument is supported by Rijsberman [2] who states that the global analysis of water scarcity is of very limited use in assessing whether individual or communities are water secure. To this effect, Rijsberman [2] states:

> The river basin is more and more adopted as the appropriate scale to understand the key processes with increasing water scarcity as human use goes up to the point where basins "close".

One such technique that has emerged supreme in monitoring changes in stored water at river basin scales, is the Gravity Recovery and Climate Experiment (GRACE) satellites [55] discussed in details in Sect. 5.3.3; see details in [47, 48, 51, 52, 90].

Timely and precise information on the changes in stored water at smaller (localized) scales of economical values, e.g., urban consumption, agriculture, industries, and mining to within 10–14 days (so far achievable by GRACE satellite) will enhance sustainable conservation and management of this precious *dwindling* resource.

> The availability of techniques that delivers information on the changes in stored water at a more local scale, is the first step towards realizing an efficient water society.

Water resource managers are able to make decisions based on timely and accurate knowledge; thereby saving considerable resources that are often spent as a penalty of inefficient decisions based on lack of information. In the south-western wheat belt of Australia, for example, accurate knowledge of changes in stored water will be beneficial to the sustainable utilization of water, while at the same time realizing the economic contribution of wheat farming to the overall Gross Domestic Product (GDP). A blind focus on the GDP's growth without paying attention to the state of salient contributors such as water stored in aquifers is detrimental, since a fall in the amount of the available water in such areas would definitely mean reduced yields.

Since the entire system of stored water is coupled within the hydrological cycle (Fig. 5.3), hydrologists will be in a position to better understand their local hydrological cycle, thanks to information at localized levels. Hydrologists will also be able to use such information to refine and calibrate local-scale models, e.g., rainfall runoff models [56], for further improvement in their hydrological cycles. This will also contribute to our understanding of the impacts of climate change on regional and global hydrological cycles.

Environmental studies have a chance of greatly benefiting from information about changes in stored water. It is widely acknowledged that stored water (surface and groundwater) plays a key role in sustaining natural biodiversity and the functioning of the environment as a whole. Knowledge of the changes in water level is therefore essential for the very survival of the entire ecosystem, which could be adversely affected by extreme change in stored water. In wetlands, for example, some vegetation and ecosystems have been known to respond to water level fluctuations [57].

Accurate monitoring of changes in stored water at smaller wetland scales will thus help in the preservation and conservation of such wetland ecosystems. Changes in water level also brings with it environmental phenomena such as salinity, compacting of aquifers due to the removal of water causing land subsidence, and changes in the properties of the top 5 cm of soil. Information on changes in stored water thus contributes enormously to the environmental conservation and protection. Remote sensing and GIS [47, 48] can be used to monitor the water quality. This is possible by employing multispectral, multi-temporal image data and analyzing parameters such as the distribution of suspended sediment, turbidity and chlorophyll. These indicators can be determined through regression analysis.

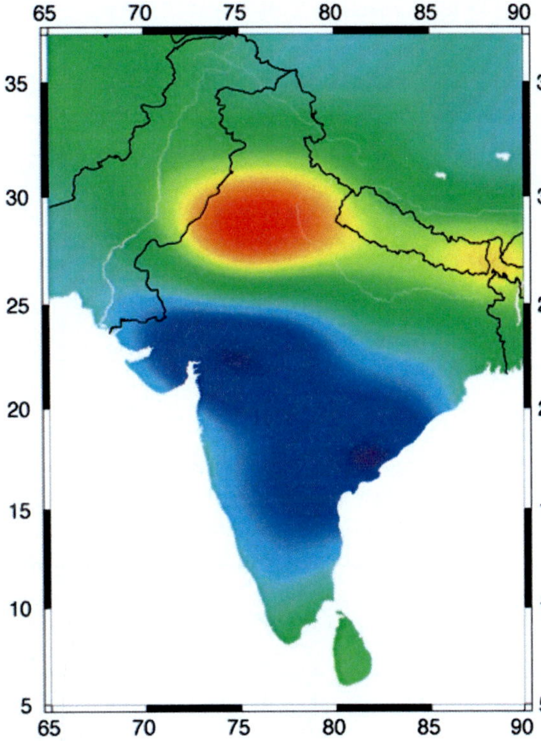

Fig. 1.2 Groundwater changes in India (2002–2008). Groundwater recharge is indicated by blue while depletion is indicated in red. *Source* NASA (I. Velicogna/UC Irvine)

1.2.2 Monitoring of Stored Water at Basin Scales

In Awange and Kiema [47, 48, 51, 52], the *geoid* is introduced as a fundamental physical surface to which all observations are referred to if they depend on *gravity*, and whose shape is influenced by inhomogeneous mass distribution within the interior of the Earth [58, p. 29]. In Chap. 5, the concept of *gravity field variations* is related to hydrological processes. Measurements of the time-varying gravity field by LEO (low earth orbiting) satellites, e.g., GRACE discussed in Sect. 5.3.3 are the key to the contribution of space monitoring of changes in water levels at basin scales, see e.g., [59]. Such techniques now enable the monitoring of groundwater recharge, see e.g., [60, 61, 90], which is the most important element in groundwater resources management and could also be applicable to monitoring salinity management measures at the catchment level. For example, in 2009, GRACE satellites showed that north-west of India's aquifers had fallen at a rate of 0.3048 m yr^{-1} (a loss of about 109 km^3 per year) between 2002 and 2008, see Fig. 1.2.[1]

[1]The Economist, September 12th 2009, pp. 27–29: Briefing India's water crisis.

1.3 Why Monitor Lake Victoria?

Lake Victoria (Fig. 1.3), the world's second largest freshwater lake, and the largest in the developing world, is a resource shared by the three East African countries: Kenya, Uganda and Tanzania. It is a source of water for irrigation, transport, domestic and livestock uses, and supports the livelihood of more than 42 million people who live around it [62, 93]. Its fish products, (i.e., Tilapia and Nile Perch) are exported the world over [62]. Its role as an indicator of environmental and climate change on long-term scales together with its global significance are documented, e.g., in Nicholson and Yin [65] and Awange and Ong'ang'a [62]. Since the 60s, the lake level has exhibited fluctuations as pointed out by Nicholson [63, 64] . The sharpest rise in the lake water level occurred during the El'Nino rains of early 60s and 1997/1998. Some reports, e.g., [66] suggest that the lake level rose by 2.5 m following the 1960s floods. Kite [67] attributed this rise to over-lake precipitation.

Although the lake has continued to attract worldwide attention due to its significance and other environmental phenomenon such as water hyacinth, in the last decade, and perhaps most threatening, Lake Victoria water level receded at an alarming rate causing concerns as to whether the lake was actually drying up as it happened in the pre-historic time, see e.g., [10]. According to Kull [68], the lake levels

Fig. 1.3 Lake Victoria basin (*Source* Kayombo and Jorgensen [70]) and weather stations (black triangles) in the Kenyan part of the region

dropped to more than 1.1 m below the 10-year average. Water levels have remained above average for more than 40-years, but current water levels are below normal and the lowest level since September 1961. The socio-economic impacts of this drastic fall in Lake Victoria water level have been reported in Awange et al. [10, 69]. At the time of writing this book (2020), the lake level has risen to a point where people are wondering whether it is retracing its shorelines of the 60s. Continuous monitoring of the lake, therefore, is paramount to its sustainable utilization, policy formulation and management.

1.4 Objectives and Aims of the Book

As part of the solution to water scarcity problem, quick fix approach has revolved around supply management approach where infrastructures, e.g., dams have been constructed or expanded to increase the available water supply. Rijsberman [2] argue, however, that although the quick fix approach has largely succeeded in producing cheap food, water supply, and sanitation to a large number of people, many people still do not have access to safe and affordable drinking water despite huge investments. Rijsberman [2] states:

> ...close to half the world population lacks access to sanitation, many rural poor do not have access to water for productive purposes, groundwater levels in key aquifers are falling rapidly, many rivers are no longer reaching the sea, etc.

As a shift from this school of thought, Rijsberman [2] points to the emergence of integrated water resource movement that has brought about organizations such as the World Water Council and the Global Water Partnership that is pushing for a demand management approach seeking;

1. to involve users more in the management of water, often through the establishment of forms of water user associations;
2. to price water and/or make it a trade-able commodity; and
3. to establish river basin authorities that integrate the usually fragmented government responsibilities for water into a single authority responsible for a hydrographically defined area, the river basin.

It is in support to item (3) above that satellites have been recognized as having the potential to provide space-based estimate of changes in terrestrial water storage. In essence, they are tools that assist water managers in conserving and controlling the utilization of dwindling water resources in a sustainable way. Water is arguably one of the most precious resource in the world, therefore, it is logical to try to monitor its distribution as efficiently as possible, and space techniques offer such opportunity [10, 71–74]. This is because one of the environmentally important signals detected by satellites such as GRACE is the temporal, e.g., gravity field variation induced by changes in the distribution of water on and below the Earth's surface, i.e., hydrology, e.g., [74]. Satellite altimetry provides the possibility of monitoring sea or lake surface heights as was demonstrated for Lake Naivasha [75]. Other studies undertaken with respect to use of GRACE to monitor hydrology include, e.g., [53, 60, 76–78, 90].

The aim of this book, therefore, is to provide a space (satellite) view of Lake Victoria. Owing to its large basin [94, 258,000 km^2], "boots on the ground" monitoring of Lake Victoria in terms of the fluctuation of its level, impacts of climate change and anthropogenic factors, the behaviour of vegetation within its basin, and the extremes, e.g., droughts, is practically impossible. The most viable approach is to turn to space techniques to remote sense the lake [95], what is demonstrated in the present book.

References

1. Molen I, Hildering A (2005) Water: cause for conflict or co-operation? ISYP J Sci World Aff 1(2):133–143
2. Rijsberman FR (2006) Water scarcity: fact or fiction. Agric Water Manag (Elsevier) 80:5–22
3. Jury WA, Vaux HJ Jr (2007) The emerging global water crisis: managing scarcity and conflict between water users. Adv Agron (Elsevier Inc) 95:77. https://doi.org/10.1016/50065-2113(07)95001-4
4. Anyah RO, Forootan E, Awange JL, Khaki M (2018) Understanding linkages between global climate indices and terrestrial water storage changes over Africa using GRACE products. Sci Total Environ 635:1405–1416. https://doi.org/10.1016/j.scitotenv.2018.04.159
5. Vörösmarty CJ, Green P, Salisbury J, Lammers RB (2000) Global water resources: vulnerability from climate change and population growth. Science 289(5477):284–288. https://doi.org/10.1126/science.289.5477.284
6. Phillips T, Nerem RS, Fox-Kemper B, Famiglietti JS, Rajagopalan B (2012) The influence of ENSO on global terrestrial water storage using GRACE. Geophys Res Lett 39(L16705):2012. https://doi.org/10.1029/2012GL052495
7. Schewe J, Heinke J, Gerten D, Haddeland I, Arnell NW, Clark DB, Dankers R, Eisner S, Fekete BM, Colón-Gonzlez FJ, Gosling SN, Kim H, Liu X, Masaki Y, Portmann FT, Satoh Y, Stacke T, Tang Q, Wada Y, Wisser D, Albrecht T, Frieler K, Piontek F, Warszawski L, Kabat P (2013) Multimodel assessment of water scarcity under climate change. PNAS 111(9):3245–3250. https://doi.org/10.1073/pnas.1222460110
8. Hurkmans R, Troch PA, Uijlenhoet R, Torfs P, Durcik M (2009) Effects of climate variability on water storage in the Colorado river basin. J Hydrometeorol 10:1257–1270. https://doi.org/10.1175/2009JHM1133.1
9. Ndehedehe CE, Awange J, Kuhn M, Agutu N, Fukuda Y (2017d) Climate teleconnections influence on West Africa's terrestrial water storage. Hydrol Process 31(18):3206–3224. https://doi.org/10.1002/hyp.11237
10. Awange JL, Sharifi M, Ogonda G, Wickert J, Grafarend EW, Omulo M (2007) The falling Lake Victoria water level: GRACE. TRIMM and CHAMP satellite analysis. Water Resour Manag. https://doi.org/10.1007/s11269-007-9191-y
11. Van Loon AF, Stahl K, Di Baldassarre G, Clark J, Rangecroft S, Wanders N, Gleeson T, Van Dijk AIJM, Tallaksen LM, Hannaford J, Uijlenhoet R, Teuling AJ, Hannah DM, Sheffield J, Svoboda M, Verbeiren B, Wagener T, Van Lanen HAJ (2016) Drought in a human-modified world: reframing drought definitions, understanding, and analysis approaches. Hydrol Earth Syst Sci 20(9):3631–3650. https://doi.org/10.5194/hess-20-3631-2016
12. Konar M, Evans TP, Levy M, Scott CA, Troy TJ, Vörösmarty CJ, Sivapalan M (2016) Water resources sustainability in a globalizing world: who uses the water? Hydrol Process 30(18):3330–3336. https://doi.org/10.1002/hyp.10843

13. Hall JW, Grey D, Garrick D, Fung F, Brown C, Dadson SJ, Sadoff CW (2014) Coping with the curse of freshwater variability. Science 346(6208):429–430. https://doi.org/10.1126/science.1257890
14. Alsdorf DE, Lettenmaier DP (2003) Tracking fresh water from space. Science 301(5639):1491–1494. https://doi.org/10.1126/science.1089802
15. Alsdorf D, Lettenmaier D, Vrsmarty C (2003) The need for global, satellite based observations of terrestrial surface waters. Eos Trans Am Geophys Union 84(29):269–276. https://doi.org/10.1029/2003EO290001
16. Vörösmarty C, Askew A, Grabs W, Barry RG, Birkett C, Dll P, Goodison B, Hall A, Jenne R, Kitaev L, Landwehr J, Keeler M, Leavesley G, Schaake J, Strzepek K, Sundarvel SS, Takeuchi K, Webster F (2001) Global water data: a newly endangered species. EOS Trans Am Geophys Union 82(5):54–58. https://doi.org/10.1029/01EO00031
17. Gleeson T, Wada Y, van Beek MFPBLPH (2012) Water balance of global aquifers revealed by groundwater footprint. Nature 488:197–200. https://doi.org/10.1038/nature11295
18. UNEP (2002) A world of salt: total global saltwater and freshwater estimates. http://www.unep.org/dewa/assessments/ecosystems/water/vitalwater/freshwater.htm. Accessed 25 Aug 2010
19. Hofman AR (2004) The connection: water and energy security. http://www.iags.org/n0813043.htm. Accessed on 25 Aug 2010
20. Gleick PH (1993) Water and conflict: fresh water resources and international security. Int Secur 18(1):79–112. https://doi.org/10.2307/2539033
21. Gleick PH (2000) The world's water: biennial report on freshwater resources 2000–2001. Island Press, Washington
22. Agola NO, Awange JL (2014) Globalized poverty and environment 21st century challenges and innovative solutions. Springer, Berlin
23. Kuylenstierna JL, Bjrklund G, Najlis P (1997) Sustainable water future with global implications: everyone's responsibility. Nat Resour Forum 21(3):181–190. https://doi.org/10.1111/j.1477-8947.1997.tb00691.x
24. Vörösmarty CJ, Douglas EM, Green PA, Revenga C (2005) Geospatial indicators of emerging water stress: an application to Africa. AMBIO
25. Duraiappah A (1998) Poverty and environmental degradation: a review and analysis of the nexus. J Hydrol 113(1–4):297–306
26. Awange JL, Khandu Forootan E, Schumacher M, Heck B (2016) Exploring hydro-meteorological drought patterns over the Greater Horn of Africa (1979–2014) using remote sensing and reanalysis products. Adv Water Resour. https://doi.org/10.1016/j.advwatres.2016.04.005
27. Awange JL, Ferreira VG, Forootan E, Khandu Andam-Akorful SA, Agutu NO, He XF (2016) Uncertainties in remotely sensed precipitation data over Africa. Int J Climatol 36(1):303–323. https://doi.org/10.1002/joc.4346
28. Omondi P, Awange J, Ogallo Ininda J, Forootan E (2013) The influence of low frequency sea surface temperature modes on delineated decadal rainfall zones in Eastern Africa region. Adv Water Resour. https://doi.org/10.1016/j.advwatres.2013.01.001
29. Omondi P, Awange J, Ogallo LA, Okoola RA, Forootan E (2012) Decadal rainfall variability modes in observed rainfall records over East Africa and their relations to historical sea surface temperature changes. J Hydrol 464–465:140–156. https://doi.org/10.1016/j.jhydrol.2012.07.003
30. IRIN humanitarian news and analysis, UN office for coordination of humanitarian affairs (2006) Global: the global water crisis: managing a dwindling resource. http://www.irinews.org. Accessed 25 Sept 2011
31. Alcamo J, Doll P, Kaspar F, Siebert S (1997) Global change and global scenarios of water use and availability: an application of WaterGAP 1.0. University of Kassel, CESR, Kassel, Germany
32. Alcamo J, Henrichs T, Rosch T (2000) World water in 2025: global modeling and scenario analysis. In: Rijsberman FR (ed) World water scenarios analyses. World Water Council, Marseille, France

33. Wallace JS (2000) Increasing agricultural water efficiency to meet future food production. Agric Ecosyst Environ 82:105–119
34. Wallace JS, Gregory PJ (2002) Water resources and their use in food production. Aquat Sci 64:363–375
35. Agola NO, Awange JL (2014) Globalized poverty and environment. Springer, Berlin
36. Yang H, Reichert P, Abbaspour K, Zehnder AJB (2003) A water resources threshold and its implications for food security. Environ Sci Technol 37:3048–3054
37. Rekacewicz P (2006) Increased global water stress. Vital water graphics 2. Le monde diplomatique. http://www.grida.no/publications/vg/water2/page/3289.aspx. Accessed 15 Apr 2012
38. Neff RA, Parker CL, Kirschenmann FL, Tinch J, Lawrence RS (2011) Peak oil, food systems, and public health. Am J Public Health 101(9):1587–1597
39. Payne WA (2010) Farming systems and food security in sub-Saharan Africa. In: Lal R, Stewart BA (eds) Food security and soil quality. Taylor and Francis, New York
40. World Bank (2003) Water resource and environment. In: Davis R, Hirji R (eds). Technical note G.2, Lake Management
41. Carolina BF (2002) Competition over water resources: analysis and mapping of water-related conflicts in the catchment of Lake Naivasha (Kenya). MSc Thesis, ITC
42. Birnie P, Boyle A (1993) International law & the environment. Camb Law J (Cambridge University Press) 52(3):540
43. Gleick PH (1999) The human right to water. Water Policy 1:487–503
44. Schelton D (1991) Human rights, environmental rights, and the right to environment. Stanf J Int Law (Heinonline) 28
45. Zehnder AJB, Yang H, Schertenleib R (2003) Water issues: the need for action at different levels. Acquat Sci: Res Boundaries 65(1):1–20. https://doi.org/10.1007/s000270300000
46. Hanjra MA, Qureshi ME (2010) Global water crisis and future food security in an era of climate change. Food Policy 35(5):365–377
47. Awange JL, Kiema JBK (2013) Environmental geoinformatics. Monitoring and management. Springer, Berlin
48. Awange JL, Kiema JBK (2018) Environmental geoinformatics. Extreme hydro-climatic and food security challenges: exploiting the big data, 2nd edn. Springer International Publishers, Berlin
49. Johnson LE (2009) Geographic information systems in water resources engineering. CRC Press Taylor & Francis Group, Boca Raton. ISBN 978-1-4200-6913-6
50. Taylor CJ, Alley WM (2001) Ground-water-level monitoring and the importance of long-term water-level data. U.S. Geological Survey Circular 1217, Denver, Colorado
51. Awange JL (2012) Environmental monitoring using GNSS. Global navigation satellite system. Springer, Berlin
52. Awange JL (2018) GNSS environmental sensing. Revolutionizing environmental monitoring. Springer, Berlin
53. Rieser D, Kuhn M, Pail R, Anjasmara IM, Awange J (2010) Relation between GRACE-derived surface mass variations and precipitation over Australia. Aust J Earth Sci 57(7):887–900. https://doi.org/10.1080/08120099.2010.512645
54. IPCC (Intergovernmental Panel on Climate Change) (2007) Contribution of working group I to the fourth assessment report
55. Tapley BD, Bettadpur S, Ries JC, Thompson PF, Watkins MM (2004) GRACE measurements of mass variability in the Earth system. Science 305:503–505. https://doi.org/10.1126/science.1099192
56. Ellett KM, Walker JP, Rodell M, Chen JL, Western AW (2005) GRACE gravity fields as a new measure for assessing large-scale hydrological models. In: Zerger A, Argent RM (eds) MODSIM 2005 international congress on modelling and simulation. The Modelling and Simulation Society of Australia and New Zealand, December 2005, pp 2911–2917. ISBN: 0-9758400-2-9
57. Casanova MT (1994) Vegetative and reproductive responses of charophytes to water-level fluctuations in permanent and temporary wetlands in Australia. Aust J Mar Freshw Res 45:1409–1419

58. Leick A (2004) GPS satellite surveying, 3rd edn. Wiley, New York
59. ILEC (International Lake Environment Committee) (2005) Managing lakes and their basins for sustainable use. A report for the lake basin managers and stakeholders. International Lakes Environmental Committee Foundation, Kusatsu, Japan
60. Awange LJ, Gebremichael M, Forootan E, Wakbulcho G, Anyah R, Ferreira CG, Alemayehu T (2014) Characterization of Ethiopian mega hydrogeological regimes using GRACE, TRMM and GLDAS datasets. Adv Water Resour 74:64–78. https://doi.org/10.1016/j.advwatres.2014.07.012
61. Hu K, Awange JL, Khandu Forootan E, Goncalves RM, Fleming K (2017) Hydrogeological characterisation of groundwater over Brazil using remotely sensed and model products. Sci Total Environ 599–600:372–386. https://doi.org/10.1016/j.scitotenv.2017.04.188
62. Awange JL, Ong'ang'a O (2006) Lake Victoria-ecology. Resource of the Lake Basin and environment. Springer, Berlin
63. Nicholson SE (1998) Historical fluctuations of Lake Victoria and other lakes in the Northern Rift Valley of East Africa. In: Lehman JT (ed) Environmental change and response in East African lakes. Kluwer, Dordrecht, pp 7–35
64. Nicholson SE (1999) Historical and modern fluctuations of lakes Tanganyika and Rukwa and their relationship to rainfall variability. Clim Chang 41:53–71. https://doi.org/10.1023/A:1005424619718
65. Nicholson SE, Yin X, Mamoudou BA (2000) On the feasibility of using a lake water balance model to infer rainfall: an example from Lake Victoria. Hydrol Sci J 45(1):75–95. https://doi.org/10.1080/02626660009492307
66. Aseto O, Ong'ang'a O, Awange JL (2003) Poverty. A challenge for the Lake Victoria basin. OSIENALA Series 5. Printed by Africa Herald Publishing House, Kendu Bay, Kenya
67. Kite GW (1982) Analysis of Lake Victoria levels. Hydrol Sci J 27(2):99–110. https://doi.org/10.1080/02626668209491093
68. Kull D (2006) Connections between recent water level drops in Lake Victoria, dam operations and drought. Available at: http://www.irn.org/programs/nile/pdf/060208vic.pdf. Accessed on 25 Sept 2011
69. Awange JL, Aluoch J, Ogallo L, Omulo M, Omondi P (2007) Frequency and severity of drought in the Lake Victoria region (Kenya) and its effects on food security. Clim Res 33:135–142. https://doi.org/10.3354/cr033135
70. Kayombo S, Jorgensen SE (2006) Lake Victoria: experience and lessons learned brief. International Lake Environment Committee
71. Awange JL, Forootan E, Kuhn M, Kusche J, Heck B (2014) Water storage changes and climate variability within the Nile basin between 2002 and 2011. Adv Water Resour 73:1–25. https://doi.org/10.1016/j.advwatres.2014.06.010
72. Awange JL, Anyah R, Agola N, Forootan E, Omondi P (2013) Potential impacts of climate and environmental change on the stored water of Lake Victoria basin and economic implications. Water Resour Res 49:8160–8173
73. Awange JL, Fleming KM, Kuhn M, Featherstone WE, Heck B, Anjasmara I (2011) On the suitability of the $4° \times 4°$ GRACE mascon solutions for remote sensing Australian hydrology. Remote Sens Environ 115:864–875. https://doi.org/10.1016/j.rse.2010.11.014
74. Awange JL, Sharifi MA, Baur O, Keller W, Featherstone WE, Kuhn M (2009) GRACE hydrological monitoring of Australia. Current limitations and future prospects. J Spat Sci 54(1):23–36. https://doi.org/10.1080/14498596.2009.9635164
75. Awange J, Forootan E, Kusche J, Kiema JKB, Omondi P, Heck B, Fleming K, Ohanya S, Gonçalves RM (2013) Understanding the decline of water storage across the Ramser-Lake Naivasha using satellite-based methods. Adv Water Resour 60:7–23. https://doi.org/10.1016/j.advwatres.2013.07.002
76. Andersen OB, Seneviratne SI, Hinderer J, Viterbo P (2005) GRACE-derived terrestrial water storage depletion associated with the 2003 European heat wave. Geophys Res Lett 32(L18405):2–5. https://doi.org/10.1029/2005GL023574

77. Forootan E, Awange J, Kusche J, Heck B, Eicker A (2012) Independent patterns of water mass anomalies over Australia from satellite data and models. Remote Sens Environ 124:427–443. https://doi.org/10.1016/j.rse.2012.05.023
78. Khandu, Forootan E, Schumacher M, Awange J, Miler Schmied H (2016) Exploring the influence of precipitation extremes and human water use on total water storage (TWS) changes in the Ganges-Brahmaputra-Meghna River basin. Water Resour Res 52(3):2240–2258
79. Othieno H, Awange JL (2016) Energy resources in Africa. Distribution, opportunities and challenges. Springer International Publishing AG, Berlin. https://doi.org/10.1007/978-3-319-25187-5
80. Agutu NO, Awange JL, Zerihun A, Ndehedehe CE, Kuhn M, Fukuda Y (2017) Assessing multi-satellite remote sensing, reanalysis, and land surface models' products in characterizing agricultural drought in East Africa. Remote Sens Environ 194:287–302. https://doi.org/10.1016/j.rse.2017.03.041
81. Mpelasoka F, Awange JL, Zerihun A (2018) Influence of coupled ocean-atmosphere phenomena on the Greater Horn of Africa droughts and their implications. Sci Total Environ 610–611:691–702. https://doi.org/10.1016/j.scitotenv.2017.08.109
82. Awange JL, Hu K, Khaki M (2019) The newly merged satellite remotely sensed, gauge and reanalysis-based multi-source weighted-ensemble precipitation: evaluation over Australia and Africa (1981–2016). Sci Total Environ 670:448–465. https://doi.org/10.1016/j.scitotenv.2019.03.148
83. IPCC (2018) Global warming of 1.5°C. An IPCC special report on the impacts of global warming of 1.5°C above pre-industrial levels and related global greenhouse gas emission pathways, in the context of strengthening the global response to the threat of climate change, sustainable development, and efforts to eradicate poverty. In: Masson-Delmotte V, Zhai P, Pörtner H-O, Roberts D, Skea J, Shukla PR, Pirani A, Moufouma-Okia W, Péan C, Pidcock R, Connors S, Matthews JBR, Chen Y, Zhou X, Gomis MI, Lonnoy E, Maycock T, Tignor M, Waterfield T (eds)
84. IPCC (2014) Climate change 2014: synthesis report. Contribution of working groups I, II and III to the fifth assessment report of the intergovernmental panel on climate change. In: Core Writing Team, Pachauri RK, Meyer LA (eds). IPCC, Geneva, Switzerland, 151 p
85. Hartmann DL, Klein Tank AMG, Rusticucci M, Alexander LV, Brönnimann S, Charabi Y, Dentener FJ, Dlugokencky EJ, Easterling DR, Kaplan A, Soden BJ, Thorne PW, Wild M, Zhai PM (2013) Observations: atmosphere and surface. In: Stocker TF, Qin D, Plattner G-K, Tignor M, Allen SK, Boschung J, Nauels A, Xia Y, Bex V, Midgley PM (Hrsg) Climate change 2013: the physical science basis. Contribution of working group I to the fifth assessment report of the intergovernmental panel on climate change. Cambridge University Press, Cambridge, United Kingdom and New York, NY, USA
86. Saji NH, Goswami BN, Vinayachandran PN, Yamagata T (1999) A dipole mode in the tropical Indian Ocean. Nature 401:360–363
87. Trenberth KE (1997) The definition of El Niño. Bull Am Meteorol Soc 78:2771–2777
88. Awange JL, Mpelasoka F, Goncalves R (2016) When every drop counts: analysis of droughts in Brazil for the 1901–2013 period. Sci Total Environ 566–567:1472–1488. https://doi.org/10.1016/j.scitotenv.2016.06.031
89. Mpelasoka F, Awange JL, Goncalves RM (2018) Accounting for dynamics of mean precipitation in drought projections: a case study of Brazil for the 2050 and 2070 periods. Sci Total Environ 622–623:1519–1531. https://doi.org/10.1016/j.scitotenv.2017.10.032
90. Agutu NO, Awange JL, Ndehedehe C, Kirimi F, Kuhn M (2019) GRACE-derived groundwater changes over Greater Horn of Africa: temporal variability and the potential for irrigated agriculture. Sci Total Environ. https://doi.org/10.1016/j.scitotenv.2019.07.273
91. Hu K, Awange JL, Kuhn M, Saleem A (2019) Spatio-temporal groundwater variations associated with climatic and anthropogenic impacts in South-West Western Australia. Sci Total Environ. https://doi.org/10.1016/j.scitotenv.2019.133599
92. Agutu NO, Awange JL, Ndehedehe C, Mwaniki MW (2020) Consistency of agricultural drought characterization over Upper Greater Horn of Africa (1982–2013): topographical, gauge density,

and model forcing influence. Sci Total Environ 709. https://doi.org/10.1016/j.scitotenv.2019.135149

93. Awange JL, Saleem A, Sukhadiya RM, Ouma YO, Kexiang H (2019a) Physical dynamics of Lake Victoria over the past 34 years (1984–2018): is the lake dying? Sci Total Environ 658:199–218. https://doi.org/10.1016/j.scitotenv.2018.12.051
94. Morgan B, Awange JL, Saleem A, Kexiang H (2020) Understanding vegetation variability and their "hotspots" within Lake Victoria basin for the 2003–2018 period. Appl Geogr
95. Alsdorf DE, Rodrguez E, Lettenmaier DP (2007) Measuring surface water from space. Rev Geophys 45(2):RG2002. https://doi.org/10.1029/2006RG000197

Chapter 2
Lake Victoria's Water Resources

Lake Victoria is a living Lake. From Kisumu (Kenya) to Mediterranean, millions of people are being fed by it!

—Pastor Martin Mbandu, former pastor, Kisumu Pentecostal Church

2.1 Features of the Lake and Its Environs

2.1.1 The Origin

It is important to know, for historical and scientific purposes, the possible origins and age of Lake Victoria (Fig. 1.3). How old is Lake Victoria? How did it come about? Has it changed much during its existence? If so, how? Why should we be interested in knowing the age of Lake Victoria? These are some of the pertinent questions, which have occupied the attention of scientists for centuries; and rightly so. The origins and age of the Lake are still shrouded in deep mystery. Although scientists have grappled with these questions, no definitive answers have been forthcoming. As it is, the origins of Lake Victoria are still the subject of scientific dispute [1]. It seems likely that it is much more recent in its origins than the other great Lakes of Eastern Africa. But there is an interesting aspect of the direction of flow of waters into the Lake. There is a scientific contention that many of the rivers now flowing east into Lake Victoria (including Kagera) once flowed west.

Hickman et al. [2, p. 36] suggests that Lake Victoria was formed due to warping and tilting of the land. They note that lower reaches of some river valleys have been raised, so that they flooded, and the directions of the rivers changed at least in the Miocene, Pliocene, and part of the Pleistocene eras; within the past 2 million years, possibly eventually into the Nile system. A more recent upthrust of the western side of the basin is thought to have however reversed these rivers, and caused Lake

J. Awange, *Lake Victoria Monitored from Space*,
https://doi.org/10.1007/978-3-030-60551-3_2

Victoria to form by flowing eastwards [1]. Although Fuggle [3] suggest that the Lake formed sometimes during the last 400,000 years, Aseto and Ong'ang'a [1, 4] suggest the possibility of the Lake having formed as recently as 25,000–35,000 years ago and dried up completely between 10,000 and 14,000 years. Samples of Earth taken from the bottom of the Lake corroborate this suggestion by indicating that the Lake may have temporarily but completely dried up 12,000–14,000 years ago. But this suggestion raises a host of other questions. For instance, *what happened to make the Lake dry up? What happened to make the dry Lake become a freshwater Lake? Could it dry up again in the future?* In attempting to address the issue of the drying of the Lake, Johnson et al. [5] suggest that the Lake probably dried up entirely about 15,000 years ago due to climatic phase of extreme dryness. Be that as it may, some scientists maintain that tectonic movement that swept across the Eastern African plateau about 14,600 years ago caused the formation of shallow basins, which filled with waters to become what is today ***Lake Victoria***. However, Some unresolved issues requiring further research include:

- Resolving the age of Lake Victoria since the conflicting explanations as to the age of the Lake still persists.
- Finding out whether the Lake can die under stress, as was supposed to have happened thousands of years ago?
- Continuous monitoring of fluctuation of its water level to establish the trend as to whether the lake is dying.
- Establishing whether the Lake is getting deeper or shallower with the passage of time and also whether it is expanding or decreasing, thus increasing or reducing its area of coverage.
- The Lake's parameters, e.g., length, area, etc., vary according to various authors as will be seen in Sect. 2.1.3. Use of satellite techniques discussed in Chap. 5 can resolve this once and for all as demonstrated by Awange et al. [6].

2.1.2 The Name "Lake Victoria"

We elaborate in this section how the name Lake Victoria was coined. Is the name Lake Victoria a proper name for the Lake? What is in a name? Does a name matter? If so, how and why? From time immemorial, these questions have been asked over and over again, and in different contexts. The general answer is that giving a name to an entity is an extremely important decision that people can make, in that, it is a means of providing an identity. A good name is given to a good thing; and a bad name is usually given to a bad thing. In the Luo language in Kenya, it is taken axiomatically that "Nying' ema ichiemogo" which is translated to mean "one eats by the fame of one's name". It is for this reason that human beings name their children after important persons, or give their children the names of their parents and fore parents to maintain their genealogy, or after important events, or in honour of important events or entities. It is also the reason why human beings give names to animals, hills, valleys, and stars. And even at an advanced age, individuals can acquire new names. Some are given names such as "sibuor" (Luo name for a lion) in recognition of one's bravery, etc.

The inhabitants around the Lake have followed suit; they always had different names for the Lake. The Lake is also known as Victoria Nyanza. This appears to have been derive from the Basuba (a tribe in Kenya) name for the Lake: "Nyancha" or "Nyanja" by certain communities in Uganda. The Luos on the Kenya side call it "Nam (Namb) Lolwe" (endless waters). The Baganda in Uganda name it "Nalubale", the Sukuma in Tanzania call it "Sukuma Lake". This being the case, why then is the Lake called Lake Victoria?

The name *Victoria* is "foreign" in that it was named in honour of an individual in a foreign land. "Foreigners" had different names for the Lake. The Arabs called it *"Ukerewe"*. But an Englishman, explorer John H. Speke, stumbled onto its southern shore and proclaimed he had discovered the fabled source of River Nile in 1858. He built a pillar at the site (Fig. 2.1) and being the first European to sight it, he decided to honour Queen Victoria of England by naming it after her. One recalls that during 1890s, Great Britain colonized Kenya and Uganda while the Germans had a stronghold of Tanzania's mainland in 1899. In 1875, an American journalist, Henry Morton Stanley, seeking to confirm Speke's claim and looking for the British missionary Dr. David Livingston, circumnavigated the Lake. He spent two weeks spinning tales of God and England to curry favour with Mtesa, king of Buganda and ruler of the northern Lake region. Then Stanley sent word back to England, calling for missionaries. They came with soldiers and traders. He popularized the name Lake Victoria in the Western world. And the name Lake Victoria has continued to-date. Nevertheless, African countries around the Lake have not been content with the present name of Lake Victoria. Thus, soon after the independence of the three surrounding East African states of Uganda, Kenya and Tanzania, there were attempts to rename the Lake so as to give it a proper African name; but such attempts have not yet succeeded, and the name Lake Victoria has been retained. It may be just a matter of time before the Lake is given its rightful appropriate African name that would give its true African identity and character. After all, changing the name of the Lake from a 'foreign' tag to an African name should not be a problem. Several examples of name change abound in East Africa and elsewhere in the continent and in the world. In the case of Kenya, African names have replaced foreign names. For instance, upon the country's independence on December 12, 1963, all landmarks, monuments and streets in Kenya that bore foreign names were given local, or to put it bluntly, African names. Lord Delamere Avenue in Nairobi was renamed Kenyatta Avenue. Queen's Way was renamed Mama Ngina Street. Government Road became Moi Avenue. And Duke Street became Ronald Ngala Street. Similar examples abound in the country. Thus changing the name of the Lake from a foreign one to an African name is in order and is consistent with the tradition of name changing initiated since independence.

Could the name "***Nyanza***" be a good candidate for the changed name, given that originally, the natives in the Suba District called it by that name? Already, the communities around the Lake have recently expressed their desire to rename the Lake. At a cultural conference held in Mwanza, Tanzania, in 2001, the cultural

Fig. 2.1 Pillar signifying the source of the White Nile erected by John H. Speke in 1858 in Lake Victoria

leaders selected from major ethnic communities around Lake Victoria questioned the rationale for retaining the name Victoria. Following extensive discussions on the matter, the leaders recommended that the Lake be renamed. Their argument was that the current name makes their Lake sound foreign and detached from their cultures.

2.1.3 Lake Victoria Basin: Physical Description

Here, only a brief background of lake Victoria is presented. For more details we refer the reader to the books and article of [1, 6–8]. Lake Victoria is the largest freshwater Lake in Africa (also largest tropical Lake) and the second largest in the world after Lake Superior in the United States. The Lake is also the largest in the developing world. It is located at Longitude 31° 39'E–34° 53' E and Latitude 0° 20'N–3° S, and has a surface area of c.a. 69,295 km^2 [6]. Its greatest length is about 400 km and its breadth is about 320 km.[1] Using a high spatio-temporal resolution Sentinel-2 remotely sensed images (see Chap. 8), Awange et al. [6] established the 2018 values

[1]Other sources, e.g., Britannica Concise put the length at 337 km and width at 240 km.

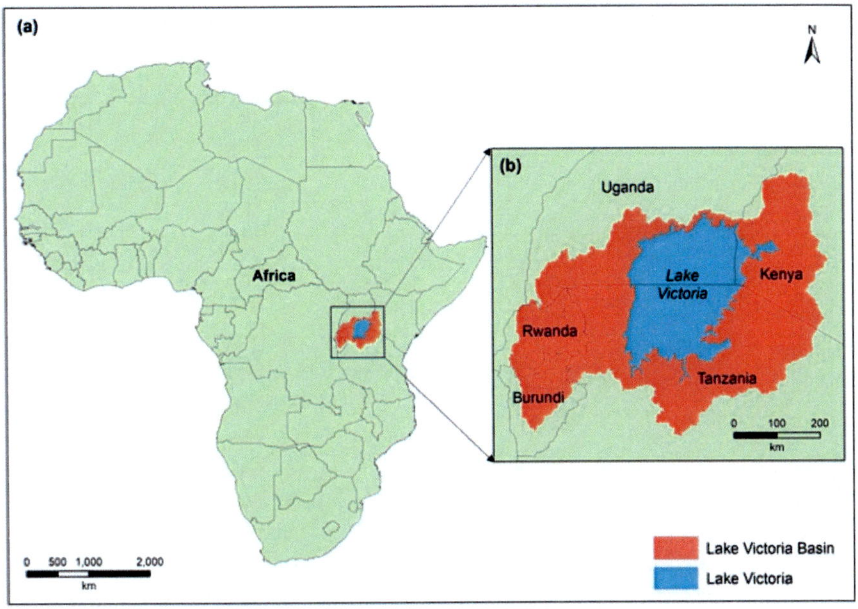

Fig. 2.2 Lake Victoria Basin (LVB), the study area. *Source* Morgan et al. [11]

to be 386 km (length) and 362 km (width).[2] It contains about 2,760 cubic kilometers of water and is situated at an altitude of 1135 m above mean sea level. The shoreline is very irregular and totals some 3,300 km in length. Hughes and Hughes [9] have put the shoreline close to 3,500 km in length, which is less than the 2018 value 4.572 km established by Awange et al. [6]. The shoreline on the Kenyan side is 760 km. Much of the Lake is relatively shallow, reaching a maximum depth of about 80 m (Fuggle [3] puts the figure to 85 m), and an average depth of about 40 m; the deepest zone (60–90 m)[3] lies toward the shore. There is little annual variation in water temperatures, the mean surface being about 24 °C, and that of deeper waters to be 23 °C.

Its basin (Fig. 2.2) spreads from Longitude 30° E–35° 50' E to Latitude 0° N–3° 20'S. The ***shoreline*** is comprised as follows: 17% in Kenya, 33% in Tanzania and 50% in Uganda. Of the total catchments area of 193,000 km^2, Burundi accounts for 7.2%, Kenya 21.5%, Rwanda 11.4%, Tanzania 44.0% and Uganda 15.9%. The entire drainage basin covers an area of 258,000 km^2. Awange et al. [6] show the lake's mean surface area of to be 69,295 km^2 (i.e., 812 km^2 or 1.2% more than that of the 37 publications that they reviewed) and its 2018 value to be 69,216 km^2 (i.e., ~733 km^2 (1.1%) more than the mean of the 37 publications). Details of this study are presented in Chap. 8. The Lake is also the source of the river Nile (i.e., the source of the White Nile while the Blue Nile originates from Ethiopia), arguably the world's

[2]The mean values of Awange et al. [6] from the 1984, 2002, 2017 and 2018 images were 388 km for length and 364 km for width respectively.

[3]82 m—Britannica Concise.

Table 2.1 Morphoedaphic characteristics of Lake Victoria [6, 10]. The mean width*, length* and area* are with respect to the Landsat (1984, 2002, 2017) and Sentinel-2 (2018) images

Characteristic	Measure
Position: Latitude	00° 20' N to 03° 00' S
Longitude	31° 39' E to 34° 53' E
Altitude [m above mean sea level]	1134
Catchment Area [km^2]	184,000
Lake Surface Area in 2018 [km^2]	69,216
Mean Area* [km^2]	69,295
Lake Area as Percentage of Catchment	37
Shoreline in 2018 [km]	4,572
Maximum Length in 2018, North-South [km]	386
Maximum Width in 2018, East-West [km]	362
Mean Width* [km]	364
Mean Length* [km]	388
Maximum Depth [m]	84
Mean Depth [m]	40
Volume [km^3]	2,760
Inflow [km^3/yr]	20
Outflow [km^3/yr]	20
Precipitation [km^3/yr]	114
Annual Fluctuations in level [m]	0.4-1.5
Flushing time [yrs]	138
Residence Time [yrs]	21

longest river. The Lake's main physical parameters have been summarized by [6, 10] in Table 2.1. Because the Lake is shallow, its volume is substantially less than that of other Eastern African Lakes with much smaller surface area. The volume of Lake Victoria's water is only 15% of the volume of Lake Tanganyika, even though the latter has less than half the surface area. The Lake straddles the Equator and touches the Equator in its northern reaches. It is shared by Kenya (6%), Uganda (45%) and Tanzania (49%). The Lake's shoreline is long and convoluted, enclosing innumerable small, shallow bays and inlets, many of which include swamps and wetlands, which differ a great deal from one another and from the Lake itself.

Lake Victoria contributes much to the regional rainfall and subsequently the hydrological cycle. Of particular importance to lake Victoria are the South-East trade winds, which picks moisture from the lake while passing. They move to the northern and western shores of lake Victoria where they eventually condense to give heavy rainfall. Uganda benefits more due to the fact that winds carry more moisture to the *Northwest* shores of Lake Victoria, with major rainfall season normally between March, April and May (MAM), with low rainfall season occurring between September, October and November (SON). There exist no remarkable dry months.

The climate of Lake Victoria catchment area is mild with small variations in monthly average air temperature between 19 °C and 25 °C throughout the year. Daily temperature fluctuates more widely, ranging from 15 °C to 30 °C. Rainfall in this catchment area averages about 1,300 mm annually, varying from 2,000 mm in highlands to 1,000 mm in the north, southwest and lowlands along the Lakeshore.

2.2 Lake Victoria: Benefits and Challenges

Of all the tropical's Lakes, Lake Victoria stands tall as the greatest fresh water body. In the entire world, it comes only second to Lake Superior. Within its surrounding, it directly supports a population of more than 42 million East Africa's inhabitants (i.e., $\frac{1}{3}$ of the combined population) [6], while globally it is the source of Tilapia (*Oreochromis niloticus*) and Nile Perch (*Lates niloticus*) [1]. So important is the Lake such that Egypt and Sudan entered the Nile treaties of 1929 and 1959 for exclusive use of its waters. Just as oil is an important resource of the world, Lake Victoria and its basin plays a major role in the world such that it appears at the top of a number of lists as pointed out by [12]. Specific roles played by this Lake are [1, 7, 8, 13]:

- Agricultural (rain-fed as well as irrigated) potential. Its surface water could be used for irrigation whereas its modulating of the regional hydroclimate contributes to rainfall that is essential for rain-fed agriculture.
- It is a source of fresh water that supports livelihood of people living within its shores. The water is used for drinking, domestic use, in industries, for transportation and sports among others.
- It contributes to the river Nile, the lifeline of Egypt and Sudan.
- It is the habitat of Tilapia (*Oreochromis niloticus*) and Nile Perch (*Lates niloticus*), the two types of fish that are in great demand the world over together with other species of fish who reside in its waters.
- Its wetlands provide homage to various bird species, insects and thus preserving biodiversity.
- It helps fight poverty by creating employment in fish industry and transportation for people living within its environs.
- It acts as eco-tourism center.

The list could go on and on but we have chosen to mention only but a few.

Although the Lake and its basin boasts of such rich endowment, the fact of the matter is that it is a sick giant [3]. This has necessitated the establishment of several measures to help address the challenges faced by this Lake. Such measures include the establishment of Lake Victoria Management Environmental Programme (LVEMP) and other institutions. The challenges facing the Lake include:

- Having as its dependents people who are faced with the health threat such as HIV/AIDs.

Fig. 2.3 A huge floating island in Lake Victoria clogging a turbine in a hydroelectric power station. *Source* https://www.bbc.com/news/world-africa-52286296

- Being infested with the deadly water hyacinth, which is depriving it of oxygen and thus suffocating it, and posing threats to hydropower plants (see Fig. 2.3).
- Faced with people living in its surrounding languish in abject poverty although it is endowed with great resources.
- Consequences of decision and policies made in far parts of the world, i.e., ecological power, and by global economic structure [3].

In general, the Lake faces four critical and mutually reinforcing problems. The *first* is the *obsolete Nile treaties* that lock up its potential for irrigated agriculture. If these treaties could be done away with, and the lake used sustainably, then its contribution to irrigated agriculture could go along way in alleviating food insecurity over the entire Greater Horn of Africa (GHA). *Second* is *productivity*. The most important commercial species caught is Nile Perch. A predatory species that has over time eaten to near extinction many species endemic to the Lake that were once important sources of food and livelihood for the local communities. The scramble for the catch has become the source of conflict between fishers from the different countries. Overfishing is attributed to the use of inappropriate gear and insufficient enforcement capacity. Open access and inappropriate exploitation increases the fishing effort against the diminishing reproduction rate. In addition, territorial boundaries tend to concentrate the fishing effort on a limited area especially in Kenya. Authorities lack the resources nor the will to effectively police the Lake.

The *third* set of challenges are *environmental*, namely resource degradation emanating from catchment degradation and discharge of human, industrial and agricultural effluent. The *fourth* set of challenges is *market related* and includes price competitiveness, quality assurance and reliability of supply. The principal export product Nile Perch, is under intense price pressure from aquaculture, in particular the new tilapia varieties. While the fishery has made substantial progress in meeting European (EU) standards, it has not yet attained the eco-labelling certification that is increasingly required by the retailers [14].

Owing to these overwhelming evidence of the importance of Lake Victoria together with its emerging threats, a continuous monitoring of its status is inevitable. A comprehensive coverage of Lake Victoria's resources and related challenges are presented in Awange and Ong'ang'a [8]. Awange [13] looks at the feasibility of weighing the Nile River Basin from space while Awange [7] presents the hydroclimate of Greater Horn of Africa and considers the potential of irrigated agriculture, e.g., from Lake Victoria's surface water. The current book looks at the feasibility of monitoring this lake using the latest state of the art space-based techniques.

2.3 Population and Demographic Features

Lake Victoria Basin (LVB) had a population of approximately 21 million in 1997, 26 million in 1999 (above 30 million if Rwanda and Burundi are considered), with a growth rate of around 3% per annum with considerably higher rates in urban areas. Interesting results are presented by [15] who puts the 2010 population to over 42 million and projects the population to increase to over 60 million in 2020, over 76 million in 2030 and tripple to over 113 million in 2050. On the commonly used figure of 30 million population of LVB, [15] writes:

> Interestingly, our population estimates derived for the LVB reveal that the total population that most stakeholders for the basin, including the Lake Victoria Basin Commission, were commonly citing - about 30 million inhabitants likely underestimated by more than 10 million people. While the number of 30 million people was being used in reports and other communications about the threat of population growth, we could identify no original source for this estimate.

The region is increasingly characterized by growing variability of ethnic composition. However the larger ethnic groups dwelling around Lake Victoria are the Luo, Samia, and the Suba people in Kenya; the Buganda and Busoga in Uganda; In Tanzania the distribution is as follow; in Mwanza region are the Sukuma, Kerewe, Jita, Kara, and Zinza. Sukuma constitute the overwhelming majority in this region. In Mara region are the Jita, Luo, Kuria, Zanaki, and Ruli. In Kagera there are Haya, Hangaza, Nyambo, Subi, and Sukuma. This means that there is a very great cultural diversity in the basin requiring a complex analytical framework as regards harmonization towards better and sustainable use of the natural resources of the basin for improving people's livelihoods.

The five countries in the Lake Region (i.e., Kenya, Uganda, Tanzania, Rwanda and Burundi) cover a total land area of 170,270 km^2, supporting one of the densest and poorest rural populations in the world. Women constitute just over 50% of the population in the region. In the Lake Victoria region as a whole, compared to the 2010 population of over 42 million, Bremner et al. [15] estimate the population increase by 43% in 2020, 80% in 2030 and 167% in 2050. About 50% of the population in the region comprise youth who need more gainful employment opportunities to avoid the risks of social insecurity. The region already experiences some insecurity because of internally displaced persons as well as refugees from neighbouring countries.

In practically all the East African countries, increasing attention is being directed toward the status of women in society. Programmes aimed at mainstreaming gender issues in development activities are at various stages of development. To date, however, the socio-economic status of women is lower than that of men in a number of ways. For instance, women have limited access to productive resources such as land and capital. The literacy levels are lower among women than among men and the number of girls in secondary schools is considerably lower than that of boys. At the primary school level, the numbers are about the same. Women bear a heavier burden of the widespread poverty. Commercial fishing and processing factories employs more men than women and has impacted negatively on women who were previously engaged in fish processing and marketing. Moreover, when men migrate from rural to urban areas in search of employment, they leave behind women and children to bear the brunt of agricultural activities. The heavy work load women bear on farms, in their reproductive and community roles is a major source of poor health. Due to lack of money, women have limited access to health facilities [16, p. 8].

There is a strong cultural dimension to environmental degradation in the region. The dictates of cultural practices of sons inheriting their fathers' land and wives owning land to cultivate are reinforcing the need to subdivide land into small units, which are uneconomic for meaningful farming. Such practices continue to generate a population of landless youth who must migrate elsewhere to earn a living and the cycle of poverty created continues to cause further environmental degradation. Population pressure on limited land leads to rapid land degradation. It is, therefore, imperative that conservation measures are adopted on a massive scale, if not then, it may not be possible to control the rate of environmental degradation.

Livelihood standards of the area have deteriorated as already noted by [17], not just because of the consequences of the population increases, but through a host of limiting factors and livestock production that lead to declining land productivity. There are also the people of Rwanda and Burundi in the wider Lake catchment region. Population density in the Lake basin is above the national averages found in all three riparian countries and the populations within these riparian communities grow at rates that are among the highest in the world. Population density in the region ranks among the highest in the world for rural areas where the majority of the people live.

2.3.1 *Historical Perspective of Early Settlements*

East Africa is largely considered the cradle of mankind, since the first paleontologic traces of humans have been found there. Most of the present Bantu-speaking inhabitants of the LVB started entering the basin around the beginning of the second century. The communities presently settled around Lake Victoria were immigrants from other parts of Africa. As early as the 13th and 14th centuries, Lake Victoria was well surrounded by organized and settled communities, which had and still enjoy a lot of interactions. The Luos arrived via the River Nile from the North of Africa and settled in the Lake Victoria basin in Uganda before moving to Kenya. When the Luos arrived, they forced out other tribes like Elgon Masai and Bantus that had earlier settled in the Lake basin. They interacted with ethnic groups that lived in Uganda. Some of the Luos later moved to North Mara in Tanzania in search of grazing ground when drought destroyed their crops. The Bantus who now occupy the Lake Victoria basin in Uganda and some parts of Tanzania trace their origins in Central Africa. Their settlement developed into indigenous kingdoms in the 14th century. Among them were the Baganda, Banyoro, Toro, Ankole and Basoga. The Baganda people created a dominant kingdom that could not be penetrated until the 19th century (see Fig. 2.4 for the Buganda Kingdom's palace[4]). The Kabaka king Mutesa II reigned until his overthrowal in 1966 by Milton Obote who himself was overthrown in 1971 by the butcher of Uganda (Idi Amin, see Fig. 2.5).

The other most important group was the Basoga who later established their kingdom around Jinja. Before colonization, the larger tribes that occupied the southern shores of Lake Victoria in Tanzania were the Wasukuma and Wanyamwezi. Currently their population runs into millions. Other smaller tribes like the Wanazaki, Wajita, Wakerewe of Tanzania; the Banyala and Suba of Kenya and the Samia of Uganda are some of the most important ethnic groups that live on the shores of Lake Victoria. We have among the communities the Asians who, although constituting a minority, were brought by the colonialists to build the then Kenya-Uganda railway. The group is among town dwellers and still controls major businesses in the Lake Victoria port towns.

The earlier settlers brought with them distinct cultures into the basin. Through interactions, intermarriages amongst the different cultures became imminent. The cross transfers of communities were enhanced among the neighbourhood communities, through trade. The barter system amongst the communities became prominent as canoes and dhows transported goods. Trade and market centers were developed and later small towns were established in the Lake ports. This was the origin of the municipalities and cities that are now well established. The major towns include Kampala, Entebbe and Jinja in Uganda, with Kampala being the capital city of Uganda; Kisumu, Homa Bay in Kenya, with Kisumu being the third largest city in Kenya after Nairobi and Mombasa; and Mwanza, Musoma and Bukoba in Tanzania. These towns' populations run into millions. The peoples of the LVB share a relatively common history characterized by intensive interaction, extensive trading, intermarriage and welfare. In other cases, assimilation occurred, e.g., the Kenyan Suba were assimilated into Luo and now speak and practise Luo culture. Young Subas can hardly speak their native Suba language but Dholuo (Luo language). In essence, the Subas are Luos for all intent and purpose. The regime of former president Moi attempted to divide the two communities through the 1988 census but the Subas and Luos remained a single

[4]http://www.buganda.com/crisis66.htm.

Fig. 2.4 Kabaka palace. **a** main office, **b**, the compound with Kabaka's flag **c** Kabaka's lake where he could go and bathe, and **d**, remnant of Kabaka's vehicle when he fled during President Obote's take over in 1966

entity. Furthermore, the communities within LVB also share the historic experience of external interventions in the area, such as the slave trade and colonialism.

In 1885, Tanzania, Rwanda, and Burundi were placed under the rule of the Imperial German Government. However, after the first World War, Tanzania was placed under the British mandate until the establishment of the Trusteeship system under the Charter of the United Nations, while Rwanda and Burundi came under Belgium rule. In 1887 Kenya was placed under the Imperial British East Africa Company, but the territory was transferred in 1895 to Britain. Likewise, the British established the Uganda Protectorate in 1894. However, after 1945, various nationalist movements emerged in this region, and began a sustained and successful effort to emancipate themselves from European domination. In 1961, Tanzania became independent followed by Uganda, Rwanda, and Burundi in 1962 while Kenya gained her independence in 1963. The delay in Kenya's independence is argued to be due to Jaramogi Oginga Odinga's (a native of Luo of Siaya County within Lake Victoria basin) refusal to assent to the throne unless Jomo Kenyatta (Kenya's first president) was released from prison.

Fig. 2.5 Historical sites within the lake Victoria surrounding. **a** Path to Uganda's torture chambers during Idi Amin's regime, **b**, **c** and **d**, inside the torture chambers

2.3.2 Impacts of Colonialism

The establishment of colonial rule over the people of the region brought with it resource management structures that removed the power from traditional leaders to the central governments of larger territories of Kenya, Uganda, and Tanganyika. This meant that people who had no interest in the Lake could be given the responsibility to manage resources such as fisheries. The ownership of the resources shifted with time. The population influx and lack of properly planned infrastructural development has brought many problems to the urban centers. These problems include non-functional sewerage systems, industrial pollution and shortage of water supply. These have become a serious headache for the citizens. County governments alone cannot address these problems. The citizens have to join them with 'own key' solutions through organized groups such as Non-Governmental Organizations (NGOs) and Community-Based Organizations (CBOs). Kenya changed its constitution in 2010 to allow for the devolved system and is proposing the current building bridges initiative (BBI) to increase county revenue from the current 15% to 35%. This if properly managed could see huge development within the Kenyan side of the LVB. However, rampant corruption and mismanagement of the resources that bedevilled the first county governments of Kisumu, Siaya, Migori, Homabay and Busia in the Kenyan side of LVB could not let this happen. Hopefully, the future county governments will learn from the mistakes of their predecessors. The current (2020) Kisumu county government under Professor Anyang Nyongo seems to have hit the right pedal.

2.4 Concluding Remarks

This chapter has discussed the origin of Lake Victoria, its name and some elements of demographic features of the region. It is a region where the population growth rate is very high and average population density also high. These two phenomena have a direct bearing on the environment, leading to serious environmental degradation, soil erosion, pollution of the Lake and the decline in soil fertility. It is therefore imperative that measures be taken to address the issues concerning population pressure and environmental degradation. Whereas these measures are necessary, it is more important that the lake waters themselves be continuously monitored. However due to its size, ground-based observation methods by themselves are insufficient hence the need of space-based methods discussed in this book.

Lake Victoria is a riparian resource shared by 3 East African countries: Kenya, Uganda and Tanzania. It serves as a source of water for irrigation, which if properly utilized, could significantly enhance GHA's food security. As argued previously, however, the caveats imposed by obsolete Nile treaties of the 1920s and 1960s have to be removed. Furthermore, its modulation of the regions climate leading to rainfall within its basin is significant for rain-fed agriculture. Moreover, it is vital for transport, domestic and livestock uses and a source of livelihood for over 42 million people living around the lake. Its fish (tilapia and Nile perch) are exported the world over. Its role as an indicator of environmental and climate change on long-term scales has been documented in [18]. The vital role of this lake and its surroundings as a source of food has been documented e.g., in Awange [7, 13] and Awange and Ong'ang'a [8]. The lake and its environment is under threat from environmental pollution and more recently from declining water levels. In 2020, the floods were back leading to speculations that the lake was retracing its 1960s level.

References

1. Aseto O, Ong'ang'a O (2003) Lake Victoria (Kenya) and its environs: resource, opportunities and challenges. Africa Herald Publishing House, Kendu Bay
2. Hickman GM, Dickins WHG, Woods E (1973) The lands and people of East Africa. Longman, Essex

3. Fuggle RF (2004) Lake Victoria: a case study of complex interrelationship. In: United nations environmental program: Africa environmental outlook case studies
4. Ong'ang'a O (2002) Poverty and wealth of fisherfolks in the Lake Victoria basin of Kenya. Africa Herald Publishing House, Kendu Bay
5. Johnson TC, Kelts K, Odada EO (2000) The Holocene history of Lake Victoria. AMBIO: A Journal of the Human Environment 29(1), 2–11. https://doi.org/10.1579/0044-7447-29.1.2
6. Awange JL, Saleem A, Sukhadiya RO, Ouma YO, Hu K (2019) Physical dynamics of Lake Victoria over the past 34 years (1984–2018): is the lake dying? Sci Total Environ 658:199–218
7. Awange JL (2021) Greater Horn of Africa's hydrometeorology. Potential for agriculture amidst extreme drought. Springer Nature International, Berlin
8. Awange JL, Ong'ang'a O (2006) Lake Victoria-ecology. Resource of the lake basin and environment. Springer, Berlin
9. Hughes RH, Hughes JS (1992) A directory of African wildlife. With a chapter on Madagascar, IUCN/UNEP/WCMC
10. Balirwa SJ (1998) Lake Victoria wetlands and the ecology of the Nile Tilapia Oreochromis niloticus. Linne. Ph.D. Dissertation, Balkema Publishers, Rotterdam, Netherlands
11. Morgan B, Awange JL, Saleem A, Hu K (2020) Understanding vegetation variability and their "hotspots" within Lake Victoria Basin (LVB: 2003-2018), 122. https://doi.org/10.1016/j.apgeog.2020.102238
12. Pitcher TJ, Hart PJB (1995) The impact of species changes in African lakes. Fish and fisheries series, vol 18. Chapman and Hall, London, p 601
13. Awange JL (2021) Nile waters. Weighed from space. Springer Nature International, Berlin
14. KLI (2004) Market based fisheries regulation strategy for Lake Victoria. In: A report of the regional inception workshop, June 6–8, 2004, Lake Naivasha Country Club, Kenya, 50pp
15. Bremner J, Lopez-Carr D, Zvoleff A (2013) Using new methods and data to assess and address population, fertility, and environment links in the Lake Victoria basin, population and the environment. Session 247. https://www.iussp.org/en/event/17/programme/paper/6136
16. Ehlin U (1997) Lake Victoria basin: natural resource under stress, Stockholm. Department of natural resource and environment, Sida
17. Aseto O, Ong'ang'a O, Awange JL (2003) Poverty. A challenge for the Lake Victoria basin. OSIENALA series, vol 5. Printed by Africa Herald Publishing House, Kendu Bay
18. Nicholson SE (1998) Historical fluctuations of Lake Victoria and other lakes in the Northern Rift Valley of East Africa. In: Lehman JT (ed) Environmental change and response in East African lakes. Kluwer, Dordrecht, pp 7–35

Chapter 3
Challenges: Sustainability and Obsolete Treaties

This treaty only benefits Egypt. We cannot sit back while we have water we cannot use to irrigate our land. Why should we preserve our water for Egypt – Raila Odinga (then Energy Minister's address to the Kenyan Parliament), and
Waste in the Lake basin is a two-lane highway: flowing down into the Lake is the waste and pollution generated upstream. In the opposite direction, leaving the basin, flows the region's wealth in different forms: exploitation of fish, and the rich gold mines in Macalder and Kitere. – The late Prof. Oyugi Aseto

3.1 Summary

Through its links with Lakes Tana and Victoria and all the rivers that feed them, the great Nile system belongs to the peoples of Rwanda, Burundi, Tanzania, Uganda, Kenya, the Sudan, Ethiopia, and Eritrea. Yet, when the present Nile Treaty was "negotiated" and signed in 1929, these peoples were never consulted. It was a treaty purely between the various British colonial regimes that lorded it over most of these African countries.

That the pact overwhelmingly favoured Egypt is no wonder. Egypt was far more important to Britain than the other colonies. Through the Suez Canal, it was a valid econo-strategic link with the riches of the Middle East, India and the Far East. But more relevant than that, the Egyptian section of the Nile Basin was (and remains) a veritable treasure trove of agricultural products, especially the raw cotton that continues to feed the maws of Lancashire's textile industries. And Egyptian farming - the great irrigation works that have maintained that country ever since Pharaoh Menes-Narmer united it in 3100 BC—have depended entirely on extraordinarily fertile alluvium, which the two Niles scoop yearly from Eastern Africa. That was why the treaty banned all the riparian countries south of Egypt from using the Nile water for their own irrigation without Cairo's consent. Even after independence,

J. Awange, *Lake Victoria Monitored from Space*,
https://doi.org/10.1007/978-3-030-60551-3_3

Cairo has clung tenaciously to this blatant injustice. The benefits have been so great that Egypt has adamantly rejected all calls to democratize the international use of the river's rich resources. But Egypt claims to belong to the comity of African nations and it is good that it has finally seen the sense of sitting down with the other "stake holders" to negotiate the issue afresh.

The need remains, nevertheless, to use the Nile waters with the greatest rationale. Egypt is right to be anxious about wastage. For the fact remains that, for that country, the Nile is the only lifeline. Other interested countries do not depend on it so completely and so desperately. Therefore, they should exploit the river with enough prudence not to harm Egypt's interests. What is to be done? One possible solution, suggested by members of the East African Legislative Assembly, is to sell the Nile waters to Sudan and Egypt, just as Egypt has been selling the Nile waters to Israel.[1]

3.2 Introductory Remarks

The greatest development challenges facing Lake Victoria and its basin are the socio-economic and ecological problems, which are mainly related to the inter-linkage between poverty and environmental degradation. These are further exacerbated by the lack of capacity among the concerned institutions to manage the resources of the Lake Victoria Basin (LVB), both human and natural, in a sustainable manner. Similarly, the judicial and institutional frameworks that govern the socio-economic activities have so far been inappropriately conceived and enforced, and in an uncoordinated manner.

Sustainable growth is one prerequisite for poverty alleviation in any country. Considering the fact that population growth in the Lake Victoria basin is in the region of 3% and that 50% of the population live under ***poverty line***, a substantial growth is required in order to alleviate poverty to any significant degree. The growth that has been seen in the region over the last few decades has mainly been based on the exploitation of natural resources. Some of it is linked to finite resources, such as mining activities (mainly diamonds and gold), other parts of it linked to agriculture and fisheries.

Critical questions on sustainable growth in the region relates to how the benefits of the exploitation of natural resources are used and to whether the non-finite resources are exploited in a sustainable way. The answers are clear enough. There are indications that a large portion of the proceeds from economic activities in the region is not reinvested in the region. Furthermore, It is evident that most present practices are unsustainable.

With the degradation of natural resources follows rising levels of poverty. Small-holder farmers in the LVB are forced to engage in increasingly desperate and unsustainable use of the natural resources. This includes cultivating marginal and fragile areas such as wetlands, and clearing vital forests to open up new arable land or to

[1]East African Standard Newspaper, June 17th, 2003.

get access to gathering fuel wood, leading to intensified soil erosion. The destruction of the basin's wetlands to create space for cultivation and grazing of cattle (see Fig. 13.2), has resulted in loss of biodiversity and a decreased filtering effect on the water entering the Lake via water courses.

Urbanization in the Lake Victoria basin in combination with widespread poverty has led to proliferation of informal "squatter" settlements in major towns. Such informal settlements lack garbage collection as well as sanitary facilities, leading to the prevalence of diseases such as diarrhea, malaria, typhoid and amoebas, among others. Existing sewage treatment facilities in all major towns have generally poor coverage and are in very poor shape. Raw sewage is discharged into small rivers or streams or directly into Lake Victoria, contributing significantly to pollution. Industries in the region have very poor waste treatment, if any, or are discharging their waste waters to the existing poor municipal waste water systems or directly into the streams and the Lake.

Given the above considerations, it is clear that the Lake receives an increased load of nutrients, organic material and other pollutants, which contribute to a rapidly increasing eutrophication resulting in strong vegetative growth (algae and plants), oxygen consumption and anaerobic conditions in the Lake's deep water and a change in the water quality. Water quality deteriorates and quantity and quality of fish decrease.

A dramatic illustration of how poverty, in combination with lack of enforcement of laws and institutional capacity, aggravates environmental degradation in the LVB is the revelations regarding poison fishing, the use of illegal nets and overfishing. A dramatic increase in fish export, lack of alternative employment opportunities, in particular for women, and decline in food security have served to perpetuate poverty, which in turn has enticed local fishermen to use unconventional fishing methods such as poisoning or illegal nets, with adverse health and environmental consequences. Furthermore, these practices threaten the future of the fish population and consequently the economy and well being of the communities surrounding the Lake.

The deterioration of Lake Victoria's ecology is demonstrated in the rapid spread of the water hyacinth, which over periods cover bays and vast Lake surface areas along the shores blocking access to fish landing sites as well as water intakes for water supply facilities [6]. Areas covered with the weed are perfect breeding places for different kinds of organisms, which leads to an increased health hazard from diseases such as malaria and bilharzia. The decreased water quality constitutes a great risk for the part of the population using Lake water directly as drinking water and causes extra costs for the municipalities using Lake water for tap water production.

The general health situation in the Lake Victoria region is alarming. The mortality in diseases such as cholera/diarrhoea, malaria, tuberculosis and HIV/AIDs is very high. The AIDs disease with its impact on society aggravates the region's economic problems. A study in 1994 by FAO reveals that both cash income and labour in farming households are partly diverted to cope with HIV/AIDs. This leaves less labour for agriculture and less income for the purchase of agricultural inputs and for off-farm activities. The disease also leaves many orphaned children, which put more strain on extended families, already under considerable stress. Of immediate concern, however, is the Corona virus, which is now (2020) posing increased health and economic challenges to the region.

Problems facing the Lake are compounded by the fact that the relevant municipalities surrounding it, e.g., Kampala in Uganda, Kisumu in Kenya, and Musoma in Tanzania lack the capacity to implement sustainable development policies within a regional context of high urbanization rate and weak national and regional economies [7]. UN–Habitat through its urban management programme initiated the Lake Victoria Region City Development Strategies Programme in early 2002 in a bid to strengthen the capacity of the three centers.

3.3 Management Issues Facing Lake Victoria

Lake Victoria faces several management problems. These include management issues like reducing conflicts regarding the usage of the shores of the Lake for fish breeding or for papyrus production for domestic usage; the question of allocation of the Lake's fishery benefits between the riparian populations and export to earn foreign exchange; the problem of minimizing overcapacity of fish production and avoiding economic waste; and the issue of reducing threats of existing fish catches from overfishing or activities other than fishing. There is evidence that total fish catches have increased in recent years, but these catches are concentrated on some individual species. That is, whereas some species are decreasing, other species are increasing. This situation causes uncertainty about the future trends in total catch from areas such as the Nyanza Gulf. What kind of decision should there be in this uncertain trend?

Lake Victoria fisheries, especially in the Nyanza Gulf (Kenya), are being exploited at a very high rate, thanks to the remarkable improvements in the capacities in fisheries such as an increase in the number of boats, improvement of communication network around the Lake, handling shades, provision of cold storage and other preservation methods such as smoking and sun drying of fish. It has been estimated that, on the Kenyan side of the Lake, there are about 21,000 fishermen operating over 5,000 boats. *This is the highest concentration of fishermen in the Lake. However, despite this large numbers, there is no knowledge of the magnitude of fish stocks and the maximum sustainable yields that should be exploited.* In spite of this uncertainty, there are warning signs of a declining fishery in Lake Victoria.

Several factors have been identified as the possible causes of the decline in fishes. These include the cutting of the papyrus swamps for making mats, reclamation for agriculture, dislodge by floating islands and other uses. Pollution is another factor. It emanates from industrial and domestic wastes from large cities like Kisumu in Kenya, Mwanza in Tanzania, and big towns like Jinja in Uganda. In addition, there is a growing threat from increased use of fertilizers and pesticides in the agricultural areas within the Lake basin. These are washed down the rivers by rainfall and eventually find their way into the Lake. Finally, there is the problem posed by the Nile Perch, which has devoured some other fish species in the Lake, thus drastically reducing the number of some species to near extinction. What kinds of management decisions should be taken in these conflicting factors?

The foregoing are some of the issues that management decisions have to encompass. More specifically, the management decisions must include the following considerations.

3.3.1 *Management of Lake Victoria Resources*

The region needs a lot more than fresh injections of funding. Ultimately, the people of the Lake can only solve the Lake basin's problems themselves. Uganda seems to have taken this sentiment to heart in its campaign to protect and boost the wetland ecologies that fringe the Lake and serve as a crucial natural filter.

Launched in 1989, the key to the Uganda National Wetlands Program's success has been community involvement, and specifically women, because the government recognizes that they are the guardians of water and fire in the community.[2] They know best how to manage those resources.

The program also offers a five-week course in wetland management, trains communities in making a range of products from sustainably harvested wetland plants such as papyrus, rattan cane and hyacinth, and is in the process of quantifying the economic contribution of the wetlands' natural services.

The establishment of colonial rule over the people of the region brought with it resource management structures that removed the power from traditional leaders to the central governments of larger territories of Kenya, Uganda and Tanganyika. This meant that people who had no interest in the Lake could be given the responsibility to manage resources such as fisheries. The ownership of the resources shifted with time.

The population influx and lack of properly planned infra-structural development has brought many problems to these cities. Such problems like sewer system, industrial pollution, shortage of water supply are a serious headache to the citizens. County governments and municipal councils alone cannot address the problems. The citizens have to join them with 'own key' solutions through organized groups such as Non-Governmental Organizations (NGOs).

The economic benefits of the Lake's resources has spread to other communities thus increasing pressure on the resources. Farming and industrial activities in the basin and the surrounding highlands has also intensified, resulting into drastic change in land use and increased pollution into the Lake. All these activities put a lot of stress on Lake Victoria's resources. The number of fish species have drastically decreased to just about 3 of commercial value at present; incidences of water borne diseases have also significantly increased; aquatic weeds began to proliferate at alarming rates. The Lake is gradually dying and could no longer support the millions of people

[2]see https://africa.wetlands.org/en/news/uganda-national-wetlands-policy-review-23-years-making-finally-kicks-off/.

who depended on it. There is therefore dire need to restore the Lake to its original supportive qualities.

The situation cannot be directly reversed to the pre-independence state. The most progressive intervention is to bring these diverse cultures together to better manage and utilize the Lake's resources. The communities have therefore formed East African Communities Organization for the Management of Lake Victoria (EcoVic) and Lake Victoria Regional Local Authorities Cooperation (LVRLAC) [2]. The two organization have formalized the operational plan to work together to improve the environmental situation of Lake Victoria.

3.3.2 Ownership of Lake Victoria: Who Owns the Lake?

If you ask fishermen on the Kenya side of the Lake, 'Who owns the fish?' they will tell you that it belongs to the government. Such attitudes indicate the extent to which many feel that they have lost out or lack a stake in managing their own resources. It doesn't help either, that with the many disasters that have stalked the region in the recent past, armies of aid agencies and NGOs have also come, doing little more than taking the place of *Mama na Baba*, the government.

All of these opportunities, bright spots flickering on the horizon, are fragile, easily extinguished. And they won't amount to much overall if the fundamental problem of the LVB is not addressed: resource ownership. It is not for lack of laws - whether old or new—*that fishermen poison fish, that industries flush their untreated waste down rivers, that municipal councils endanger the lives of their citizens by emptying raw sewage directly into the Lake*. These acts of irresponsibility are nurtured in an environment where entire communities have been disenfranchised and the *extraction* and *exportation* of wealth have become the dominant trends. Until these basic issues are addressed, the region's economic potential will be something we'll be talking about forever.

3.3.3 Jurisdiction and the Political Environment

In terms of the surface area, Kenya, Tanzania and Uganda, now partner states in the East African Community (EAC), respectively have control over 6, 49 and 45% of the total Lake area.

Overwhelmingly, the politics of management and ownership of the Lake fall into the larger context of the establishment of the East African Community.[3] Within the Community, two institutions on Lake Victoria have been established. These are the Lake Victoria Fisheries Organization, which is specific to fisheries, and the

[3]The Treaty for the Establishment of the East African Community. Arusha, Tanzania.

Lake Victoria Development Program (covering general development matters of the Basin) [1, 2].

The EAC Partner States recognize three important and convergent issues relating to management of the shared waters. These are; *firstly*, that they share an interest in the well being of the Lake and its living resources and in the rational management and sustainability of these resources. *Secondly*, they recognize the need to develop Lake Victoria region as an Economic Growth Zone. *Thirdly*, they agree that management decisions relating to any portion of the Lake, within the territorial limits of any one of the Partner States, will affect the others, and hence there is the concomitant necessity that management decisions take such into account.

Fragmentation of the Lake management institutions in Kenya is a big problem. Kenya does not have appropriate institutional mechanisms for integrated management of Lake Victoria and its basin. The management responsibility of the Lake cuts across jurisdictional, administrative and national borders, making it difficult to establish a sound management framework. Moreover, active stakeholder involvement has been lacking in the establishment of many institutions responsible for the management of the Lake and its basin.

Managing the Lake Victoria and its basin should be guided by a common long-term vision. For the Lake to continue to provide benefits into the future, a comprehensive management approach involving all stakeholders and covering all activities affecting the water resources throughout the watershed is required. To work effectively, management plans must be developed at the community level, involve the participation of all groups who benefit directly and indirectly from the Lake, and have clear and transparent procedures for resolving conflicts.

3.3.4 Development Challenges

After several decades of marginal economic growth, increasing poverty amidst escalating environmental degradation, the East African countries are confronted with a series of critical transitions to attain sustainable socio-economic growth. These transitions include:

- Demographic transition toward an optimal population size, structure and distribution in relation to the environment and natural resources.
- Social transition toward a more equitable sharing of development opportunities and benefits with priority to the poor majority.
- Gender transition toward expanded rights and participation of the vulnerable, particularly women in the development process.
- Economic transition toward equity-led growth with priority to the poor and to protecting the environment and natural resources needed for future development.
- Agricultural transition toward sustainable land use and land management options to enhance food security, improved livelihoods and conservation of land water and vegetation resources.

- Energy transition toward efficient and less polluting sources of energy. Priority should be to the development of renewable sources and affordable alternatives to fuel wood for the poor majority.
- Technological transition toward accelerated industrial development with priority to technologies that produce less waste and are more energy and resource efficient.
- Institutional transition toward effective national and regional institutional arrangements with priority to integrating economic, equity and environmental imperatives in planning and decision-making within and among different ministries and countries.
- Governance transition toward greater public accountability and participation with priority to new partnerships among governments, industries and NGOs.
- Capacity building transition toward national and regional self-reliance with priority to accelerated development and use of local know-how, technology and expertise.
- Development budget transition from aid dependence to self-reliance. Peace and security transition toward a new era of regional co-operation and integration with priority to the peaceful settlement of disputes.

3.4 The River Nile Treaties

In the previous sections, we noted the issues concerning Lake Victoria and its environs. For a complete and comprehensive understanding of the Lake, it is important to address the question of the Nile Treaties. These treaties were signed by Egypt and other interested parties to safeguard the interest of Egypt. These treaties controversies have arisen in recent years, with other riparian states demanding the abrogation of the treaties. This chapter presents an overview of the Nile Treaties and their implications for the economic development of the riparian states upstream.

3.4.1 The Origins of River Nile

Most history books credit the "discovery" of the source of the Nile to the 19th century explorer John Speke (see Fig. 2.1) who captivated the Western world with this news in 1862.[4] This claim, however, no longer holds. In fact, a little-known German explorer, Bruckhart Waldekker, proved Speke wrong.[5] It is now accepted that the Nile originates from two distinct geographical zones—the basin of the "White Nile" and the "Blue Nile". The White Nile originates from the Great Lakes Region and is fed by the Bahr-el -Jebel water system to the North and East of the Nile Congo rivers divides. The Blue Nile originates in the Highlands of Ethiopia and Eritrea, so do the other major tributaries of the Nile, Atbara and the Sobat. Recent research has

[4]Britannica Concise.

[5]John Mbaria, "Revoke Obsolete River Nile Treaty" in the Daily Nation, Thursday, March 29, 2002.

indicated that the Ethiopian Highlands contribute about 85% of the total volume. Lake Victoria, though, is a major source of River Nile. The Lake contributes less than 15% of Nile water. About 84% of the water in the Lake comes from the seven rivers in Kenya, even though Kenya owns only 6% of the Lake.

3.4.2 Defects of Past Treaties: Legal Implications

Most of these consist only of an article in the treaties and agreements about colonial boundaries and economic territories. In chronological order [5]:

1. The Anglo Italian protocol signed on 15th April 1891.
2. The treaty between Britain and Ethiopia of 15th May 1902.
3. The agreement between Britain and the government of the independent state of the Congo signed on 9th of May 1906.
4. The 1901 agreement between Britain and Italy over the use of the River Gash.
5. The Tripartite (Britain–France–Italy) Treaty of December 13, 1906.
6. The 1925 exchange of notes between Britain and Italy concerning Lake Tanner.
7. The agreement between Egypt and Anglo Egyptian Sudan dated 7th May 1929.
8. The 1959 Nile Waters Agreement (between Egypt and Sudan).

Let us closely examine these treaties.

The Anglo-Italian Protocol of April 15, 1891: Only Article III of this treaty refers to the Nile water. The remaining articles define the colonial territorial claims of Great Britain and Italy in East Africa. Article III states that the Italian government engages not to construct on the Atbara River, in view of irrigation, any work, which might sensibly modify its flow into the Nile. This article of the treaty is curious in light of the fact that neither this river flowed in the territory claimed by Italy nor was Italy colonizing a country near the Atbara River, in order to have a claim over the river. However, the reference to the Atbara River on the part of Britain made some sense since the Sudan and Egypt, through which the Atbara flows, were within the British colonial territory. The reason for Italy to sign such an agreement foregoing its irrigation development without receiving any benefit in return is unclear. Moreover, for Great Britain to be interested in including this reference in a treaty with a country at a distance of some thousands of kilometers from the Atbara River makes the essence of the agreement more irrelevant.

It appears that the intent of the treaty was not the use of the Nile water, but to establish a colonial boundary. Given this context, the treaty cannot be seen as an agreement over property rights to the river. Even if, say, the treaty is assumed to define use rights to the river, what was meant by the term "sensibly modify the flow into the Nile"? The volume of Atbara water used upstream to be considered, as a sensible modification by downstream users of Britain colonies is undefined. The language used is too vague to provide the parties with clear property rights and guarantees for water use. It is not unreasonable, therefore, for the remaining riparian states to see no reason for this treaty to provide a historical base for binding present and future cooperation on Nile water use.

The Treaty between Great Britain and Ethiopia of May 1902: The aim of this treaty was to establish the border between Ethiopia and the Sudan. One of its articles, Number III, relates to the use of Nile water. The English version, as reviewed by Britain and later by the Sudan, reads as follows: "*His Majesty the Emperor Menilik II, King of Kings of Ethiopia, engages himself towards the Government of His Britannic Majesty not to construct or allow to be constructed any work across the Blue Nile, Lake Tana, or the Sobat, which would arrest the flow of their waters except in agreement with His Britannic Majesty's Government and the Government of Sudan*" The Amharic version, however, gave a different meaning and understanding to Ethiopia and "was never ratified by this country". The treaty was understood by Ethiopia as follows: "*…Securing and maintaining the prior agreement of Britain before construction of any work on the Nile tributaries; not to stop (arrest) the flow of the Nile rivers did not mean not to use; and, that the treaty was made between Britain (colonizer of the Sudan) but not with the Sudan as it was under the colonial power of Britannia. As Britain is no longer ruling the Sudan, this agreement does not hold at present.*"

The 1902 agreement has been the most controversial treaty in the history of Nile agreements as both parties claimed that their own understanding of the treaty was correct. And, not only has the claim remained controversial but it has also been the cause of disputes, which threaten the socio-political and economic dynamics of the basin environment and efforts toward future cooperation. This is because, first, referring to this agreement, Sudan has argued that Ethiopia should not use the Nile water without the permission of the Sudan. Second, the Sudanese claim has been supported by Egypt with the possibility of military retaliation if Ethiopia used the Nile water. Third, there is the possibility that this threat has played some role in Ethiopia's extremely poor record in food production, though poor agricultural policies have no doubt played the major role.

This controversy remained a threat to present and future cooperation over the Nile waters. By forbidding the Nile water from being arrested by Ethiopia, the treaty inadvertently advocated the principle of sustainable development. Whether or not its intent was out of concern over the impact of development of Nile waters on the environment is subject of much debate. Stopping a flow of a river creates an artificial Lake that destroys ecological systems and often results in the relocation of people. However, it is unrealistic to believe that this treaty was predictive enough to anticipate the impact of water impoundment on biophysical and social systems in the basin, looking at its objectives and the time it was constituted.

The Agreement between Britain and the Government of the Independent State of the Congo on 9th May 1906: This was an agreement on the colonial boundary of the Congo between Britain and Belgium. The Congo being called an 'independent state' when the treaty was signed by the Government of Belgium on behalf of this country was hypocritical. Article III of the agreement was about the Nile waters and it stated that: "*The Government of the independent state of the Congo undertakes not to construct, or allow to be constructed, any work over or near the*

Semliki or Isango River which would diminish the volume of water entering Lake Albert except in agreement with the Sudanese Government".

Belgium signed this unfair agreement on behalf of the Congo, despite the agreement entirely favoring the downstream users of the Nile waters and restricting the people of the Congo foregoing its Nile water use. The agreement did not require downstream users to consult the upstream countries for anything that they might do to the Nile waters. In this treaty, it is difficult to find any incentive for riparian states that might enhance their future cooperation since it involved neither the principle of equitable water use nor the approach of integrated water development.

The 1901 Agreement between Britain and Italy over the Use of the River Gash: The agreement states that "*The Government of Erythraea, while recognizing all its rights on the waters of the Gash and having regard to the requirements of the Colony, sees no difficulty in declaring that, in so far as the regime of the waters of that river are concerned, it will regulate its conduct in accordance with the principles of good neighbourship*". Evidence is scarce, however, on whether or not this treaty bound the parties to the agreement. Nevertheless, of all the treaties and agreements made during the colonial period, it could be said that this agreement was the most equitable. Because of difficulty in ensuring equitable water use, the agreement was defined and reinforced later by the "the Anglo-Egyptian Exchange of Notes" with subsequent detailed arrangements of 1925.

The exchange of notes included technical provisions suitable for practical implementation as follows: *"Quantified allocation, to each party, of water from the river Gash, Flow regime terms and conditions for water allocation, and the Amount of annual payment by the Sudan to Eritrea as a proportion of Sudanese revenues from irrigated cultivation at Kassala"*. This treaty is held by some of the riparian states as not being binding because the colonial signatory governments are no longer present in the Nile basin. Nevertheless, it can be used as a basis for the effort being undertaken to establish cooperation among the riparian countries.

The Tripartite (Britain–France–Italy) Treaty of December 13, 1906: Article 4 (a) of this treaty dealt with the use of the Nile water in Ethiopia's sub-basin. It states: "To act together …to safeguard; …the interests of Great Britain and Egypt in the Nile Basin, more especially as regards the *regulation of the waters of that river and its tributaries (due to consideration being paid to local interests) without prejudice to Italian interests*". This treaty denied "**the absolute sovereignty**" of Ethiopia over its water resource. It resulted in Ethiopia immediately notifying its rejection of the agreement by indicating that no country had the right to stop it from using its own water resources. Neither Ethiopia's military power nor its international political and economic influence was strong enough to protect Ethiopia's sovereign rights over its water resource. Ethiopia's rejection of this agreement was a revision, if not retraction, of the May 15, 1902 treaty signed between Ethiopia and Britain.

The 1925 Exchange of Notes between Britain and Italy Concerning Lake Tana: Britain and Italy had signed an agreement in 1919 over Lake Tana, of Ethiopia, which read in part as follows: "In view of the predominating interests of Great Britain in respect of the control of the waters of Lake Tana, Italy offers Great Britain her support, in order that she may obtain from Ethiopia the concession to *carry out works of barrage in the Lake itself …*".

In 1925 it was expanded to say that:

> *…Italy recognizes the prior hydraulic rights of Egypt and the Sudan …not to construct on the head waters of the Blue Nile and the White Nile (the Sobat) and their tributaries and influence any work which might sensibly modify their flow into the main river*".

Ethiopia opposed this agreement and notified the parties of its objections as follows: To the Italian government, Ethiopia said:

> *The fact that you have come to an agreement, and the fact that you have thought it necessary to give us a joint notification of that agreement, make it clear that your intention is to exert pressure, and this in our view, at once raises a previous question. This question, which calls for preliminary examination, must therefore be laid before the League of Nations.*

And, to the Britannia government Ethiopia said:

> *The British Government has already entered into negotiations with the Ethiopian Government in regard to its proposal, and we had imagined that, whether that proposal was carried into effect or not, the negotiations would have been concluded with us; we would never have suspected that the British Government would come to an agreement with another Government regarding our Lake.*

When an explanation was required from the British and the Italian governments by the League of Nations, they denied challenging Ethiopia's sovereignty over Lake Tana. This notwithstanding, however, there was no explicit mechanism enforcing the agreement. A reliable and self-enforcing mechanism that can protect the property rights of each stakeholder is essential if the principle of economically and ecologically sustainable international water development is to be applied".

The Agreement between Egypt and Anglo-Egyptian Sudan of 7th May 1929: This agreement included the following terms:

(a) Egypt and Sudan utilize 4 billion cubic meters of the Nile flow per year, respectively;
(b) The flow of the Nile during January 20 to July 15 (dry season) would be reserved for Egypt;
(c) Egypt reserves the right to monitor the Nile flow in the upstream countries;
(e) Egypt assumed the right to undertake Nile river related projects without the consent of upper riparian states.
(f) Egypt assumed the right to veto any construction projects that would affect her interests adversely.

It is important to note that Egypt was still under British influence in 1929 and neither Sudan nor the remaining riparian states, aside from Ethiopia, were independent. The roles of both referee and player were taken, in the process of this agreement, by Britain in the name of its colonial territories in order to favor one, Egypt, by limiting the rights of the Sudan, and by rejecting those of the remaining riparian states. The agreement became the basis for the next agreement, called *The 1959 Nile Water Agreement*, which opened a door for Egypt and the Sudan to acquire rights to Nile water resources and for the full utilization of these waters by developing the Aswan high Dam, with its huge impact on the biophysical and social (disposition of human settlement) environment of the basin.

In May 1929, the Egyptian Prime Minister and the British High Commissioner to Egypt, signed the Nile River Treaty. Although the 1929 agreement was concluded between the British High Commission in Cairo and the Egyptian Government, while the 1959 treaty was between Egypt and the Sudan governments, *the pacts bind Uganda, Tanzania, and Kenya to date. They bar the three countries from using Lake Victoria waters without Egypt's permission*. The 1929 treaty was signed at a time when Lake Victoria was thought to be the source of river Nile. The treaty was a culmination of previous agreements made in 1889, 1891 and 1902—all which merely satiated British selfish interests against the Italian and (later) the Ethiopian Government. All ended up securing, for the 96% Egyptians crowded along the Nile, the use of 48 billion cubic meters of the water each year while Sudan got 4 billion cubic meters. These agreements were not sensitive to the water needs of the other eight countries in the basin (Tanzania, Uganda, Kenya, Rwanda, Burundi, Ethiopia, and the Democratic Republic of Congo).

The treaty acknowledged Egypt's natural and historical rights to the Nile waters. "Without the consent of the Egyptian Government, no irrigation or hydro-electric works can be established on the tributaries of the Nile or their Lakes, if such works can cause a drop in water level harmful to Egypt", so says a section of the treaty.

The original agreement concluded in 1929 heavily favored Egypt's "historic rights" and gave Egypt and Sudan 100% use and control of the waters from Lake Victoria. Even in the event of floods, other riparian states had no permission to use this water. Thus, in 1949, Owen Falls Treaty prohibited Uganda from using Lake Victoria water for irrigation in Karamoja.

The Nile Treaty of 1959: This was an agreement between the Sudan and Egypt for full utilization of Nile waters. In the 1950s, Egypt was planning the Aswan High Dam project to collect the entire annual flow of the Nile water. The objective of the 1959 Agreement was to gain full control and utilization of the annual Nile flow. One of the financiers of the project, the International Bank for Reconstruction and Development (IBRD) required a secure water allocation for Sudan and compensation for the population to be dislocated due to the project. In 1956 Sudan had become an independent country and wanted previous agreements, which it saw as being unfair, to be changed, so that it could pursue another agreement on the use of Nile water with Egypt. At the beginning of the talks, both Sudan and Egypt claimed large areas of irrigable land and amounts of Nile water: Sudan claimed 44 billion cubic meters of Nile water to irrigate 2.22 MH, while Egypt claimed even more water than Sudan that irrigates 7.1 MH. The debate over the claims delayed the agreement, but whether or not Sudan agreed, the construction of the Aswan High Dam was seen as a development priority for Egypt. Although neither the Sudan nor Egypt were contributors to the Nile water but only users, the agreement for the full utilization of the Nile Waters was signed in 1959, between Sudan and Egypt. The agreement contained the following main points:

1. The controversy on the quantity of average annual Nile flow was settled and agreed to be about 84 billion cubic meters measured at Aswan High Dam, in Egypt.
2. The agreement allowed the entire average annual flow of the Nile to be shared between the Sudan and Egypt at 18.5 and 55.5 billion cubic meters, respectively.
3. Against this backdrop in the sharing of the waters of the Nile, which disregarded the contribution of the upper riparian states, changing the status quo has been an issue of hairsplitting interpretation. This is why sound water resources management is very crucial.

The general observation from the treaties discussed so far is that, countries have tried to make interpretations in their own words and contexts. It is, however, important for countries to adopt new ways that will help them forge ahead. This will be achieved through respect of the object and the purpose of treaties as established by the International Law. According to the Vienna Law of treaties, a treaty shall be interpreted in good faith in accordance with the ordinary meaning to be given to the terms of the treaty in that context and in the light of its object and purpose.

The object and the purpose of any agreement therefore must serve the interests of all the parties. In conclusion, as discussed above, none of the treaties and agreements dealing with the use of Nile waters signed during the colonial period involved all the riparian countries and they did not deal equitably with the interests of these riparian states. Also they did not take in to account the impact of water development on the basin's social and biophysical environment.

Some Important Terms of the 1959 Nile Treaty:

Egypt and Sudan signed a Treaty in Cairo on November 8, 1959. Its title was "*United Arab Republic and Sudan Agreement (with annexes) for the Full Utilization of the Nile Waters*". The agreement came into force on December 12, 1959. The 1959, agreement set Egypt's share at 55.5 bcm per year and Sudan's 18.5 bcm per year.

Agreement on a Unified View:

1. The Treaty said in part: "If it becomes necessary to hold any allegations with any riparian state, outside the boundaries of the two republics, the governments of the Sudan Republics and the United Arab Republic (Egypt) shall agree on a unified view after the subject is studied by the said technical commission. The said unified view shall be the basis of any negotiations by the commission with the said states".
2. The Sudan–Egypt treaty also notes: "If the negotiations result in an agreement to construct any works on the river, outside the boundaries of the two republics, the joint technical commission shall, after consulting the authorities in the governments of the States concerned, draw all the technical execution details and the working maintenance arrangements".
3. The treaty also gives Egypt and Sudan powers to "supervise the carrying out of technical agreements" by other riparian states.
4. According to the agreement, any other country that wishes to lay claims on the Nile waters, and that includes Lake Victoria, will only get a share, if Egypt

and Sudan agree. The treaty reads: "As the riparian states, other than the two republics, claim a share in the Nile Waters, the two republics have agreed that they shall jointly consider and reach one unified view regarding the said claims".[6]

Deductions in Equal Parts:

1. The Treaty further says: "And if the said consideration results in the acceptance of allotting an amount of Nile water to one or the other of the said states, the acceptance amount shall be deducted from the shares of the two republics in equal parts, as calculated at Aswan. Even then, any water taken out will have to be monitored by a technical commission (which) shall make the necessary arrangement with the states concerned".
2. Little is known about an annex document in which Sudan gave a "water loan" to Egypt. The annex reads: "The Republic of the Sudan agree in principle to give water loan from the Sudan's share [to Egypt] in order to enable the latter to proceed with her planned programs for agricultural expansion".

Sharp focus has been put on the Nile water agreement but it is the 1959 treaty signed by Egypt and the Sudan that has shocking clauses. According to the treaty, none of the Lake Victoria nations can use the Lake's water without supervision by a "technical commission" appointed by Egypt and Sudan. Further, they can only be allocated a certain quota by the two. In one clause, the two nations agree, "if the other riparian states claim a share in the Nile waters, the two republics shall jointly consider and reach one unified view. And…the accepted amount shall be deducted from shares of the two". The treaty authorizes a technical commission to "make the necessary arrangement with the states concerned to ensure that their water consumption shall not exceed the amounts agreed upon".

Consequences of the Nile Treaty: The implementation of the Nile Treaty means the following consequences:

According to the treaty, Egypt's approval must be obtained if any project is to be undertaken on the Nile River. Archival information shows that as early as 1949 when the Owen Dam was constructed, the British Government wrote to the Uganda Electricity Board telling it that the project would only go ahead "*If the Egyptian Government approved it*". To date, there is an Egyptian resident engineer at the dam.

A document titled "Exchange of Notes Constituting An Agreement Between the Government of Great Britain and Northern Ireland and the Government of Egypt regarding the Construction of the Owen Falls Dam, Uganda signed at Cairo on December 5, 1949' is one of the many agreements on the issue. The pack was signed by Britain and Egypt and it binds the former colonies.

Military Coup against Dissenting Foreign Governors: When President Gamal Abdel Nasser begun building the Aswan High Dam, contrary to the agreement, Sudan withdrew from the treaty. Later, there was a military coup in Sudan. The new Sudanese Government renegotiated the treaty resulting in the 1959 agreement, which increased Egypt's share to 55.5 billion cubic meters (or 82% of the annual flow), while Sudan's share was increased to 18.5 billion cubic meters (or 18%). Egypt is today accused

[6]Daily Nation newspaper, March 28, 2002.

of being behind the coup that followed the standoff and saw Mr. Jaffer Numeiri take power in Sudan. Once again, Egypt and Sudan ignored Kenya and all the other basin countries and went ahead to irrigate 6 million acres and 2.75 acres, respectively.

Threat of Usage of Force: To sustain this selfishness, the two countries have used the threat of force to intimidate other countries from utilizing the waters flowing into the Nile. For instance, in May 1978, the then Egyptian Minister of Irrigation and Land Reclamation, Mr. Abd a-Azim Abu al-Ata warned countries in the Horn of Africa that any attempt to control the source of the Nile was "an act of aggression".

Such has been the sensitivity with which Egypt treats any attempt by upstream countries to use the Nile waters for their needs. Even Kenya, which contributes two-thirds of the water flowing into Lake Victoria, is not spared from this threat. "The day that Kenya decides to use water from Lake Victoria, we'll have less water in Egypt. One litre of water used for their irrigation will be reduced from water received in Cairo", a senior official in Egypt was quoted by the Financial Times of London in January 1988. The then Egyptian Ambassador to Kenya, Dr Rifaat al-Ansary reacted that his country had nothing to do with this and there was no controversy. But as the Ambassador denied any controversy, other countries were not happy with the fact that his country, and Sudan, in 2001 insisted that the treaty could only be renegotiated, if other Nile Basin countries first recognized it as binding.

Nevertheless, the agreement goes against the internationally recognized riparian states' rights of Kenya and the other eight Nile Basin countries. This law states that a riparian state has sovereignty over the stretch of any river within its international boundaries and may use such water as it deems fit so long as it does not infringe upon the rights of other riparian states on the same river. It could be with this law in mind that Tanzania disregarded the treaty in 1962 while Uganda constructed a second hydro-electric power generation dam on the Bujagari Falls on the Nile. Kenya seems to have settled for the renegotiation option.

The Near War with Ethiopia over the Use of Nile Water: In June 1980, Egypt nearly went to war with Ethiopia after the latter opposed moves by Egyptian President Anwar Sadat to divert the Nile waters to the Sinai Desert, which is outside the Nile Basin. After making peace with Israel, Mr Sadat had promised Israelis that he would irrigate the desert. On its part, Ethiopia had threatened to obstruct the Blue Nile, a notion that made Egypt prepare for war. Egypt has also placed its hydrologists along the 5,584 km-long course of the Nile ostensibly to monitor its volume. Nonetheless, Ethiopia is in the process of completing the biggest dam in the continent; the Grand Ethiopian Renaissance Dam,[7] which is currently (2020) causing tension between it and Egypt.[8]

The eight countries need to disregard the treaty as already done by Uganda and Ethiopia if they are to meet their rising water needs. Today, their combined population stands at 202 million against 98 million Sudanese and Egyptians. Though a substantial section of the eight countries' population may have alternative sources of water, this has not been adequate for the rising population.

[7] https://www.internationalrivers.org/campaigns/grand-ethiopian-renaissance-dam.

[8] https://www.bbc.com/news/world-africa-50328647.

The Riparian States and Egyptian Concept on use of River Nile: Meanwhile countries sharing Lake Victoria cannot use its water for development projects. Egypt and Sudan must agree that Kenya and indeed the other basin countries have as much right over the use of the waters as they do. While there is merit in the Tanzanian President's statement in 1998 that the Nile resource should be "a formidable unifying factor instead of a source of division and conflict", the eight countries should handle the controversy with more boldness. Collectively, they should disregard the treaty so that Egypt and Sudan can approach the issue in a spirit of give-and-take. The eight Basin countries should not continue honoring a treaty that neglects others' interests.

Demands to Revoke the River Nile Treaty: Sudan was the first country to demand a review of the 1929 treaty. The action brought a brief standoff between Khartoum and Cairo after Sudan got independence from Britain in 1956. In April 2002, there was a heated debate in the Kenyan Parliament about the Nile Treaties signed in 1929 and 1959, respectively. The Members of Parliament questioned the legality of those treaties. Consequently, there was a heated debate as to whether Kenya should or should not honor the treaties. Surprisingly, the debate, just like the treaty, was pegged on the wrong assumption that Lake Victoria is actually the only source of river Nile and the MPs, during their contribution, argued from this point of ignorance.

As East African politicians question the legality of a treaty that gives Egypt and the Sudan absolute rights over river Nile waters, one issue is quickly coming to the fore: How will the two nations react given their enormous economic and agricultural interests? In Kenya, the then Energy Minister Raila Odinga and then minister for roads and public works and Kisumu East Town MP Gor Sungu questioned the legality of the 1929 and 1959 Nile waters Treaties while the same issue have cropped up at the new East African Legislative Assembly. Recently, Mr. Odinga threw the first salvo in Kenya's parliament, calling upon the government to review the treaty and denouncing it as "*obsolete*", accusing Egypt of planning to "Export" the water to Sinai via a tunnel [1]. The story of the tunnel is little known, although it is part of a huge land reclamation project in Sinai Desert called the North Sinai Agricultural Development project. Since 1987, the project has been diverting Nile water to agricultural developing plots west of Suez Canal,[9] with allegations that the water may end up in Israel.

Debate on a 1929 treaty that bound countries in the upstream of River Nile from "touching" its waters has been simmering for a while now. The treaty gave Egypt and Sudan absolute right to use 100% of the river's water. The debate re-surfaced in April 2002, when a legal consultant with the United Nations Environmental Program, Prof. Charles Okidi, implored the three East African countries to ignore the treaty and go ahead and utilize the waters of Lake Victoria. He asked East African countries to ignore the 1929 treaty.[10] Yet, unlike this treaty, which was binding on British colonies, the 1959 one is between two independent nations—Egypt and Sudan—without reference to other concerned countries. The then Energy Minister Raila

[9] see https://earthobservatory.nasa.gov/images/1495/egypts-north-sinai-agricultural-development.

[10] Daily Nation newspaper, March 29, 2002.

Odinga saying the treaty was "unfair" had earlier made a similar call. By and large, the Kenya government seems reluctant to be drawn into the controversy. It is not clear why Government officials have continued to refer to this flawed water use arrangement as an "Agreement".

For one, it was made before Kenya was a state, as we know it. "This treaty only benefits Egypt. We cannot sit back while we have water we cannot use to irrigate our land. Why should we preserve our water for Egypt"? Mr. Odinga asked Parliament? Experts predict minimal impact on the Nile if East African countries harvest Lake Victoria water for local use. Parliament was right in demanding a renegotiation of the 1929 Nile Water Agreement between Britain and Egypt over the use of Nile's waters. A treaty they did not sign in some distant past should not bind Kenyans.[11] In June 2003, members of the East African Legislative Assembly (EALA) called for a renegotiation on the Nile Treaty with a view to selling waters of the Nile to Sudan and Egypt.

Source of Acrimony: With approximately 125 million people depending on the Nile for survival, the river is seen today as possible source of acrimony in East Africa. At the Regional Regulative Assembly, Uganda member Yona Kanyomozi questioned the importance of the 1949 Owen Fall Dam Agreement and wondered why Egypt was using more water than agreed. He also wonders why Egypt does not participate in conservation of Lake Victoria. He said, "What bothers me is that when Uganda developed a scheme to divert some of the Nile waters to Karamoja for irrigation, the plan was opposed by Egypt, yet for them they can do anything with the Nile waters".[12]

How the treaty binds other riparian countries is doubtful. But since then Egypt—the most powerful military power in the region - has been warning that it could go to war over Nile waters. In 1979, Egyptian President Anwar Sadat said: "The only matter that could take Egypt to war again is water". And in 1988, then Egyptian foreign minister Boutros-Ghali, who latter became UN Secretary-General, went a step further. He said: "The next war in our region will be over waters of the Nile, not politics".[13]

Such sentiments have for the years caused Nile nations eager to renegotiate the Nile water treaties to tread with caution. They are aware that for Egypt, in particular, Nile waters are a matter of life and death. From as early as 1898, when a French expedition tried to control the headwaters of the of the Nile, Egypt has discussed the Nile with passion. In 1958, it sent a military expedition to a disputed territory with the Sudan, pending negotiations over Nile waters. This eventually led to a coup and installation of a pro-Egyptian government in the country. From 1978 onwards tensions have persisted between Egypt and Ethiopia after Ethiopia proposed to construct dams on headwaters of the Blue Nile. That was when President Sadat warned that Egypt's next war would be over the Nile. But can the region go to war over water? The only recorded incident of an outright war over water was 4,500 years ago in mod-

[11] Daily Nation newspaper, October 19, 2001.

[12] John Mbaria: Revoke obsolete Nile treaty. Daily Nation newspaper, March 29, 2002.

[13] John Mbaria: Revoke obsolete Nile treaty. Daily Nation newspaper, March 29, 2002.

ern southern Iraq between two Mesopotamian city-states over the Tigris-Euphrates. Scholars say that 80% of water's war consists of verbal threats and posturing by state leaders, probably aimed at their own internal constituents.

In the past 50 years, there have been only 37-recorded events in which people actually shot at each other over water along international borders. Of those, 27 were between Israel and Syria over the Jordan and Yarmouk rivers. Is the Nile meandering towards that, or is this just a cold water war?

The New Nile River Treaty: In September 2002, the Coordinator of the Uganda-based inter-state organization, the Nile Basin Initiative, informed a workshop meeting in Kisumu, Kenya, that a new legal framework was to be signed in two months' time by 10 riparian countries, which include Burundi, Democratic Republic of Congo, Egypt, Ethiopia, Eritrea, Kenya, Rwanda, Sudan, Tanzania and Uganda.

The Nile Basin Initiative was launched in 1999 following consultations among the 10 riparian countries. Although Egypt is today backing a new initiative among the Nile Basin riparian states, which might redistribute more equitably the river's water usage rights, it is not clear how far this will go. Note 3.1[14] below provides a historic perspective of the basins initiatives.

Note 3.1 *(A Brief History of the Nile Basin Initiative (NBI), 1992–2001)*

1992: In 1992 the Council of Ministers (Nile-Com) of Water Affairs of the Nile Basin States launched an initiative to promote cooperation and development in the Basin. Six of the riparian countries—the Democratic Republic of Congo (D.R.C), Egypt, Rwanda, Sudan, Tanzania, and Uganda; formed the Technical Cooperation Committee for the Promotion of the Development and Environmental Protection of the Nile Basin (TECCONILE). The other four riparian states participated as observers. Within this framework, the Nile River Basin Action Plan (NRBAP) was prepared with support from CIDA. One of the projects (Project 3), whose objective is to develop a co-operative framework for management of the Nile, was endorsed by all countries during the 3rd meeting of the Nile-COM (in Arusha, 9–11 February, 1995) and is being implemented with UNDP funding.

1995: The World Bank was asked by the COM to play a lead role in co-ordinating the inputs of external agencies to finance and implement the NRBAP.

March 1997: The request from COM to the World Bank was reiterated.

June 1997: The request was accepted by the World Bank. The World Bank proposed that it undertakes the task in partnership with UNDP and CIDA, and that a review and consultation process be launched prior to a Consultative Group-style donor meeting.

November 1997: A review of the Nile River Basin Action Plan was undertaken with the support of an International Advisory Group (IAG). The IAG meeting of international experts was held at Coolfont near Washington, D.C., USA.

January 1998: A Special Review Meeting was held in Cairo with senior officials attending from the riparian countries to discuss the Draft Review Report and move towards the definition of a priority, revised Action Program. Discussions converged on two complementary ideas which provide a structure for the revised Action Program—a shared vision, and action on the ground.

March 1998: The 6th COM Meeting, held in Arusha, Tanzania, was attended by eight riparian countries (all except Eritrea and D.R. Congo). It was a major milestone in Nile co-operation. The meeting considered the Revised Action Plan.

July 1998: The 1st meeting of the Nile Technical Advisory Committee (Nile-TAC) was held in Dar es Salaam, Tanzania, under the chairman ship of Mr. Meraji Msuya.

---

[14]Source: The Nile Basin Initiative Secretariat (Nile-SEC).

21–22 Sept 1998: The 2nd meeting of the Nile-TAC, held in Arusha, Tanzania. Meeting agrees on the Terms of Reference of the Nile-TAC, its Rules of Procedure, its Policy Guidelines and a Plan of Action.

23–24 Sept 1998: Extra-Ordinary meeting of the Council of Ministers (Nile-COM) was held in Arusha, Tanzania.

22nd Feb 1999: Extra-ordinary meeting of the Nile Basin Council of Ministers, was held in Dar es Salaam, United Republic of Tanzania. Agreed minutes were prepared which formally established the Nile Basin Initiative.

23–24 Feb 1999: 3rd meeting of Nile-TAC in Dar es Salaam, Tanzania.

4–7 May 1999: Nile-TAC held a Strategic Planning and training workshop in Sodere, Ethiopia, which initiated the preparation of projects within the Basin-wide Shared Vision Program.

10–14 May 1999: 4th meeting of the Nile-TAC in Addis Ababa, Ethiopia. The formation of Working Groups to develop Project Concept Documents for the Shared Vision Program.

12–13 May 1999: 7th meeting of the Council of Ministers for Water Affairs in the Nile Basin States (Nile-COM), Addis Ababa.

15th May 1999: First meeting of the Eastern Nile Council of Ministers.

1st June 1999th The Nile Basin Initiative Secretariat started operations in the former TEC-CONILE building in Entebbe, Uganda, with Mr. Meraj Msuya as Executive Director.

30th August–3rd Sept 1999: 5th meeting of the Nile-TAC in Entebbe and second meeting of the Working Groups to develop Project Concept Documents for the Shared Vision Program in preparation for the ICCON meeting.

3rd Sept. 1999: The official opening of the Nile Basin Secretariat offices in Entebbe, Uganda.

13–18 Dec. 1999: National experts from various sectors meet at the Nile Basin Secretariat in Entebbe to share ideas and work together in initiating project studies under the Shared Vision Program.

24–26 Jan. 2000: 6th Nile-TAC meeting is held at the Nile Basin Secretariat in Entebbe to agree on the work plan for the priority project proposals for the Nile Basin.

18 March 2000: Senior officials from the Nile Basin Initiative present their "Shared Vision" for the time to the international community during the Second World Water World Forum in the Hague, the Netherlands, attended by over 4,000 participants from all over the world.

23rd March 2000: 7th Nile-TAC meeting is held in Deft, the Netherlands to review alternative work plans and meeting schedules towards the first ICCON.

31st July–3rd August 2000: 8th Nile-TAC meeting is held in Khartoum, Sudan, as a precursor to the 8th Nile-COM meeting, to finalize proposals for the priority projects under preparation.

4–5 Aug. 2000: 8th Nile-COM meeting is held in Khartoum. It endorses the priority projects being prepared under the Shared Vision Program and instructs the Nile-TAC to complete preparation of full project documents to be submitted to the Nile-COM in the first week of December 2000. The Council confirms February 2001 and Geneva, Switzerland as the time and venue of the first ICCON.

21–25 Aug. 2000: National experts from the Nile Equatorial Lakes region meet in Entebbe to share ideas on a series of possible joint projects for the region, to be prepared under the Nile Equatorial Lakes Subsidiary Action Program (NELSAP).

20–29 January 2001: National experts from the Eastern Nile meet in Addis Ababa to share ideas about joint projects being prepared under the Easter Nile Subsidiary Action Program (ENSAP).

6th September 2002: Mr. Antoine Sendama, Co-ordinator of Nile Basin Initiative, announces that a new legal framework—a revision of the Nile Treaty—is to be signed by 10 riparian countries (nations neigbouring water masses). The countries are Burundi, Congo, Egypt, Ethiopia, Eritrea, Kenya, Rwanda, Sudan, Tanzania and Uganda.

Fig. 3.1 The disputed Migingo Island in Lake Victoria (right). GNSS receivers were used to establish that the island belongs to Kenya, thus resolving a territorial dispute between Kenya and Uganda *Source* Daily Nation, Kenya

3.5 Inter-State Conflicts

Fishermen have frequently found themselves in situations where they trespass into a neighbouring State's territory with the end results often being conflict, leading to arrests and the confiscation of boats and fishing equipment. A real-case scenario is illustrated by Migingo Island in Fig. 3.1,[15] which is an island currently disputed between Kenya and Uganda due to it being home to the dwindling Nile Perch (*Lates niloticus*) fish. Owing to uncertainty about the boundary, Global Navigation Satellite System (GNSS [3, 4]) was used by a team of surveyors from both countries to mark the boundary and establish that the disputed island belongs to Kenya.

3.6 Concluding Remarks

Two main challenges face the lake and its basin; (i) sustainability issue, which entails utilization of the resources with an eye on the future, and in so doing, balancing the social, economic and environmental pillars [3, 4], and (ii), the question of adhering to obsolete treaties that prevent the full exploitations of the resources in a sustainable way. This Chapter has expounded on these issues.

References

1. Aseto O, Ong'ang'a O (2003) Lake Victoria (Kenya) and its environs: resource opportunites and challenges. Africa Herald Publishing House, Kendu Bay, Kenya
2. Awange JL, Ong'ang'a O (2006) Lake Victoria: ecology resource and 571 environment. Springer, Berlin
3. Awange JL (2012) Environmental monitoring using GNSS. Global navigation satellite system. Springer, Berlin, New York
4. Awange JL (2018) GNSS environmental sensing. Revolutionizing environmental monitoring. Springer, Berlin, New York

[15]Source: http://www.nation.co.ke.

5. Bullock A, et al (1995) Report on LVEMP tasks 11, 16 and 17. FAO, Rome
6. Ongore C, Aura C, Ogari Z, Njiru J, Nyamweya C (2018) Spatial-temporal dynamics of water hyacinth, Eichhornia crassipes (Mart.), other macrophytes and their impact on fisheries in Lake Victoria, Kenya. J Great Lakes Res 44:1273–1280. https://doi.org/10.1016/j.jglr.2018.10.001
7. UN–Habitat (2005) Cities development strategies for improved urban environment and poverty reduction in Lake Victoria region. Kampala, Kisumu and Musoma. A document of the process, achievements and lessons. Urban management programme—working paper series

Chapter 4
Lake Level: Dam Operations Versus Droughts

"The secrecy of hydrologic and dam operations data for Lake Victoria and the Victoria Nile is worrying."–D. Kull [1]

4.1 The Water Level Drop: A Summary

The period 2002–2006 saw a sharp decrease in Lake Victoria's water level by more than 1.1 m below the 10-year average. In 2004 it was claimed that the dams at Owen Falls (Nalubaale and Kiira) were responsible for a portion of the lake's drop, while others insisted that the drop was due only to the droughts at that time.

A study by D. Kull in 2005 sought to determine what factors contributed to what extent to the 2002–2006 drops in Lake Victoria. The data had for some time been kept out of the public eye, but released reports as well as on-line resources allowed for rough analyses of the situation. At the same time, the implications for the designs and benefits of the Owen Falls and Bujagali dams of the 2002–2006 level and outflow changes of Lake Victoria, as well as past observed hydrology, was analysed.

The major conclusions of the study were:

1. Severe drops in Lake Victoria (2004–2005) were approximately 45% due to drought, and 55% due to over-release from the Owen Falls dams (Nalubaale and Kiira).
2. The Owen Falls dams had not been adhering to the Agreed Curve for operations, releasing more water than dictated.

[1] This is an invited Chapter from Mr. Daniel Kull, currently a Senior Disaster Risk Management Specialist with the World Bank Group. This article was written in 2005 while Mr. Kull was an independent specialist, originally published online by the International Rivers Network (see, e.g., https://www.internationalrivers.org/sites/default/files/attached-files/full_report_pdf.pdf).

J. Awange, *Lake Victoria Monitored from Space*,
https://doi.org/10.1007/978-3-030-60551-3_4

3. Based on the Lake Victoria hydrology, as well as observations from the past 100+ years, the Owen Falls dams are likely over-dimensioned.
4. The current hydrology, long-term observations and non-adherence to the Agreed Curve for Owen Falls dam operations must be considered in the cost-benefit analysis of the proposed Bujagali dam.
5. The lack of public information on dam releases, dam operations and river flows is disturbing and makes it difficult for outsiders to soundly judge implemented and proposed hydroelectric projects on the Victoria Nile.

Future climates [13], which will likely involve "drier conditions, lower lake levels and lower downstream river flows" [22], will exacerbate conclusions 3 and 4, making it increasingly more difficult for Victoria Nile dams to produce their projected power, and thus challenge hydropower on the Victoria Nile as a viable energy alternative for Uganda.

4.2 The 2002–2006 Severe Drop in Lake Victoria Level

Since late 2003, Lake Victoria's water level dropped by over 1.1 m from its 10-year average (Fig. 4.1). As of December 27, 2005, it was approximately 10.69 m, reaching the lowest level since 1951 [25].

It should be noted that all Lake Victoria water levels in this chapter are given with reference to the Jinja gauge. The cited water levels are thus not elevations. The 0 m, or datum, of the Jinja gauge is 1122.86 m above mean sea level, such that the actual

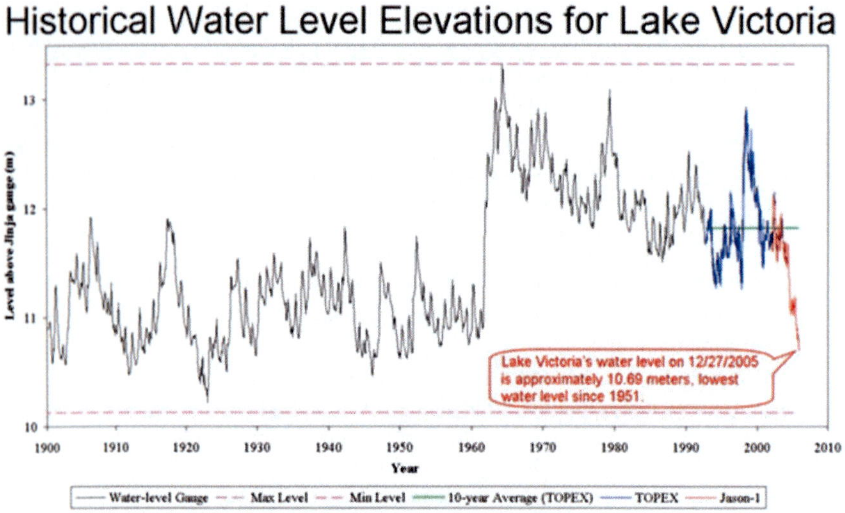

Fig. 4.1 Historical water levels of Lake Victoria *Source* [25]

elevation above mean sea level of the lake can be computed by adding 1122.86 to the Jinja gauge value [23]. For instance, a lake level of 11 m is equal to $11 + 1122.86 = 1133.86$ m above sea level.

4.2.1 Nalubaale Dam: Turning Lake Victoria into a Reservoir

Since 1959, the outflow of Lake Victoria, the second largest freshwater lake in the world has been under human control, through the Nalubaale dam (originally called Owen Falls Dam), located at Jinja, Uganda. The construction of this hydropower dam effectively transformed Lake Victoria from a natural lake to a reservoir, controlling the lake's outflow to the Victoria Nile (which eventually becomes the White Nile) [27].

Originally, the outflow of Lake Victoria, while driven by inflow from tributaries, rainfall on the lake, and evaporation from the lake, was controlled "hydraulically" by Ripon Falls. Ripon Falls acted as a natural weir and constriction, allowing a certain flow of water to exit the lake depending on the level of water in the lake. The Nalubaale Dam submerged Ripon Falls, which were also excavated in preparation for the Dam, thus assuming hydraulic control over the lake.

4.2.2 Agreed Curve Mimics Natural Flows

An "Agreed Curve" (based on agreements in 1949, 1953 and again in 1991 between Uganda and Egypt) was developed for the operation of Nalubaale Dam to dictate how much water should be released from Lake Victoria, based on the water level in the lake, shown in Fig. 4.2. This operating rule was developed in a way to retain the original (natural) pre-Nalubaale Dam relationship between lake level and outflow. Dam operators adjust the outflow based on a water balance of the lake computed every ten days [21].

By developing such an Agreed Curve, the Nalubaale dam's operators acknowledged the importance of keeping Lake Victoria levels in sync with hydrologic developments and natural conditions. Allowing the lake level to dictate the outflow meant that fluctuations in rainfall and evaporation determined how much water would flow out of the lake, as would occur in a natural state. In effect, hydropower demands took a back seat to the requirement of mimicking natural conditions.

Figure 4.3 shows observed water levels in Lake Victoria since 1800. It can be seen that from the late 1800s to about 1960, lake levels averaged between 11 and 12 m, indicating on the Agreed Curve in Fig. 4.2 an average outflow range of 600–1100 m^3/s. However since about 1961, water levels and thus outflow has maintained a higher level than before.

The Agreed Curve is not considered to be a significant factor in determining lake level [21], meaning that by mimicking the natural system, when the dams are

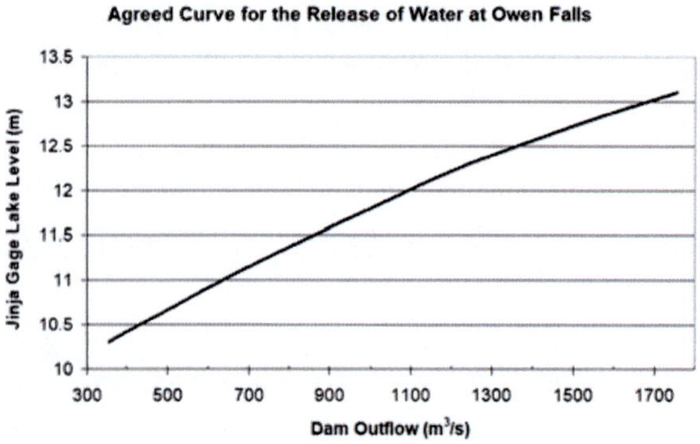

Fig. 4.2 The "Agreed Curve" dictating how the Owen Falls dams are operated

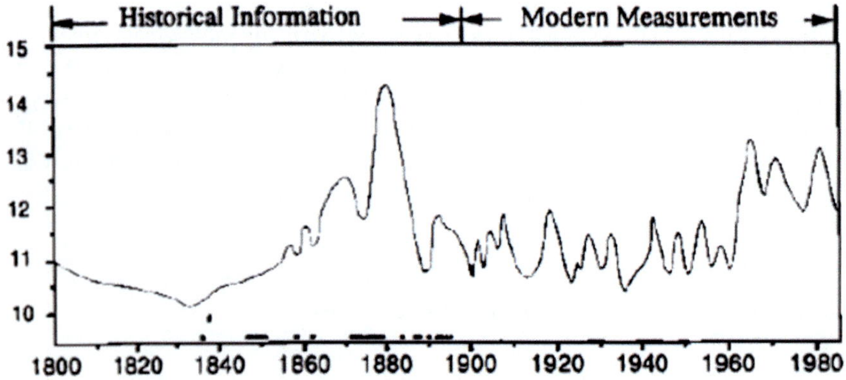

Fig. 4.3 Historical water levels of Lake Victoria from [9]

operated according to the Agreed Curve, they allow nature to "run its course". In this way, lake inputs (direct rainfall and tributary flows) and outputs (evaporation and "natural" outflow) determine the lake level, as they would have in the natural state without the dams. This implies, since the historical data used to develop the Agreed Curve included previous droughts, dam operations according to the Agreed Curve would not lead to unnatural extreme drops in lake levels.

4.2.3 Kiira Dam: Extending the Owen Falls Hydropower

Not all the water being released by Nalubaale was being utilised for hydropower production. Water release management at Nalubaale, with the aim of adhering to the Agreed Curve, involved balancing turbine and sluice gate flows, with only the sluices providing direct control. Some of the water flowed through the dam's turbines, while some flowed through the sluice gates down a spillway. By regulating spillway flow, the sluice gates also determined how much water passed through the turbines.

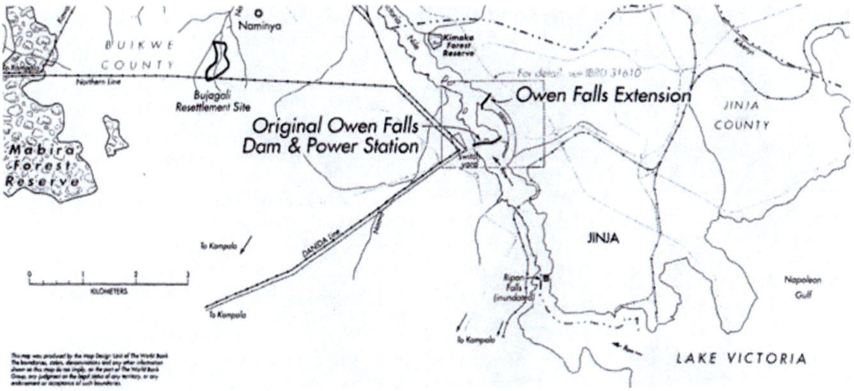

Fig. 4.4 Location of Nalubaale and Kiira dams (from [18])

The need for more power in Uganda resulted in a second dam being added one kilometer from the existing Nalubaale dam [5]. The Owen Falls extension, called Kiira, was built to utilise the "excess" water being spilled by the sluices of Nalubaale, thus generating more electricity. Work started on the Kiira project in 1993 and major construction was completed in 1999. In July 2000 the first 2 turbine and generator units (officially "Units 11 and 12", continuing from the original 10 turbines at Nalubaale), were commissioned. The 3rd Kiira turbine, Unit 13, was commissioned in 2002. Design and project management was by Acres International (now Hatch Acres) of Canada.

The turbines of Kiira Dam are a few meters lower than those of Nalubaale, therefore utilizing the same "head" (water drop) from Lake Victoria plus some additional head, resulting in increased relative energy capacity. A 1.3 km canal above Nalubaale diverts water to Kiira in such a way that the two dams now in combination control the Lake Victoria water level and outflow. Figure 4.4 shows a map of the two dams.

With the opening of Kiira's turbines, the aforementioned balancing of turbine and sluice gate flows at Nalubaale to adhere to the Agreed Curve became more difficult, if not impossible, as additional water was released through the new dam [15].

4.3 2002–2006 Severe Drops in Lake Victoria: The Cause

In order to estimate what the impacts of both drought and dam operations on Lake Victoria's water levels were, the annual water balance[2] of the lake was analysed under different scenarios. Two average annual water balances, provided in [17] and [26], were averaged in order to obtain a "consensus" water balance.

A non-drought year was analysed along with mild, strong and severe drought year scenarios. Table 4.1 shows the assumptions used to develop the drought scenar-

[2]A water balance represents the relationship between the lake's input (direct rainfall and tributary flows), output (evaporation and outflow), and the resultant change in water stored in the lake and thus change in lake level for a given period of time.

Table 4.1 Drought scenarios used for lake water balance analysis

	No drought (%)	Mild drought (%)	Strong drought (%)	Severe drought (%)
Direct rain & Tributary flows	100	90	80	70
Evaporation	100	105	110	115

ios, with 100% values equalling average annual direct rainfall, tributary flows and evaporation.

A reduction by 10% of annual rainfall and tributary flow may not appear to be much of a drought. It must be considered, however, that if all of the missing rainfall was concentrated during the March–May wet season, it would represent a loss of 25% of the rains for those 3 months. If it was concentrated in the December–February dry season, it would represent a 55% rainfall deficiency for the season. The additional assumption of increased evaporation, based on the premise that during a drought there are less clouds and thus higher radiation and evaporation, further strengthens the drought scenarios. As a major component of the water balance, a 5% increase in annual evaporation is equivalent to a 3.6% decrease in annual lake input (rainfall and tributary flow).

The water balance and resultant changes in lake level were computed under five assumptions of Nalubaale and Kiira combined dam operations: according to the Agreed Curve and average annual releases of 700, 900, 1100 and 1300 m^3/s. It was assumed that the lake level at the beginning of the year was equal to the recent 10-year average. The results are show in Fig. 4.5.

It can be seen in Fig. 4.5 that both drought and dam operations have strong impacts on lake levels. Under the mild drought scenario, operations according to the Agreed Curve would result in about half the lake level drop caused by average releases of 1300 m^3/s. The dynamic nature of the Agreed Curve can also be seen as drought intensity increases, less water is released, reducing the relative drop in lake level as compared to the steady release scenarios. If the analysis were to assume a lower starting lake level (as was the case when the study was undertaken), the Agreed Curve would further dampen the impact of the drought relative to the steady releases.

Using the same water balance model, 2002–2006 changes in the level of Lake Victoria were simulated. 2004 and 2005 were analyzed by estimating departures from average monthly rainfall to compute yearly difference from average rainfall in the Lake Victoria basin, based on data and graphics from [2] and [10]. Despite some periods of less than normal rainfall, for example April–July 2004 and the last months of 2005, the rough estimates of total departure from average annual rainfall are −10% for 2004 and −15% for 2005.

Prevailing reports indicated that for the period of October 2005 to 25 January 2006, rainfall in parts of the Lake Victoria Basin were 45–70% of normal [3]. This is equivalent to 8–14% of annual rainfall, indicating that the −15% estimate used

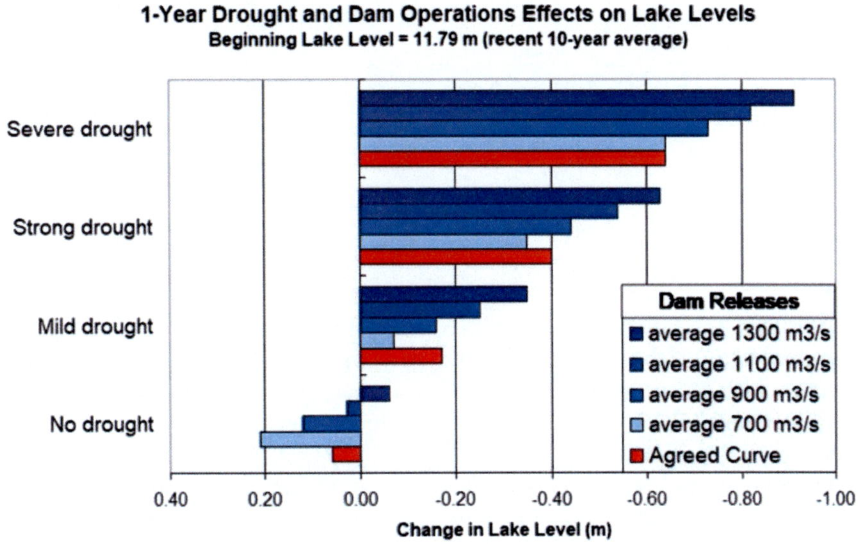

Fig. 4.5 The impact of droughts and dam operations on Lake Victoria levels

Table 4.2 Water balance simulation results for 2004 and 2005. Asterix (*) refers to [25]

	Observed Lake level begin (m)*	Observed Lake level end (m)*	Observed Lake level drop (m)*	Lake drop (m) expected with Agreed Curve	Cause of Lake level drop (drought based on Agreed Curve) on (%)	Cause of Lake level drop (additional factor) (%)
2004	11.59	11.13	0.46	0.20	43	57
2005	11.13	10.69	0.44	0.21	48	52

for the 2005 analysis was well within range. Owing to a lack of data, it was assumed that evaporation was 10% greater than normal for both years.

The analysis assumed that dam operations adhered to the Agreed Curve. As it represents the natural hydrologic functioning of the lake, any lake drop occurring with Agreed Curve operations represents natural drought effects. Therefore, any observed drops in lake level in excess of that expected from the Agreed Curve were caused by an additional factor. Results are given in Table 4.2.

It can be seen that although the droughts of 2004 and 2005 contributed to the lowering of the lake level, if dam operations had adhered to the Agreed Curve, the lake levels would be around 50 cm higher. There was obviously an additional factor in the reduction of the lake level.

Table 4.3 Estimated average annual combined releases prescribed by the Agreed Curve, verses those necessary for the observed lake drop

	Agreed curve average flow (m^3/s)	Needed for observed drop average flow (m^3/s)	Percent difference (%)
2004	857	1387	162
2005	752	1114	148

Assuming this additional factor was dam releases not in adherence to the Agreed Curve, an analysis was performed to estimate what the average annual combined (both dams) releases would have had to have been in order for the observed drop to occur, shown in Table 4.3. The Agreed Curve annual average flows shown in Table 4.3 are computed based on the observed lake level at the beginning of 2004, followed by Average Curve releases for the two years. This assumption results in the lake level being about 0.45 m higher than observed at the end of 2005. This differs slightly from the expected lake drops shown in Table 4.2 (0.41 m) because these were computed individually per year, while the computations in Table 4.3 utilise the cumulative impacts of two years of non-Agreed Curve operations.

It can be seen that for 2004 and 2005, more than 50% over the prescribed Agreed Curve flow has been released from Nalubaale and Kiira. If the two years are analysed separately, thus using the observed water level at the beginning of 2005 as opposed to the higher water level that would have resulted from Agreed Curve flows in 2004, the expected Agreed Curve flows for 2005 average 646 m^3/s, such that the over-release approaches 75% of the prescribed flow for that year. Based on the results of this analysis (Tables 4.2 and 4.3), it must be concluded that the ***severe Lake Victoria drops that occurred in 2004 and 2005 were about 45% due to drought, and 55% due to over-releases from Nalubaale and Kiira***. Similar conclusion are also arrived, e.g.., [28, 29] who employ Gravity Recovery and Climate Experiment (GRACE) mission discussed in Chaps 5 and 9.

Anecdotal evidence supports this conclusion: a dry period from June–September 2004 was followed by heavy rain during October to December of the same year. Experts assured that this wet period would raise the lake level "back to normal", but this did not occur [11].

4.3.1 Sample Analysis: Dam Releases Above the Agreed Curve

Technical Report 7 of the Study on Water Management of Lake Victoria reviews a study on the river hydraulics between Lake Victoria and Nalubaale and Kiira [23]. It investigated primarily the impacts of the river channel and Lake Victoria levels on hydropower production.

Table 4.1 of Technical Report 7 lists the Nalubaale and Kiira outflows for August 19–21, 2004. These are provided to indicate the data used for the calibration and validation of a numerical simulation model discussed in the preceding section. Dam operations data was otherwise not publicly available. Monthly flow data was available from the Global Runoff Data Centre for 1946–1970 and 1973–1982, but no later nor higher resolution (daily, hourly, etc.) data was publicly available. In many other countries, such data is public domain, which is appropriate considering water and thus major lakes and rivers are public goods. The secrecy of hydrologic and dam operations data for Lake Victoria and the Victoria Nile is worrying.

During the 3 days of provided data, the estimated lake level at Jinja was about 11.24 m [25]. According to the Agreed Curve, when the lake is at this level, about 740 m^3/s should be released. Although fluctuating between 530 and 882 m^3/s over the three days, Nalubaale was releasing an average 712 m^3/s, which is not far from the prescribed 740 m^3/s. These fluctuations are somewhat surprising, as the levels of Lake Victoria, which as discussed according to the Agreed Curve should dictate the dam outflows, could not have changed so quickly.

However, in addition to the 712 m^3/s from Nalubaale, during the same time, Kiira was releasing an average 658 m^3/s, such that for the lake level of 11.24 m, a total of average 1370 m^3/s was being released. This was a full 630 m^3/s over the Agreed Curve (an over-release of 85%). In other words, this was almost double the prescribed release for the given lake level. This over-release is very close to the average annual flow of 1387 m^3/s estimated earlier for 2004 (see Table 4.3), further supporting the conclusion that over-releases contributed to lake level drops of 2002–2006.

On March 1, 2005, a Ugandan engineer, Hilary Onek (who is also an MP for Lamwo County, Kitgum) claimed that over 1400 m^3/s was being released from Nalubaale and Kiira dams combined [11]. On that date, the observed lake level was about 11.05 m [25], which according to the Agreed Curve, require an outflow of 653 m^3/s. The total releases from the dams were thus on this day more than double the agreed release for the observed lake level.

The Lake Victoria Policy Brief of December 2005, as quoted in [7], indicates that during November 2005 the average combined dam releases were about 1100 m^3/s. During that month, the average Lake Victoria level was approximately 10.8 m [25], indicating an Agreed Curve prescribed release of about 550 m^3/s. During November, 2005, combined Nalubaale and Kiira dams releases were thus about double the prescribed releases, and are again, extremely close to the average estimated for 2005 (1114 m^3/s) in Table 4.3.

It is clear that at least for some of the time since Kiira came on-line, the combined outflows from Nalubaale and Kiira dams were far above that prescribed by the Agreed Curve. Dam operations have therefore, no longer mimicking the natural Lake Victoria outflows.

According to the WREM Hydraulic Model Study [23], at lake levels around 11.1 m, combined (both dams) maximum hydropower production would result in total outflow of about 1460 m^3/s. It would thus appear, based on the analysed flow data for August 2004, March 2005, and November 2005, that dam operators were maximizing hydropower production, which was in violation of Uganda's agreement with Egypt.

4.3.2 Mixed Messages on Plans and Operations

The East African media (in Uganda and Kenya, primarily) had conflicting messages about this controversy. Numerous articles made references to the lake level dropping because of drought. But a few articles stated that the lake's level was being negatively affected because of the operation of the dams.

For example, an article in The Sunday Vision (Uganda) on 4 Jan. 2006 stated that "The (Study on Water Management of Lake Victoria) report heaped the blame for the continued falling water levels on the over leasing of water to generate electricity at the two existing dams, Kiira and Nalubaale" [14].

In the same article, Dr. Frank Sebbowa of the Electricity Regulation Authority denied the dams were at fault, blaming instead global warming. He said: "*The new dam (Kiira) is supposed to replace the old dam (Nalubaale), which has become obsolete*" [14]. The article reported that the sluice gates at Nalubaale had been closed, and all flows were now going through Kiira, "which then rules out the arguments of over leasing of water."

That was a mis-representation: the reality was that Nalubaale was still releasing water through its turbines, but the water that used to pass through its sluice gates (thus not producing electricity) was now passing through Kiira. On 5 January 2006, Maj. Gen. Kahinda Otafire, Minister of Water, Lands and Environment (Uganda) was reported to have said that the Government was considering closing one of the dams in order to maintain the water levels [8]. This indeed confirmed that both Nalubaale and Kiira dams had been in operation.

Dr. Sebbowa's statements were also in conflict with the goal of Kiira to utilise the spare hydro capacity being spilled by the sluices of Nalubaale. If Kiira was designed to replace Nalubaale, as Dr. Sebbowa stated, it can be assumed that reference to this would have been made in World Bank Project Appraisal Documents for the Uganda 4th Power Project, which was not the case [19]. Indeed, references to Uganda's future hydropower generation included inputs from both Nalubaale and Kiira.

4.3.3 Disputed Hydrology

The hydrology of Lake Victoria, especially the outflow into the Victoria Nile, has long been a topic of disagreement among hydrologists and engineers. Between 1960–1964, Lake Victoria experienced a massive 2.5 m increase in lake level, which many experts have attributed to a period of excessive rainfall, but the precise cause of which is not agreed.

Computed averages for lake outflow depend greatly on the period of record used. Since the 1960–1964 rises in lake level, outflow has been greater than before 1960. Table 4.4 shows some of the many different computed values for average annual flow from Lake Victoria.

Table 4.4 Average outflow from Lake Victoria into the Victoria Nile at Jinja. The values with * are based on monthly data from the Global Runoff Data Centre (GRDC), station Jinja 1946–1970, station Owen Reservoir 1973–1982. Data for 1971 and 1972 not available

Time period	Flow (m^3/s)	Source
Before 1960	600	[19]
1900–1960	650	[21]
1925–1959	668	de Baulny and Baker (1970) in [26]
1901–2001	838	[21]
1912–1982	884	[4]
1948–1970	956	[4]
1945–1984	983	Flohn (1983) in [26]
1946–1982*	985	Author's analysis
"Average flow"	1004	[19]
1940–1995	1007	[12]
1950–1979	1092	Flohn and Burkhardt (1985) in [26]
1956–1978	1092	Piper et al. (1986) in [26]
1960–2001	>1100	[19]
1956–1978	1144	Howell et al. (1988) in [26]
1956–1978	1144	in [26]
1964–2001	1164	[21]
1960–1982*	1218	Author's analysis
1970–1974	1245	Kite (1982) in [26]
1961–1970	1352	[4]

The impact of the period of record used to compute average flow is quite apparent in Table 4.4. Those using data only after 1960 are of the opinion that the higher water levels and thus outflows from 1960 are relatively permanent, representing the modern state of the lake.

Surface water modelling is based on stochastic hydrology, and the longest available and sound period of record should always be used for hydrologic statistical analysis. The longer the time-series of observations, the better dynamic physical processes, natural cycles, extremes and changes are captured. "As the number of samples grows, the accuracy typically improves" [6, 30].

This has been recognised by the World Bank: "The river is susceptible to some hydrological cycles but these are included within the available record length of 100 years" [20].

4.4 Implications for the Owen Falls Complex

The original Kiira design and cost-benefit analysis were based on the higher average flows experienced after 1961. Acres International assumed a 99% probability for the continuation of the higher flows (thus implying a 1% probability for reversion to earlier low flows) for their investment risk analysis. This assumption was supported by a break-even analysis that showed that only a 61% probability for the high flows to continue was needed for Kiira to produce a net benefit. It was acknowledged that Kiira would not be economic if the low-flow observed before 1961 returned and continued [16].

The World Bank again later recognised the challenges that Lake Victoria outflows and levels could pose to the viability of the proposed additional Kiira Units 14 and 15, admitting the flows were a function of an uncertain hydrological regime. It was, however, determined that the flow "available in the short-term (in particular during 2003-5) was the most critical" [19].

The design, therefore, emphasized the higher flows of 1965–1998, neglecting the extremely high 1961–1964 flow, which resulted in an average flow of 1004 m^3/s. The higher starting level of the lake (as observed since 1961) and the assumption that "from a hydrological viewpoint it is more likely that the higher post 1965 inflow pattern will continue as opposed to a reversion to the older lower flows" are given as reasons for this decision [19].

It is unclear what hydrological viewpoint, when taken in context of the full hydrologic record and the inherent variability of climate and hydrology, would assume a high likelihood of continued high lake levels and flows. Although by omitting the 1961–1964 flows, the utilised flow is lower than figures computed for 1960–2000, it can still be considered optimistic in relation to the full historical record, and particularly with today's new reality.

The situation has now changed, with the water level dropping severely and outflow reduced. In 2002, the World Bank's investigation panel drew a conclusion that is beginning to ring true in light of the 2002–2006 situation: "If (average flow is) 650 (m^3/s), then the Owen Falls complex (Nalubaale and Kiira dams) is over-designed and incapable of full capacity" [21].

Figure 20a of the Hydraulic Model Study on Water Management of Lake Victoria [23] shows the combined maximum power generation capacity of the two dams. It can be seen that at the water levels of about 10.69 m (corresponding to an elevation of 1133.55 m), the proposed turbines 14 and 15 would be useless.

Section 14.4.1.2 of the USER Manual: Lake Victoria Decision Support Tool (LVDST) of the Study on Water Management of Lake Victoria [24] shows the turbine power curves of the 10 original Nalubaale units. For the full 180 MW of Nalubaale to be realised, a flow of at least 850 m^3/s is needed, depending on the lake level.

Section 4.5.2 of [24] shows that at lake water elevation of 1134.0 m (equivalent to Jinja gage of 11.14 m), in order to produce combined (both Nalubaale and Kiira operating) 200 MW of power, a lake outflow of about 1130 m^3/s is needed. This assumes optimising Kiira's turbines as a priority over Nalubaale turbines, as the

Table 4.5 Flow and lake level needed for power production found in [24]

Minimum condition	Minimum condition	Dam output	Dam output	Dam output
Flow (m^3/s)	Lake level (m)	Nalubaale (MW)	Kiira (MW)	Total (MW)
800	11.14	–	144	144
850	all	180	–	180
1130	11.14	56	144	200

newer units are more efficient. This optimised combination (lake level 11.14 m, total outflow 1130 m^3/s), is much removed from the Agreed Curve, which dictates a total flow of 694 m^3/s for this lake elevation. The same figure shows that even flow only through Kiira, at about 800 m^3/s, is above the Agreed Curve, and produces just over 144 MW. Table 4.5 summarises the scant data available on minimum flow and lake level needed for hydropower generation as extracted from [24].

It can be concluded that under today's and the long-term averaged (100+ years) lake level and flow conditions (for instance a flow of 838 m^3/s, as defined in [21]), the average power capacity for the Owen Falls complex under sustainable operations is somewhere below 200 MW. It thus appears that the original Nalubaale plant, with its expansion to 180 MW, was well designed to capture the hydroelectric potential of the Owen Falls site in a sustainable manner.

Considering that the project target capacity of the Owen Falls complex is 380 MW, with so far 300 MW having been installed, it must be concluded that the complex, especially Kiira, has been over-designed. This scenario was noted in the project documents: "The only significant risk to economic feasibility would arise if low hydrologic regime flows of the magnitude of the pre-1961 stream flow data set were to occur. In this case, the extension would not be economic." [16].

Recent technical reports further support this conclusion: "The full installed capacity of the (existing and proposed) Kiira units (208 MW) cannot be utilized until 1137 m lake level is realized, which, of course, is undesirable from a lake level management standpoint." [23].

4.5 Implications for Bujagali Dam

The prospect of the flow and thus hydropower output of Bujagali dam meeting cost-benefit expectations is a bit more positive, as the project design appears to have been based on the full hydrologic record [20].

It is interesting to note, however, that hydrologists from the dam's proponents, Acres International [1] (also involved in the Kiira projects), disagreed with hydrologists from the Institute of Hydrology (IOH—United Kingdom), Electricity de France (EdF) and Knight Piesold (United Kingdom). The Acres' hydrologists believed that the pre-1960 data was not reliable, such that the long term mean flow would be about 1100 m^3/s. The IOH, EdF and Knight Piesold hydrologists believed the 1925–1960 data to be reliable, such that their estimate of long-term mean flow was 870 m^3/s, and that "lake outflow conditions similar to the pre-1960 low flow sequence could occur again in the design life of a hydraulic structure" [20].

One can assume that Acres optimistic view of the hydrology was biased, especially when considering they were disagreeing with professional hydrologists from the internationally renowned IOH (as opposed to engineers performing hydrology on the side, as is often the case in engineering firms), in a situation where such a disagreement benefited the viability of their project.

Bujagali was designed assuming the flow released from Lake Victoria through the Owen Falls complex would be in accordance with the Agreed Curve [20]. As it is clear now that the Agreed Curve is no longer being respected and the Victoria Nile flow regime has changed, the original long-term energy output assessment for Bujagali is no longer valid.

4.6 Addendum to the 2005 Article

Thanks to an anonymous source, new data on Lake Victoria and its outflow management has become available since the original report.[3] The data consists of observed daily lake levels, associated agreed curve discharges and actual Nalubaale and Kiira Dam releases for September 2004 to the beginning of March 2006, as reported by Eskom Uganda Ltd.

4.6.1 Original 2005 Estimates Satisfactorily Accurate

In the original study it was assumed that during 2005, precipitation was 15% lower than normal, and evaporation was 10% greater than normal. Based on the new data and assuming again 10% above average annual evaporation, the Lake Victoria water balance model indicates precipitation was actually 17% lower than normal in 2005.

The original study computed that the average releases for 2005 for Nalubaale and Kiira Dams were 1114 m^3/s. The new data shows actual releases averaged 1180 m^3/s, about 6% greater than the original calculations. Considering the rough data and the estimated parameters used in the original study, this magnitude of error is acceptable.

[3](see, e.g., https://www.internationalrivers.org/sites/default/files/attached-files/full_report_pdf.pdf).

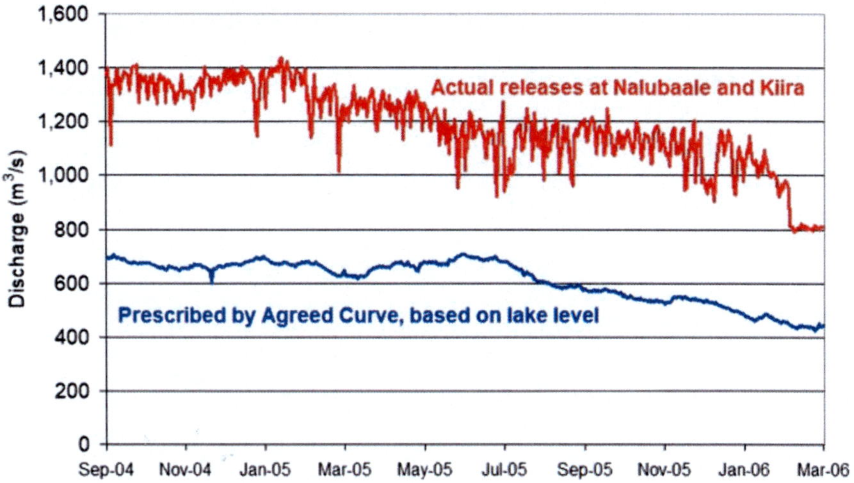

Fig. 4.6 Actual and Agreed Curve-prescribed total releases at Nalubaale and Kiira Dams for September 2004 to early March 2006

4.6.2 Negligence of Agreed Curve Greater Than Originally Reported

Figure 4.6 shows the newly available data of discharge prescribed by the Agreed Curve (based on observed lake levels), and the actual releases from the two dams.

It can be seen that the actual releases were far greater than those prescribed by the Agreed Curve. Indeed, during the period September 2004 to March 2006, the actual releases were 194% of those prescribed by the agreed curve: almost double.

For 2005, the over-release was 191% of the Agreed Curve, which accounted for approximately 0.26 m in additional lake level drop, or 51% of the total (drought and over-release induced) lake drop. The original study yielded an estimate of 0.23 m additional lake level drop and 53% of total drop due to over-releases, again a close estimate when compared to the newly available data. During the full period of data (September 2004 to March 2006), over-releases resulted in an approximate additional 0.4 m of water level drop.

4.6.3 New Release Policy Still Not Adhering to Agreed Curve

The newly available data indicates a New Release Policy was implemented on February 6, 2006, limited to a maximum total discharge of 850 m^3/s. Actual average releases for the one month of available data (Feb. 6–March 5, 2006) were less, around 806 m^3/s. In any case, with observed lake levels indicating an average Agreed Curve-prescribed release of 440 m^3/s during this period, the New Release Policy is still far above the Agreed Curve.

It is clear that the Agreed Curve is still not being adhered to, and although the New Release Policy is moving in the right direction, it continues to ignore the natural water balance of Lake Victoria. Such choices are often political, and in this case it appears that the sustainability of a regional resource of social, economic and environmental benefits is of secondary consideration.

4.7 Concluding Remarks

Because of uncertainty, a "cautious and incremental approach to the extension of Owen Falls capacity has been adopted" [19]. Unfortunately, it would seem that there has been no similar caution in recent Nalubaale and Kiira Dam operations. The Agreed Curve is no longer being adhered to, and the resultant over-release of water from Nalubaale and Kiira is contributing to the severe drop in water level in Lake Victoria.

The drops in Lake Victoria threaten the future performance of the Owens Falls dams, as well as to a lesser degree the constructed Bujagali Dam. "It is clear that future climates imply drier conditions, lower lake levels and lower downstream river flows" [22]. It is unknown if Lake Victoria will recharge to the high levels and outflow experienced during 1961–2000, and if such a recharge could occur, whether it would be in the next years or only in 100 years. Viable non-hydro, or at least hydro not on the Victoria Nile, power generating alternatives must therefore be considered for Uganda.

It is ironic that some years of what appears to be optimised hydropower and thus benefit production since Kiira opened have now resulted in the economic viability of Kiira being challenged. This would appear to be a case of environmental and economic sustainability ignored!

References

1. Acres International (2005) Owen falls generating station: additional powerhouse, project fact sheet, Report P9226. http://www.hatchacres.com
2. Famine Early Warning Systems Network (2006a) Uganda monthly reports. USAID-funded FEWS NET. http://fews.net
3. Famine Early Warning Systems Network (2006b) Africa: weather hazards assessment 26 Jan 1 Feb 2006. USAID-funded FEWS NET, report on ReliefWeb.int
4. Karyabwite DR (2000) Water sharing in the Nile river valley. Project GNV011: Using GIS/remote sensing for the sustainable use of natural resources, UNEP/DEWA/GRID, Geneva
5. Lahmeyer International (2004) Owen falls extension (Completion of Kiira Station) Uganda, project fact sheet. http://www.lahmeyer.de
6. Maidment DR (ed) (1993) Handbook of Hydrology. McGraw-Hill Inc., New York

7. Mugabe D, Kisambira E (2006) Lake Victoria levels at Jinja Dam raise eyebrows. East African Business Weekly, 16 January
8. Nyanzi N, Nandutu A (2006) Jinja power dam to close over drought. Daily Monitor, Kampala, Uganda, 6 January
9. Nicholson SE, Yin X, Ba MB (2000) On the feasibility of using a lake water balance model to infer rainfall: an example from Lake Victoria. Hydrolog Sci J 45(1):75–95
10. National Oceanic and Atmospheric Administration (2006) Africa rainfall estimate climatology (CPC ARC). National Weather Service, Climate Prediction Center, Camp Springs, MD, USA. http://www.cpc.ncep.noaa.gov/products/fews/AFR_CLIM/afr_clim.html
11. Onek H (2005) Kiira's extension had a design flaw. The new vision, opinion section, Kampala, Uganda, 1 March
12. Roskar J (2000) Assessing the water resources potential of the Nile river based on data, Available at the Nile forecasting centre in Cairo. Republic of slovenia ministry of the environment and spatial planning, hydrometeorological institute of Slovenia, UDC: 556.53(282.263.1), COBISS: 1.01
13. Phoon SY, Shamseldin AY, Vairavamoorthy K (2004) Assessing impacts of climate change on Lake Victoria Basin, Africa. People-centred approaches to water and environmental sanitation, 30th WEDC international conference, Vientiane, Lao PDR
14. Tenywa G (2006) Lake Victoria is drying up. Sunday vision, special report, Kampala, Uganda, 4 January
15. Wamaniala VN (2002) The development and management of hydropower resources in Uganda. Hydro power resources development and management course, Trondheim, Norway, 3–20 June
16. World Bank (1991) Staff appraisal report for the Uganda third power project, May 29, Washington D.C
17. World Bank (1996) Staff appraisal report for the Lake Victoria environmental management project, report no. 15429-AFR, Washington, D.C
18. World Bank (2000) Assessment of generation alternatives Uganda, 18 May, Washington, D.C
19. World Bank (2001) Project appraisal document on a proposed international development association credit in the amount of SDR 48 million (US$ 62 million equivalent) to the republic of Uganda, report no: 22318-UG, Africa energy team Africa regional office (AFTEG), 8 June
20. World Bank and IFC (2001) Project appraisal document for the Bujagali hydropower project in the republic of Uganda. Africa region energy team, world bank, and power department, international finance corporation, Washington, D.C., 14 November
21. World Bank (2002) Investigation report; Uganda: third power project, fourth power project, Bujugali hydropower project. The inspection panel, report no. 23998, Washington, D.C., 23 May
22. WREM International Inc. (2005a) Climate change impact assessment technical report 10, Study on water management of Lake Victoria, prepared by water resources and energy management international inc. for the Uganda ministry of energy and mineral development, June
23. WREM International Inc. (2005b) Hydraulic model study technical report 7, study on water management of Lake Victoria, prepared by water resources and energy management international inc. for the Uganda Ministry of energy and mineral development, June
24. WREM International Inc. (2005c) USER manual: Lake Victoria decision support tool (LVDST) technical report 9, study on water management of Lake Victoria, prepared by water resources and energy management international inc. for the Uganda ministry of energy and mineral development, June
25. United States Department of Agriculture (2005) Low water levels observed on Lake Victoria. Production estimates and crop assessment division, foreign agricultural service, USDA. http://www.fas.usda.gov/pecad/highlights/2005/09/uganda_26sep2005/
26. Yin X, Nicholson SE (1998) The water balance of Lake Victoria. Hydrolog Sci J 43(5):789–811
27. Awange JL, Forootan E, Kuhn M, Kusche J, Heck B (2014) Water storage changes and climate variability within the Nile Basin between 2002 and 2011. Adv Water Resour 73:1–25. https://doi.org/10.1016/j.advwatres.2014.06.010

28. Awange JL, Sharifi MA, Ogonda G et al (2008) The falling Lake Victoria water level: GRACE, TRIMM and CHAMP satellite analysis of the Lake Basin. Water Resour Manage 22:775. https://doi.org/10.1007/s11269-007-9191-y
29. Awange J, Anyah R, Agola N, Forootan E, Omondi P (2013b) Potential impacts of climate and environmental change on the stored water of Lake Victoria Basin and economic implications. Water Resour Res 49(12):8160–8173. https://doi.org/10.1002/2013WR014350
30. Awange JL, Saleem A, Sukhadiya RM, Ouma YO, Kexiang H (2019a) Physical dynamics of Lake Victoria over the past 34 years (1984–2018): Is the lake dying? Sci Total Environ 658:199–218

Part II
Remote Sensing Techniques

Chapter 5
Satellite Remote Sensing

To supplement ground-based observations and water resources management over large and poorly gauged areas, satellite remote sensing (SRS) plays a crucial role in monitoring the various components of the hydrological cycle. SRS can now effectively monitor almost all components of the water balance equation

–Khandu [96]

5.1 Satellite, Reanalysis and Model Data

Lake Victoria Basin (LVB)'s overall water movement expressed through the hydrological cycle is characterised by a simple water balance equation: $\triangle S = P - E - Q$, where $\triangle S$ is the basin's total water storage (TWS representing the sum of groundwater, soil moisture, vegetation, and surface water), P the basin's precipitation, E its evapotranspiration, and Q its runoff. Due to its sheer size of area of 258,000 km^2 [1], monitoring changes in $\triangle S$, P, E, Q through "boots on the ground" ground-based (in-situ) observations is practically impossible and a daunting task indeed. On the one hand, the in-situ (ground-based, see e.g., [85]) data may be inaccessible while on the other hand, the meagre accessible data where available, might be inconsistent or suffer from missing data.

To supplement in-situ observations and water resources management over large and poorly gauged LVB, ***satellite remote sensing SRS*** offers the only possibility of rigorously monitoring changes in these components ($\triangle S$, P, E, Q) of the hydrological cycle within the LVB as will be shown in the Chapters ahead. To this end, the basin's total water storage changes $\triangle S/dt$, where dt is the temporal differences, e.g., changes between consecutive months can now accurately be measured using the Gravity Recovery And Climate Experiment (GRACE, discussed in Sect. 5.3.3) mission launched in 2002, see e.g., [2]. It enables the closure of LVB terrestrial water

J. Awange, *Lake Victoria Monitored from Space*,
https://doi.org/10.1007/978-3-030-60551-3_5

budget by providing large-scale quantitative estimates of changes in TWS fields ΔS on an approximately monthly basis.

LVB's precipitation P can be remotely sensed using both infra-red (IR) and microwave (MW) sensors on board geostationary satellites and low-Earth orbiting satellites (e.g., Tropical Rainfall Measuring Mission, TRMM, Kummerow et al. [17]). TRMM satellite, jointly managed by National Aeronautics and Space Administration (NASA) of the United States and Japan Aerospace Exploration Agency (JAXA) was designed to monitor and study tropical rainfall. It has the advantage of providing four dimensional (X, Y, Z, t) rainfall and latent heat data over inaccessible areas such as the ocean, un-sampled terrains, etc. The primary rainfall parameters on TRMM are the TRMM Microwave Imager (TMI), the Precipitation Radar (PR) and the Visible and Infrared Radiometer System (VIRS) [17]. Infrared-based rainfall estimates have a longer historical record and have higher spatio-temporal resolution than the more accurate (and shorter temporal record) Microwave-based estimates but suffer from higher uncertainty due to various problems [96, 97]. Awange et al. [4] discusses uncertainties in satellite based derived precipitation over Africa while Awange et al. [3] compares the performance of merged products in Africa and Australia. In this book, various satellite based precipitation will be employed and as such, they are discussed at their point of use. Evapotranspiration E on its part can be monitored using MODerate Resolution Imaging Spectroradiometer (MODIS) onboard NASA's Terra and Aqua satellites [22], which is also useful for monitoring vegetation as will be shown in Chaps 13 and 14. Details of MODIS will be covered in those chapters where they are used. For surfaces runoff Q, useful for measuring changes in levels of the lake and rivers, satellite altimetry missions [6, 7] such as Topex/Poseidon, Jason I and II, and the European Space Agency (ESA) discussed in Sect. 5.5.1 are employed.

Besides satellite data, ***reanalyses*** have made significant contributions to many global/regional hydrological and climatic studies. With the release of many new high-resolution reanalyses in the past decade, see e.g., [35–38], their application into regional- and basin-scale studies have become increasingly valuable [96]. Yet certain elements of the analyzed fields (e.g., precipitation) remain highly uncertain at global and regional scale both in terms of trends and interannual variabilities. The reliability of reanalysis fields can considerably vary in space and time due to lack of adequate observational data, instrumental changes, changing mix of observations, biases in observations, etc., which can introduce spurious variability and trends into reanalysis fields [96]. Khandu [96] opines that since reanalysis products are increasingly used as regional climate forcing data and hydrological model inputs, it is vital to estimate their skills in a region. A reanalysis system consists of (i) a "data assimilation system" that combines available observations from various data sources (e.g., SRS-based observations), and (ii), a "forecast model" consisting of a atmospheric model at its core, which is often coupled to a land surface model and/or ocean model [35, 36, 38]. The performance of reanalyses have significantly improved with the assimilation of additional data from SRS-based observations, e.g., [37, 38, 40–42]. Reanalysis are employed in various chapters of this book to complement satellite-based remotely sensed data. In this regard, the individual reanalysis product used will be discussed in the respective chapter.

Climate models and hydrological models add knowledge to the key climate and hydrologic process and has been rigorously applied to understand the contributions of climate change (both natural and man-made) and other drivers (e.g., land use change, water use) on water resources [42, 96]. SRS-based observations are also assimilated into hydrological models that are widely used to quantify climate change impacts on various hydrological components as will be demonstrated for LVB. Climate models and hydrological models are also employed in subsequent chapters to complement satellite remote sensing and help in understanding of the LVB hydrological process.

5.2 Remotely Sensed Landsat and Sentinel-2 Products

The multi-spectral Landsat and Sentinel-2 data have a wide range of spectrum bands that includes, visible to Near-Infrared (VNIR) and shortwave Infrared (SWIR) wavelengths. Landsat satellite system collect multi-spectral images that are useful for monitoring and understanding human activities, e.g., detecting, measuring and highlighting changes in landscape patterns over time including those of coastal studies [5]. Since changes in the landscape are dynamic, these imagery can be used to study and capture extent of changes at large scale within short time frames and at no cost since the images are free. Without remotely sensed data, the task of capturing and mapping changes in landscape within LVB would be much more difficult and time consuming. The imagery from Landsat platform, which can be accessed freely from USGS website (https://earthexplorer.usgs.gov/) are powerful data source providing accurate and timely spatial information on changes in the Earth's landscape. Because of its long-term digital archive since 1972, it has been used in many land use and land cover change studies over different parts of the world, see e.g., [8].

Over the past 46 years, there have been numerous studies that have used Landsat imagery for agriculture forecasting and management (production and conservation), see e.g., [9–11]. Its other applications are reported, e.g., in [12–16, 18–21]. Landsat data, therefore, are important when there are no other accurate way of mapping and detecting changes due to cost, time, and access constraints. Landsat datasets obtained from the USGS are Level 1 products, which have been corrected for terrain effects, calibrated radiometrically, and geographically georeferenced [5].

To validate the used Landsat images whose spatial resolution is 30 m, the Sentinel-2 (MSI) with 10 m spatial resolution, officially launched by ESA on June 23, 2015, with a 5-day temporal resolution or repeat cycle [23] are employed. The corporation between ESA and the USGS provides Sentinel-2 (MSI) data in Level-1C product to end users and can be accessed from the USGS website and the official website of the Copernicus Open Access Hub for the European Space Agency (ESA) (https://scihub.copernicus.eu/). Sentinel-2 data has been processed by ESA and USGS to include radiometric and geometric corrections along with ortho-rectification to generate highly accurate geo-located products for researchers [24]. Landsat and Sentinel-2 employed in the chapters a head are discussed therein.

5.3 Remote Sensing of Gravity Variations

In the subsections that follow, it is explained how satellites (particularly the low earth observing-LEO satellites) are used to monitor variations in gravity field, which are in turn used to remotely sense the changes in stored waters of Lake Victoria. The most significant success of a LEO satellite is evidenced in the Gravity Recovery and Climate Experiment (GRACE) satellites discussed in Sect. 5.3.3. A possible use of the global navigation satellite systems (GNSS, see [25]) to measure variations in water mass is illustrated by Tregoning et al. [26] whose predictions derived from GRACE measured fields show a correlation with GNSS measured deformations, suggesting the possible use of such deformations to infer changes in stored water potential on much shorter temporal and spatial scales than GRACE provides (and with low-latency), while averaging over much larger spatial scales than afforded by multipath amplitude measurements [27]. In this book, GRACE satellites are employed to monitor changes in Lake Victoria Basin (LVB)'s stored water (see Sect. 5.1). To understand how GRACE satellites work, let us first discuss the relationship between mass variation and gravity next.

5.3.1 *Mass Variation and Gravity*

Two types of gravity field variation exists. The *first* is the long-term, also known as mean gravity field or static gravity field, which is due to the static part of the gravity field. The variation is constant over a very long time interval. To simplify the understanding of the mean gravity field, consider a case where you frequently go to a given destination and along the way, you pass a given mountain. Every time you pass by, the mountain is stationary at the same place. Come months and years, the mountain, which represents concentration of mass is stationary and does not change its position. In simple terms, one can visualize this mountain as representing the long term changes in gravity field as its mass concentration does not seem to change. Study of long-term (mean gravity field) is useful in understanding the solid structure of the Earth, ocean circulation, and in achieving a universal height measuring system. In this respect, the GOCE (Gravity field and the steady state-of-the ocean circulation explorer, Fig. 5.1) satellite[1] that completed its mission in mid-October 2011 used the state-of-the-art gradiometer with improved accuracy to provide data that are useful in mapping changes in gravity, see e.g., Hirt [28]. GOCE data is expected to benefit other studies such as those concerned with earthquakes, changes in sea level, and volcanoes.[2]

The *second* type of variation of the Earth's gravity field is associated with those processes that occur over shorter time scales, such as atmospheric circulation or the hydrological cycle. In the illustration of the mean gravity field above, we used

[1] https://www.esa.int/Our_Activities/Observing_the_Earth/GOCE.

[2] See, e.g., https://www.esa.int/esaCP/SEMV3FO4KKF_Germany_0.html.

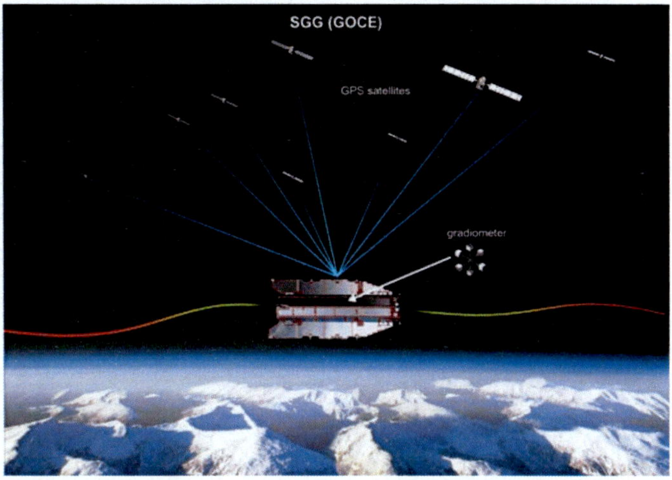

Fig. 5.1 Global Navigation Satellite System (GNSS) satellites track the GOCE satellite in space, thus contributing to the determination of its position (©ESA). The GOCE satellite's accurate determination of the static gravity field is expected to contribute towards studies of changes in sea level, earthquakes, and volcanoes. Figure modified by Rieser [33]

the mountain as an example. To visualize time variable gravity field, instead of a mountain, consider now a small pond along your frequently used route to a given destination. As you frequently pass this pond, you will notice that its water level changes depending on climatic or human induced impacts. If it rains heavily and over a long period of time, the level rises while if there is a prolonged drought of massive use by humans, the level of the pond falls. This change in pond level over time symbolizes changes in mass with time. This is known as the *time-varying gravity field* and is the component, which enables the monitoring of, for example, changes in water level of Lake Victoria as will be demonstrated in Chaps. 9 and 11.

By removing the effects of the other processes that cause changes in the time variable gravity field, changes in *terrestrial water storage* can thus be inferred from the observed temporal changes in the terrestrial gravity field. By assuming the density of water as 1.00 g/cm^3, and following the relation of [69], Ellet et al. [29] present the relationship between changes in stored water and gravity as

$$\Delta S = 0.419\Delta g, \tag{5.1}$$

where water storage change ΔS is given in units of cm of water and gravity change Δg is in units of microGal (10^{-6} cm/s^2). From Eq. (5.1), it is seen therefore, that monitoring variations in the gravity field can enable hydrological changes to be monitored.

5.3.2 *High and Low Earth Orbiting Satellites*

At the broadest conceptual level, Low Earth Observing (LEO) satellites' gravity field missions observe (either directly or indirectly) gradients in the Earth's external gravitational field. This is essentially done through differential measurements between two or more points, thus largely eliminating spatially correlated errors. When done from space, two approaches can be used, e.g., [30, 31]:

1. Satellite-to-satellite tracking (SST), or
2. A dedicated gravity gradiometer on board a satellite, coupled with SST.

The SST methods can use either low-low inter-satellite tracking (ll-SST, see Fig. 5.2, right), where two LEO satellites track one another and additional observations in terms of high precision ranges and range rates between the two satellites are taken, or high-low inter-satellite tracking (hl-SST, see Fig. 5.2, left), where high-Earth orbiting satellites (notably GPS) track a LEO satellite. The low-low mode, compared to the high-low mode, has the advantage of signal amplification leading to a higher resolution of the obtained gravity variations, up to the medium wavelength spectrum of a few hundred km in spatial extent [30]. Taking this further, a combination of ll-SST and hl-SST is conceptually better still, as demonstrated by the GRACE mission (Fig. 5.2, right) with a baseline length between the two satellites of about 220 km. This is treated in detail in the next section.

In order to detect temporal gravity field variations at smaller spatial scales, the satellite(s) being tracked must be in as-low-as-possible orbits (close to the mass source), with the satellites being as free as possible from the perturbing effects of atmospheric drag [30]. In addition, so-called de-aliasing models (for correcting short-term—6 h—variations due to atmosphere and ocean mass variations) have to be used to mitigate the propagation of unwanted signals (e.g., leakage from the oceans) into the derived gravity solutions, e.g., [32].

5.3.3 *Gravity Recovery and Climate Experiment*

The GRACE mission, launched on 17th of March 2002, and which ended its mission on mid-October 2017, consisted of two near-identical satellites following one another in nearly the same orbital plane (about 400 km altitude) separated by a distance of 220 km; the so-called tandem formation (see Fig. 5.2, right). The ll-SST was measured using K-band ranging, coupled with hl-SST tracking of both satellites by GNSS (Fig. 5.2, right). Gobal Navigation Satellite System (GNSS) receivers were placed on GRACE satellites to measure occulted signals, see [25], and also to determined the orbital parameters of GRACE satellites required in order to determine gravity changes. On-board accelerometers monitored orbital perturbations of non-gravitational origin, see, e.g., [25].

GRACE mission processes GNSS data to contribute to the recovery of long-wavelength gravity field, remove errors due to long-term onboard oscillator drift,

GNSS tracking a low satellite, e.g., CHAMP.	GNSS tracking 2-low satellites, which are tracking each other, e.g., GRACE.

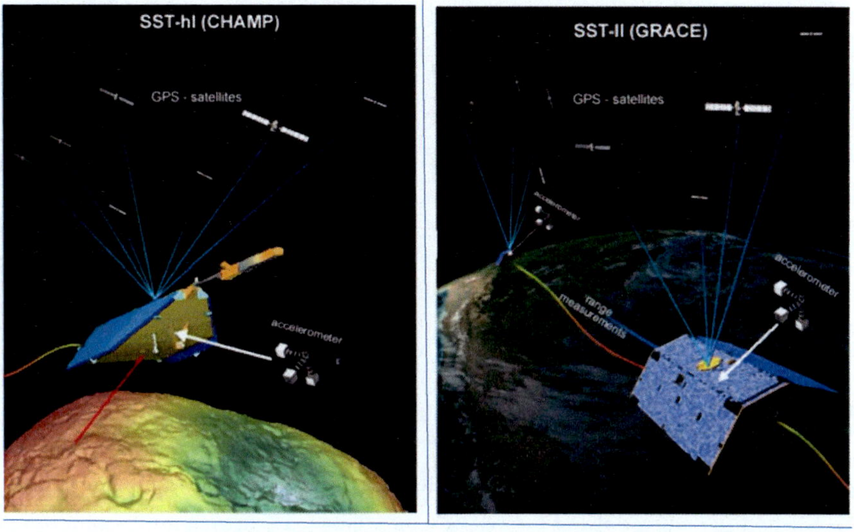

Fig. 5.2 Left: SST-hl realized with CHAMP (©GFZ Potsdam ([2.2]). Right: A combination of ll-SST and hl-SST realized with GRACE and GNSS satellites (©GRACE - CSR Texas ([2.2]). Figures modified by Rieser [33]). GNSS satellites are used in determining the positions of these satellites in space. For the GRACE satellites (right) inter-satellite distances can be computed from these positions and compared to the measured K-band distances, thus providing additional independent information

and aligns measurements between the two spacecraft [34, Sect. 8.4.4]. The timing function of GNSS for precision orbit determination, in terms of position and velocity as a function of time, enabled orbits to be determined within an accuracy better than 2 cm in each coordinate [34, Sect. 8.4.4]. The precise locations of the two satellites in orbit allowed for the creation of gravity maps approximately once a month.[3] These gravity maps, when converted to total water storage maps, are useful for monitoring changes in stored water potential, e.g., those of LVB as demonstrated in Chap. 9.

The Earth's gravity field is mapped by making accurate measurements of changes in the distance between the satellites, using GNSS and a microwave ranging system. These changes in the distances between the two satellites occur due to the effect of the gravity (mass concentration) of the Earth. As the lead satellite passes through a region of mass concentration, it is pulled away from the trailing satellite (Fig. 5.2, right). As the trailing satellite passes over the same point, it is pulled towards the lead satellite thus changing the distance between the satellites.

Time-variable gravity field solutions are obtained by the exploitation of GRACE observation data over certain time intervals, i.e., every month [39, 43], or less, e.g., [44, 45]. There are a number of institutions delivering GRACE products, each

[3]http://www.csr.utexas.edu/grace/publications/brochure/page11.html.

applying their own processing methodologies and, often, different background models. GRACE mission, which lasted for 15 years provide scientists with an efficient and cost-effective way to monitor time-varying component of the gravity field with unprecedented accuracy and in the process yield crucial information about the distribution and flow of mass within the Earth's system. The process causing gravity variations that are currently being studied by GRACE include [46];

- changes due to surface and deep currents in the ocean leading to more information about ocean circulation, e.g., [47, 48],
- changes in groundwater storage on land masses such as that of LVB, relevant to water resource managers, e.g., [46, 49–51, 84],
- exchanges between ice sheets or glaciers and the oceans, needed for constraining the mass balance of the global ice regime and sea level change, e.g., [52, 53],
- air and water vapour mass change within the atmosphere, vital for atmospheric studies, e.g., [54, 55], and
- variations of mass distribution within the Earth arising from, e.g., on-going glacial-isostatic adjustments and earthquakes, e.g., [26, 56].

Currently, river and lake basins of the order of 200,000 km^2 and above in area can be successfully studied using the GRACE products [57]. In general, to understand how the GRACE satellites monitor changes in freshwater (all groundwater, soil moisture, snow, ice, and surface waters), first, the larger effect of the mass of the Earth, i.e, the static gravity field discussed in Sect. 5.3.1, which is always a constant G_0 corresponding to nearly 99% of the total field, is computed from a static model (e.g., *GGSM01S* [43]) and removed by subtracting it from the monthly gravity field ($G(t)$) measured by GRACE at a time t [58], i.e.,

$$\Delta G(t) = G(t) - G_0, \tag{5.2}$$

to give the monthly time-variable gravity field $\Delta G(t)$. Alternatively, instead of computing the static gravity field G_0, mean gravity field over a given number of years can also be used. Changes mostly related to the atmosphere and ocean, which occur over time scales shorter than one month, are then removed using models, see e.g., Wahr et al. [59]. Remnant atmospheric and oceanographic effects that last for more than one month can be removed using atmospheric and ocean circulation models before water storage change can be analyzed. The resulting difference in Eq. (5.2), which is called the gravity field anomaly is usually due to changes in stored water. If we consider $\overline{\Delta C}_{nm}(t)$ and $\overline{\Delta S}_{nm}(t)$ to be the normalized Stoke's coefficients expressed in terms of millimeters of geoid height, with n and m being degree and order respectively, the time-variable geoid in (5.2) is then expanded in-terms of spherical harmonic coefficients (see [60]) as

$$\Delta G(t) = \sum_{n=1}^{N}\sum_{m=0}^{n}(\overline{\Delta C}_{nm}(t)cos(m\lambda) + \overline{\Delta S}_{nm}(t)sin(m\lambda))\overline{P}_{nm}\left(cos(\theta)\right), \tag{5.3}$$

where N is the maximum degree of expansion, θ is the co-latitude, λ the longitude and $\overline{P}_{nm}$ the fully normalized Legendre polynomial [60]. From the gravitational spherical harmonic coefficients (5.3), the equivalent water thickness is computed using the following steps:

1. The gravitational residual coefficients are converted into the surface density coefficient differences by [59]

$$\begin{pmatrix} \Delta\check{C}_{lm}(\mathrm{M}_j) \\ \Delta\check{S}_{lm}(\mathrm{M}_j) \end{pmatrix} = \frac{\rho_{\mathrm{avg}}}{3\rho_{\mathrm{w}}} \frac{2l+1}{1+k'_l} \begin{pmatrix} \Delta\bar{C}_{lm}(\mathrm{M}_j) \\ \Delta\bar{S}_{lm}(\mathrm{M}_j) \end{pmatrix}, \tag{5.4}$$

where k'_l is the *load Love* number of degree l, $\rho_{\mathrm{avg}} = 5517\,\mathrm{kg/m^3}$ the average density of the Earth, and $\rho_{\mathrm{w}} = 1000\,\mathrm{kg/m^3}$ the density of water.
2. The spatial variation of the surface density is then computed through

$$\Delta\sigma(\theta, \lambda, \mathrm{M}_j) = R\rho_{\mathrm{w}} \sum_{l=1}^{l_{\max}} \sum_{m=0}^{l} [\Delta\check{C}_{lm}(\mathrm{M}_j)\cos m\lambda + \Delta\check{S}_{lm}(\mathrm{M}_j)\sin m\lambda]\bar{P}_{lm}(\cos\theta), \tag{5.5}$$

where $R = 6378137\,m$ is the radius of the Earth and $\Delta\sigma$ is in $\mathrm{kg/m^2}$.
3. Finally, the changes in total water storage (TWS) are calculated by

$$\mathrm{TWS}(\phi, \lambda, \mathrm{M}_j) = \frac{\Delta\sigma(\theta, \lambda, \mathrm{M}_j)}{\rho_{\mathrm{w}}} = \frac{\Delta\sigma(\theta, \lambda, \mathrm{M}_j)}{1000} \quad [\text{meters}]. \tag{5.6}$$

The first steps in the analysis of GRACE data would provide an estimate of the changes in total water storage. In the second step, the changes can then be separated into their various components as discussed, e.g., in [46, 58] to obtain changes in the respective components (e.g., groundwater, surface water, soil moisture, and ice).

As stated previously, the GRACE satellites exceeded their planned 5 year life-span and ended their mission in mid-October 2017. However, given the excellent results that have been delivered so far, see e.g., [61] and also Fig. 1.2, GRACE follow-on mission (GRACE-FO) was launched on the 22nd of May 2018. Although GRACE-FO satellites, like their predecessor, will use the same kind of microwave ranging system giving a similar level of precision, they will also test an experimental instrument using lasers instead of microwaves, which promises to make the measurement of their separation distance at least 20 times more precise.[4] An example of GRACE's application is presented in Yang [62], where it is used to constrain recent freshwater flux from Greenland where the data show that Arctic freshwater flux started to increase rapidly in the mid-late 1990s, coincident with a decrease in the formation of dense Labrador Sea Water, a key component of the deep southward return flow of the Atlantic Meridional Overturning Circulation (AMOC). Recent freshening of the polar oceans may be reducing formation of Labrador Sea Water and hence may be weakening the AMOC [62].

[4] http://gracefo.jpl.nasa.gov/mission/.

5.4 Gravity Field and Changes in Stored Water

In the discussion that follows, the concept of *gravity field variations* discussed in Sect. 5.3 is related to hydrological processes. Measurements of the time-varying gravity field by LEO satellites, e.g., GRACE discussed in Sect. 5.3.3 are the key to monitoring of changes in water levels at basin scales. Such techniques now enable the monitoring of groundwater recharge, which is the most important element in groundwater resources management and could also be applicable to monitoring changes in Lake Victoria's water levels. For example, in 2009, GRACE satellites showed that north-west of India's aquifers had fallen at a rate of 0.3048 m yr^{-1} (a loss of about 109 km^3 per year) between 2002 and 2008 (see Fig. 1.2).[5]

5.4.1 Gravity Field Changes and the Hydrological Processes

The hydrological cycle (Fig. 5.3) refers to the pathway of water in nature, as it moves in its different phases through the atmosphere, down over and through land, to the ocean and back to the atmosphere [63]. The associated variations in gravity field are therefore caused, e.g., by

- the redistribution of water in the oceans, including e.g., El Niño and Southern Oscillation (ENSO) events,
- movement of water vapour and other components in the atmosphere,
- seasonal rainfall; snow and subsequent drying and melting,
- groundwater extraction, or
- drying and filling of lakes, rivers, and reservoirs.

5.4.2 Monitoring Variation in Stored Water Using Temporal Gravity Field

The potential of using the relationship between temporal gravity changes and hydrology (Fig. 5.3) was first recognized by Montgomery [64] who estimated specific yield through a correlation between gravity and water-level changes [65]. In 1977, Lambert and Beaumont [66] used a gravity meter to correlate groundwater fluctuations and temporal changes in the Earth's gravity field. Goodkind [67] recorded observations from seven super conducting gravimetric stations to examine non-tidal variations in gravity and noted that at one of the stations (Geysers geothermal station), much of the variation could be correlated with rainfall and seismic activity. Such measurements had not been possible before the advent of super conducting gravimeters, thus providing evidence of the existence of temporal variation in gravity.

[5]The Economist, September 12th 2009, pp. 27–29: Briefing India's water crisis.

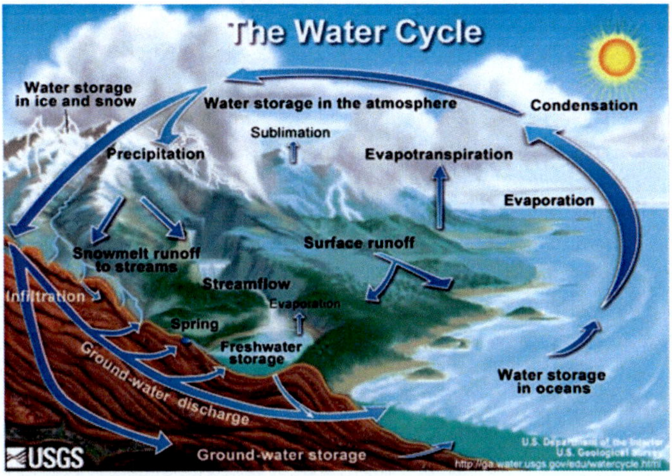

Fig. 5.3 Components of hydrological cycle that lead to temporal variations in the gravity field. *Source* US Geological Survey (USGS)

In 1995, while estimating the atmospheric effects on gravity observations around Kyoto, Mukai et al. [68] noted that changes in gravity around the station could have been caused by changes in underground water. In the same year, Pool and Eychaner [69] assessed the utility of temporal gravity-field surveys to directly measure *aquifer-storage* changes and reported gravity changes of around 100–134 μGal, equivalent to 2.4–3.2 m of water column, considering infinitely extended sheet approximation. Their results from the analysis of changes in stored water in the aquifer indicated an increase in the gravity field of 158 ± 6 μGal when the water table rose by about 17.7 m, providing further evidence of the possibility of using temporal gravity-field surveys to monitor changes in stored aquifer water. In fact, according to Bower and Courtier [70] who analyzed the effect of precipitation on gravity and well-levels at a Canadian absolute gravity site, 90% of the gravity variation was found to be due to the effects of precipitation, evapotranspiration and snow-melt.

The last decade has also recorded increased use of temporal gravity field studies in monitoring changes in stored water, see e.g., [29]. It saw the beginning of satellite missions dedicated to monitoring temporal variations in the gravity field. Smith et al. [71] investigated the ability of ground-based gravity meters to monitor changes in soil moisture storage.

Moving from local tests to regional, a different application of gravity surveys was investigated by Damiata and Lee [72], who simulated the gravitational response to aquifer hydraulic testing. The synthetic system was composed of an unconfined shallow aquifer and the purpose of the investigation was to assess the potential of the gravity measurements for detecting groundwater extraction. Drawdown due to pumping causes a decrease in mass and consequently in gravity measured at the surface. The results showed that the gravitational response to aquifer testing could be used to monitor the spatial development of the drawdown cone. For the configuration considered in the investigation, the signal was of the order of tens of μGals and could be detected up to several hundred meters away from the pumping well.

Water storage changes, such as changes in soil moisture, snow and ice cover, surface and groundwater, including deep aquifers, can be monitored either by in-situ observations or indirectly through changes in gravity [43]. While in-situ observations provide valuable localized information, they suffer from limited spatial coverage for regional to continent-wide studies [73]. Any change in water storage also manifests itself in a change in the gravity field. This property can be used to infer water-storage changes from time-variable gravity observations as demonstrated by Rodell and Famiglietti [49] for 20 globally distributed drainage basins of sizes varying from 130,000 to 5,782,000 km^2 to assess the detectability of hydrological signals with respect to temporal and spatial variations. Space borne techniques can provide time-variable gravity observations on a regional and global scale, thus allowing for large-scale water storage monitoring and the ability to close the 'gaps' between locally limited in-situ observations [74].

Since the launch of the GRACE satellite mission in 2002 (see Sect. 5.3.3) until the end of its mission in mid-October 2017, it provided a new and powerful tool for studying temporal gravity field changes as evidenced from numerous articles assessing the potential of GRACE recovering hydrological signals, see e.g., Awange et al. [30, and the references therein]. Tapley [43] provided early results of the application of the GRACE products for detecting hydrological signals in the Amazon-Orinoco basin. Following these results, numerous other authors have subsequently applied GRACE to detect hydrological signals in various situations and locations, see references in [30].

For instance, Ramillien et al. [46, 58] and Andersen et al. [76] investigated the potential of inferring inter-annual gravity field changes caused by continental water storage changes from GRACE observations between 2002 and 2003, and compared these changes to the output from four global hydrological models. It was possible to correlate large scale hydrologic events with the estimated change in the gravity field for certain areas of the world to an accuracy of 0.4 μGal, corresponding to 9 mm of water, see also [76–78].

Syed et al. [79] examined total basin discharge for the Amazon-Orinoco and Mississippi river basins from GRACE, while Rodell et al. [75] estimated groundwater storage changes in the Mississippi basin. Crowley et al. [80] estimated hydrological signals in the Congo basin, while Schmidt et al. [81] and Swenson et al. [57, 82] used GRACE to observe changes in continental water storage. Winsemius et al. [83] compared hydrological model outputs for the Zambezi River Basin with estimates derived from GRACE. Monthly storage depths produced by the hydrological model displayed larger amplitudes and were partly out of phase compared to the estimates based on GRACE data. Likely reasons included leakage produced by the spatial filtering used in the GRACE data, and the difficulty to identify the time of satellite overpass as opposed to simply averaging over the whole period. Awange et al. [2, 84] used GRACE to study the fall of Lake Victoria's water level in Africa. This last example will be elaborated upon in more detail in Chap 9. The Nile Basin and water changes in West Africa have been studied, e.g., by [87, 88] while its application to study agricultural droughts and changes in aquifer storage over the Greater Horn of Africa was pioneered in the works of Agutu et al. [89, 90].

As already discussed in Sect. 5.3.3, GRACE satellites detect changes in the Earth's gravity field by measuring changes in the distance between the two satellites at a 0.1 Hz sampling frequency. The variation in the distance between the two twin satellites caused by gravitational variations above, upon, and within the Earth all have an effect on the satellites. This variation in gravity could be due to *rapid* or *slow* changes caused, for example by the redistribution of water in the oceans, the movement of water vapor and other components in the atmosphere, the tidal effect of the Sun and Moon, and the displacement of the material by earthquakes and glacial isostatic adjustment. The data therefore must be processed to isolate these effects so as to retain only those which correspond to the process of interest, in this case, terrestrial water storage changes, see e.g., [91]. Equation (5.6) is used to compute changes in stored water.

5.5 Satellite Altimetry

5.5.1 Remote Sensing with Satellite Altimetry

Satellites altimetry (Fig. 5.4) operates in two steps:

- *First*, the precise orbit of the satellite, i.e., its position, is determined. Through this, its *height* above the Earth is obtained.
- *Second*, range measurements are made by obtaining the time an emitted signal (radar or laser) travels to the Earth's surface and reflected back to the satellite.

GNSS satellites, see [25, 86], contribute to the *first step* where height is determined through GNSS receivers onboard the space satellites that enables monitoring of ranges and timing signals from GNSS satellites. The observed GNSS ranges provide precise and continuous tracking of the spacecraft, thereby delivering its position $\{\phi, \lambda, h\}$ at any time. The height component h is useful in determining the measured height (see Fig. 5.4). Besides GNSS tracking, other approaches such as satellite laser ranging (SLR) and DORIS (Doppler Orbitography and Radio positioning Integrated by Satellite) are also used to ensure that precise orbit determination is achieved.

In the *second step*, the Earth's surface heights (e.g., lake surface, ocean surface, glaciers, and ice sheets) are measured using ranges from the space altimetry satellite to the surface of interest. Radar altimeters send microwave signals to the Earth's surface and measures the time taken by the reflected signals to travel back. Using the velocity of light, the distance from the satellite to the Earth's surface is derived. Since the signals pass through the atmosphere from and to the satellites, they are affected by the atmosphere and as such, atmospheric corrections have to be made

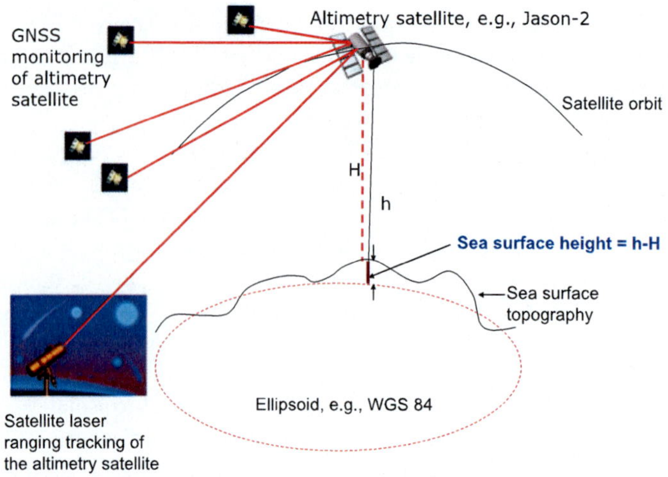

Fig. 5.4 Global Navigation Satellite System (GNSS) in support of monitoring changes in sea/lake level through the determination of the altimetry satellites' precise orbit. From the precise orbital parameters, the height component h is useful in determining changes in sea/lake level through the difference $\{h - H\}$, where H is measured by multiplying the speed of light with the time taken by the signals to travel from and to the satellite divided by 2, since the same distance is covered twice

(see Chap. 7 on how this are corrected). The sea surface height is then obtained by subtracting the measured ranges in step 2 from the GNSS-derived satellite heights in step 1 (Fig. 5.4).

Example 1 (Satellite altimetry monitoring of Lake Victoria water level) As an example, Lake Victoria water levels for the period 1993 to 2006 are plotted for both TOPEX/Poseidon altimetry-derived heights and tide gauge (in-situ) data in Fig. 5.5. The figure indicates a close relationship between the two data sets. Crétaux et al. [92] compares water levels of Lake Victoria from the Jinja tide gauge and those from Jason-1 altimetry satellite and obtain a coefficient of correlation value of 0.99 with a standard deviation of 2.7 cm for the period 2004 to 2007. This supports the fact that satellite altimetry provides useful information on changes in the lake level. Satellite altimetry are largely used for observing changes in sea level. By averaging the few-hundred thousand measurements collected by the satellite in the time it takes to cover the global oceans (i.e., 10 d for TOPEX/Poseidon), global mean sea level can be determined with a precision of several millimeters [93]. Such information is vital for mitigation of disasters related to sea level changes. Detailed satellite altimetry study on East African lakes (e.g., Fig. 5.5) are given e.g., in [92, 95].

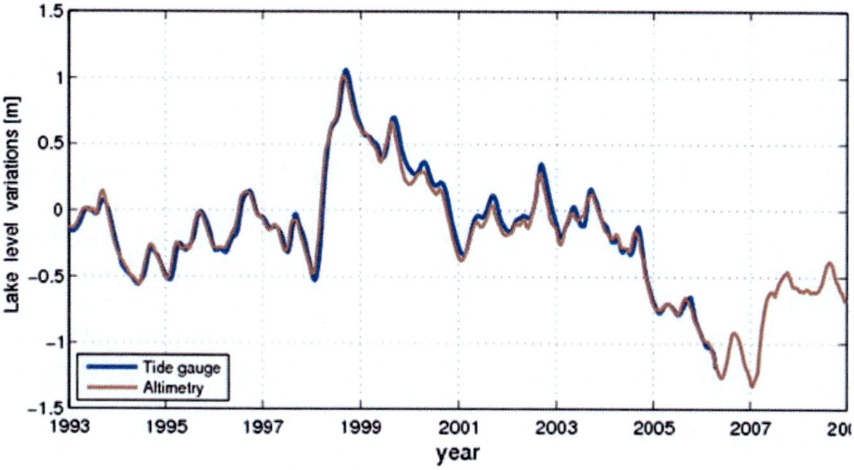

Fig. 5.5 A comparison of water gauge readings at Jinja station in Uganda (near Lake Victoria's outlet, see Fig. 13.1) and water Levels from Topex/Poseidon and Jason-1 altimetry satellites. The figure shows a close match between the tide gauge and satellite altimetry data (cf. Fig. 9.10 obtained from GRACE)

5.5.2 *Satellite Altimetry Missions*

The first altimetry mission was TOPEX/Poseidon, developed by NASA and the Centre National d'Etudes Spatiales (CNES) and launched on 10 August 1992. Its mission ended in 2006 after 13 years of operation, providing 11 years of data. It was followed by Jason-series (Jason-1 was launched on 07/12/2001 and Jason-2 on 20/06/2008). Both TOPEX/Poseidon and Jason-1 were dedicated to measuring global mean sea level from space. TOPEX/Poseidon orbited at 1336 km above the Earth and covered the global oceans every 10 days, measuring the heights of the ocean surface directly underneath the satellite with an accuracy of 2–4 cm or better when averaging over several measurements [93]. Jason-2 is expected to be replaced by Jason-3 launched on 17th of January 2016, and subsequently Sentinel-6. Sentinel-6 will continue high precision ocean altimetry measurements in the 2020–2030 time-frame using two successive, identical satellites (Jason-Continuity of Service): Jason-CS-A and Jason-CS-B, and as a secondary objective, collect high resolution vertical profiles of temperature using the GNSS Radio-Occultation sounding technique discussed in Awange [25, 86].[6] Combined, all these satellites will provide long-term series of data capable of undertaking *climatological studies* resulting from changes in lakes and sea level.

ICESat (launched on January 12, 2003) uses a 1064 nm-laser operating 40 Hz to make measurements at 172 m intervals over ice, ocean, and land [94]. It combines state-of-the-art laser ranging capabilities with precise orbit and attitude control and

---

[6] https://eospso.nasa.gov/missions/sentinel-6.

knowledge to provide very accurate measurements of ice sheet topography and elevation changes along track. It has the specific objective of measuring changes in polar ice as part of NASA's Earth Observing System.

By observing changes in ice sheet elevation, it is possible to quantify the growth and shrinkage of parts of the ice sheets with great spatial detail, thus enabling an assessment of ice sheet mass balance and contributions to sea level. Moreover, because the mechanisms that control ice sheet mass loss and gain in accumulation, surface ablation, and discharge presumably have distinct topographic expressions, ice sheet elevation changes also provide important insights into the processes causing the observed changes [94]. ICESat-2 launched on 15th September 2018 is expected to be a follow-on mission to ICESat (Fig. 5.6) with improved laser capability compared to ICESat and will have the objectives of measuring ice sheet changes, sea ice thickness, and vegetation biomass. Achieving these objectives will contribute to the following [94]:

- Contribute to the development of predictive models that capture both dynamic and surface processes.
- Since the thickness distribution of sea ice controls energy and mass exchanges between the ocean and atmosphere at the surface, and the fresh water fluxes associated with melting ice serve as stabilizing elements in the circulation of the North Atlantic waters, basin-scale fields of ice thickness are therefore essential to improve our estimates of the seasonal and interannual variability in regional mass balance,

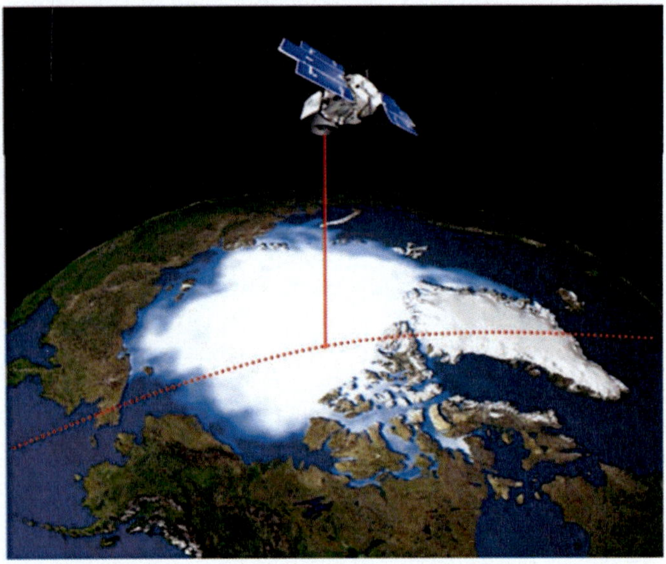

Fig. 5.6 Schematic diagram of ICESat on a transect over the Arctic. ICESat uses a 1064 nm-laser operating 40 Hz to make measurements at 172 m intervals over ice, oceans, and land *Source* Abdalati et al. [94]

the freshwater budget of the polar oceans, and the representation of these processes in regional and climate models.

- The capability of producing a vegetation height surface with 3 m accuracy at 1 km spatial resolution, assuming that off-nadir pointing can be used to increase the spatial distribution of observations over terrestrial surfaces. This sampling, combined with a smaller footprint of 50 m or less, would allow characterization of vegetation at a higher spatial resolution than ICESat, and is expected to provide a new set of global ecosystem applications.
- In addition, the atmospheric measurement capability of ICESat-2, even at near-IR wavelengths, will enable global measurements of cloud and aerosol structure to extend the record of these observations beyond those provided by the current lasers onboard ICESat.

5.6 CHAMP Radio Occultation Satellite

Radio occultation with GPS takes place when radio signals from a transmitting GPS satellite, setting or rising behind the Earth's limb, are received by a Global Positiong System (GPS) receiver aboard a Low Earth Orbiting (LEO) satellite. The meteorological products of occultation comprise the air pressure P, air temperature T and the water vapour pressure P_w. By measuring the occulting GPS signal due to the effect of the atmosphere, CHAMP satellite (LEO) is capable of providing accurate tropospheric measurements to sub-kelvin accuracy, see e.g., [2, 25, 86]. This technique is applied in Chap. 9. In Chap. 6, GNSS reflectomatry (GNSS-R) is presented as a potential future technique for complementing satellite altimetry.

5.7 Concluding Remarks

Satellite remote sensing and its application to water resource monitoring, particularly GRACE, is a new and active area of research. The data that has been collected from GRACE so far has provided information that was hitherto difficult to fathom. The new technique clearly promises to contribute significantly to water resource studies of Lake Victoria and other freshwater lakes the world over. When the life span of the various missions (e.g., GRACE) is reached, thousands of data sets will have been collected that will help to unravel some of the complex nature of the hydrological cycle.

References

1. Awange JL, Saleem A, Sukhadiya RM, Ouma YO, Kexiang H (2019) Physical dynamics of Lake Victoria over the past 34 years (1984–2018): is the lake dying? Sci Total Environ 658:199–218. https://doi.org/10.1016/j.scitotenv.2018.12.051
2. Awange JL, Sharifi MA, Ogonda G et al (2008) The falling lake Victoria water level: GRACE, TRIMM and CHAMP satellite analysis of the lake basin. Water Res Manag 22:775. https://doi.org/10.1007/s11269-007-9191-y
3. Awange JL, Hu KX, Khaki M (2019) The newly merged satellite remotely sensed, gauge and reanalysis-based multi-source weighted-ensemble precipitation: evaluation over Australia and Africa (1981–2016). Sci Total Environ 670:448–465
4. Awange JL, Ferreira VG, Forootan E, Khandu Andam-Akorful SA, Agutu NO, He XF (2016) Uncertainties in remotely sensed precipitation data over Africa. Int J Climatolol 36(1):303–323. https://doi.org/10.1002/joc.4346
5. USGS (2015) Landsat data. USGS, United States
6. Berry PAM, Garlick JD, Freeman JA, Mathers EL (2005) Global inland water monitoring from multi-mission altimetry. Geophys Res Lett 32(16). https://doi.org/10.1029/2005GL022814.
7. Shum CK, Ries JC, Tapley BD (1995) The accuracy and applications of satellite altimetry. Geophys J Int 121:321–336. https://doi.org/10.1111/j.1365-246X.1995.tb05714.x
8. Jensen JR, Rutchey K, Koch MS, Narumalani S (1995) Inland wetland change detection in the Everglades Water Conservation Area 2A using a time series of normalized remotely sensed data. Photogram Eng Remote Sens 61:199–209
9. Kauth R, Lambeck P, Richardson W, Thomas G, Pentland A (1979) Feature extraction applied to agricultural crops as seen by Landsat
10. Stow D, Tinney L, Estes J (1980) Deriving land use/land cover change statistics from Landsat-A study of prime agricultural land
11. Peterson U, Aunap R (1998) Changes in agricultural land use in Estonia in the 1990s detected with multitemporal Landsat MSS imagery. Landsc Urban Plan 41:193–201
12. Cihlar J (2000) Land cover mapping of large areas from satellites: status and research priorities. Int J Remote Sen 21:1093–1114
13. Jusoff K, Senthavy S (2003) Land use change detection using remote sensing and geographical information system (GIS) in Gua Musang district, Kelantan, Malaysia. J Trop For Sci 15:303–312
14. Porter-Bolland L, Ellis EA, Gholz HL (2007) Land use dynamics and landscape history in La Montaña, Campeche, Mexico. Landsc Urban Plann 82:198–207
15. Dewan AM, Yamaguchi Y (2009) Land use and land cover change in Greater Dhaka, Bangladesh: using remote sensing to promote sustainable urbanization. Appl Geogr 29:390–401. https://doi.org/10.1016/j.apgeog.2008.12.005
16. Kesgin B, Nurlu E (2009) Land cover changes on the coastal zone of Candarli Bay, Turkey using remotely sensed data. Environ Monit Assess 157:89–96
17. Kummerow C, Barnes W(1998) The tropical rainfall measuring mission (TRMM) sensor package. J Atmos Ocean Technol 15:809–817
18. Vittek M, Brink A, Donnay F, Simonetti D, Desclée B (2014) Land cover change monitoring using landsat MSS/TM satellite image data over West Africa between 1975 and 1990. Remote Sens 6:658–676
19. Yadav P, Kapoor M, Sarma K (2012) Land use land cover mapping, change detection and conflict analysis of Nagzira-Navegaon corridor, Central India using geospatial technology. Int J Remote Sens GIS 1:90–98
20. Otukei JR, Blaschke T (2010) Land cover change assessment using decision trees, support vector machines and maximum likelihood classification algorithms. Int J Appl Earth Obs Geoinf 12:S27–S31
21. Moran EF (2010) Land cover classification in a complex urban-rural landscape with QuickBird imagery. Photogram Eng Remote Sens 76:1159

22. Justice CO, Townshend JRG, Vermote EF, Masuoka E, Wolfe RE, Saleous N, Roy DP, Morisette JT (2002) An overview of MODIS Land data processing and product status. Remote Sens Environ 83:3–15. https://doi.org/10.1016/S0034-4257(02)00084-6
23. ESA (2017). European space agency, Copernicus open access hub. In: SERCO
24. Drusch M, Del Bello U, Carlier S, Colin O, Fernandez V, Gascon F, Hoersch B, Isola C, Laberinti P, Martimort P (2012) Sentinel-2: ESA's optical high-resolution mission for GMES operational services. Remote Sens Environ 120:25–36
25. Awange JL (2018) GNSS environmental sensing. Revolutionizing environmental monitoring, Springer, Berlin, New York
26. Tregoning P, Watson C, Ramillien G, McQueen H, Zhang J (2009) Detecting hydrologic deformation using GRACE and GPS. Geophys Res Lett 36:L15401. https://doi.org/10.1029/2009GL038718
27. Hammond WC, Brooks BA, Bürgmann R, Heaton T, Jackson M, Lowry AR, and Anandakrishnan S (2010) The scientific value of high-rate, low-latency GPS data, a white paper
28. Hirt C, Gruber T, Featherstone WE (2011) Evaluation of the first GOCE static gravity field models using terrestrial gravity, vertical deflections and EGM2008. Quasigeoid Heights J Geodesy 85:723–740. https://doi.org/10.1007/s00190-011-0482-y
29. Ellett KM, Walker JP, Western AW, Rodell M (2006) A framework for assessing the potential of remote sensed gravity to provide new insight on the hydrology of the Murray-Darling Basin. Aust J Water Res 10(2):89–101
30. Awange JL, Sharifi MA, Baur O, Keller W, Featherstone WE, Kuhn M (2009) GRACE hydrological monitoring of Australia. Current limitations and future prospects. J Spatial Sci 54(1): 23–36. https://doi.org/10.1080/14498596.2009.9635164.
31. Rummel R, Balmino G, Johannessen J, Visser P, Woodworth P (2002) Dedicated gravity field missions—principles and aims. J Geodyn 33(1):3–20. https://doi.org/10.1016/S0264-3707(01)00050-3
32. Schrama EJO, Visser PNAM (2007) Accuracy assessment of the monthly GRACE geoids based upon a simulation. J Geod 81(1):67–80. https://doi.org/10.1007/s00190-006-0085-1
33. Rieser D (2008) Comparison of GRACE-derived monthly surface mass variations with rainfall data in Australia. MSc thesis, Graz University of Technology
34. Prasad R, Ruggieri M (2005) Applied satellite navigation using GPS. GALILEO and augmentation systems, Artech House, Boston/London
35. Kalnay E, Kanamitsu M, Kistler R, Collins W, Deaven D, Gandin L, Iredell M, Saha S, White G, Woollen J, Zhu Y, Chelliah M, Ebisuzaki W, Higgins W, Janowiak J, Mo KC, Ropelewski C, Wang J, Leetmaa A, Reynolds R, Jenne R, Joseph D (1996) The NCEP/NCAR 40-year reanalysis project. Bull Am Meteorol Soc 77(3):437–470. https://doi.org/10.1175/1520-0477(1996)077<0437:TNYRP>2.0.CO;2
36. Onogi K, Tsutsui J, Koide H, Sakamoto M, Kobayashi S, Hatsushika H, Matsumoto T, Yamazaki N, Kamahori H, Takahashi K, Kadokura S, Wada K, Kato K, Oyama R, Ose T, Mannoji N, Taira R (2007) The JRA-25 reanalysis. Q J R Meteorol Soc 85(3):369–432. https://doi.org/10.2151/jmsj.85.369
37. Saha S, Moorthi S, Pan H-L, Wu X, Wang J, Nadiga S, Tripp P, Kistler R, Woollen J, Behringer D, Liu H, Stokes D, Grumbine R, Gayno G, Wang J, Hou YT, Chuang HY, Juang H-MH, Sela J, Iredell M, Treadon R, Kleist D, Delst PV, Keyser D, Derber J, Ek M, Meng J, Wei H, Yang R, Lord S, Dool HVD, Kumar A, Wang W, Long C, Chelliah M, Xue Y, Huang B, Schemm J-K, Ebisuzaki W, Lin R, Xie P, Chen M, Zhou S, Higgins W, Zou C-Z, Liu Q, Chen Y, Han Y, Cucurull L, Reynolds RW, Rutledge G, Goldberg M (2010) The NCEP climate forecast system reanalysis. Bull Am Meteorol Soc 91(8):1015–1057. https://doi.org/10.1175/2010BAMS3001.1
38. Dee DP, Uppala SM, Simmons AJ, Berrisford P, Poli P, Kobayashi S, Andrae U, Balmaseda MA, Balsamo G, Bauer P, Bechtold P, Beljaars ACM, van de Berg L, Bidlot J, Bormann N, Delsol C, Dragani R, Fuentes M, Geer AJ, Haimbergere L, Healy SB, Hersbach H, Holm EV, Isaksen L, Kållberg P, Köhler M, Matricardi M, McNally AP, Monge-Sanz BM, Morcrette J-J, Park B-K, Peubey C, de Rosnay P, Tavolato C, Thépaut J-N, Vitarta F (2011) The ERA-Interim

reanalysis: configuration and performance of the data assimilation system. Q J R Meteorolog Soc 137:553–597. https://doi.org/10.1002/qj.828

39. Luthcke S, Rowlands D, Lemoine F, Klosko S, Chinn D, McCarthy J (2006) Monthly spherical harmonic gravity field solutions determined from GRACE inter-satellite range-rate data alone. Geophys Res Lett 33:L02402. https://doi.org/10.1029/2005GL024846
40. Rodell M, Houserand PR, Jambor U, Gottschalck J, Mitchell K, Meng C-J, Arsenault K, Cosgrove B, Radakovich J, Bosilovich M, Entin JK, Walker JP, Lohmann D, Toll D, (2004) The global land data assimilation system. Bull Am Meteorol Soc 85:381–394. https://doi.org/10.1175/BAMS-85-3-381
41. Rienecker MM, Suarez MJ, Gelaro R, Todling R, Bacmeister J, Liu E, Bosilovich MG, Schubert SD, Takacs L, Kim GK, Bloom S, Chen J, Collins D, Conaty A, da Silva A, Gu W, Joiner J, Koster RD, Lucchesi R, Molod A, Owens T, Pawson S, Pegion P, Redder CR, Reichle R, Robertson FR, Ruddick AG, Sienkiewicz M, Woollen J (2011) MERRA: NASA's modern-era retrospective analysis for research and applications. J Clim 24(14):3624–3648. https://doi.org/10.1175/JCLI-D-11-00015.1
42. Döll P, Müller Schmied H, Schuh C, Portmann FT, Eicker A (2014) Global-scale assessment of groundwater depletion and related groundwater abstractions: combining hydrological modeling with information from well observations and GRACE satellites. Water Res Res 50(7):5698–5720. https://doi.org/10.1002/2014WR015595
43. Tapley BD, Bettadpur S, Ries JC, Thompson PF, Watkins MM (2004) GRACE measurements of mass variability in the earth system. Science 305:503–505. https://doi.org/10.1126/science.1099192
44. Bruinsma S, Lemoine J, Biancale R, Valès N (2010) CNES/GRGS 10-day gravity field models (release 2) and their evaluation. Adv Space Res 45(4):587–601. https://doi.org/10.1016/j.asr.2009.10.012
45. Lemoine F, Luthcke S, Rowlands D, Chinn D, Klosko S, Cox C (2007) The use of mascons to resolve time-variable gravity from GRACE. In: Tregoning P, Rizos C (eds) Dynamic planet. Springer, Berlin, pp 231–236
46. Ramillien G, Cazenave A, Brunau O (2004) Global time variations of hydrological signals from GRACE satellite gravimetry. Geophys J Int 158(3):813–826. https://doi.org/10.1111/j.1365-246X.2004.02328.x
47. Chambers D, Wahr J, Nerem R (2004) Preliminary observations of global ocean mass variations with GRACE. Geophys Res Lett 31(L13310). https://doi.org/10.1029/2004GL020461
48. Wahr J, Jayne S, Bryan F (2002) A method of inferring changes in deep ocean currents from satellite measurements of time-variable gravity. J Geophys Res 107(C12):3218. https://doi.org/10.1029/2002JC001274
49. Rodell M, Famiglietti JS (1999) Detectability of variations in continental water storage from satellite observations of the time dependent gravity field. Water Res Res 35(9):2705–2724. https://doi.org/10.1029/1999WR900141
50. Tiwari V, Wahr J, Swenson S (2009) Dwindling groundwater resources in northern India, from satellite gravity observations. Geophys Res Lett 36:L18401. https://doi.org/10.1029/2009GL039401
51. Werth S, Güntner A, Petrovic S, Schmidt R (2009) Integration of GRACE mass variations into a global hydrological model. Earth Planet Sci Lett 27(1–2):166–173. https://doi.org/10.1016/j.epsl.2008.10.021
52. Baur O, Kuhn M, Featherstone W (2009) GRACE-derived ice-mass variations over Greenland by acocunting for leakage effects. J Geophys Res 114(B06407). https://doi.org/10.1029/2008JB006239
53. Velicogna I (2009) Increasing rates of ice mass loss from the Greenland and Antarctic ice sheets revealed by GRACE. Geophys Res Lett 36:L19503. https://doi.org/10.1029/2009GL040222
54. Boy J-P, Chao B (2005) Precise evaluation of atmospheric loading effects on Earth's time-variable gravity field. J Geophys Res - Solid Earth 110(B8):4–12. https://doi.org/10.1029/2002JB002333

55. Swenson S, Wahr J (2002) Estimated effects of the vertical structure of atmospheric mass on the time-variable geoid. J Geophys Res 107(B9):2194. https://doi.org/10.1029/2000JB000024
56. Barletta V, Sabadini R, Bordoni A (2008) Isolating the PGR signal in the GRACE data: impact on mass balance estimates in Antarctica and Greenland. Geophys J Int 172(1):18–30. https://doi.org/10.1111/j.1365-246X.2007.03630.x
57. Swenson S, Wahr J, Milly PCD (2003) Estimated accuracies of regional water storage variations inferred from the Gravity Recovery and Climate Experiment (GRACE). Water Res Res 39(8):1223. https://doi.org/10.1029/2002WR001736
58. Ramillien G, Frappart F, Cazenave A, Güntner A (2005) Time variations of land water storage from an inversion of two years of GRACE geoids [rapid communication]. Earth Planet Sci Lett 235(1–2):283–301. https://doi.org/10.1016/j.epsl.2005.04.005
59. Wahr J, Molenaar M, Bryan F (1998) Time variability of the Earth's gravity field: Hydrological and oceanic effects and their possible detection using GRACE. J Geophys Res (Solid Earth) 103(B12):30205–30230. https://doi.org/10.1029/98JB02844
60. Heiskanen WA, Moritz H (1967) Physical geodesy. W.H Freeman and Company, San Francisco
61. Arras C, Jacobi C, Wickert J, Heise S, Schmidt T (2010) Sporadic E signatures revealed from multi-satellite radio occultation measurements. Adv Radio Sci 8:225–230. https://doi.org/10.5194/ars-8-225-2010
62. Yang Q (2016) Applications of satellite geodesy in environmental and climate change. Graduate Thesis and Dissertations. http://scholarcommons.usf.edu/etd/6440. accessed 26 Jan 2017
63. Brutsaert W (2005) An introduction hydrology, 4th edn. Cambridge University Press, New York
64. Montgomery EL (1971) Determination of coefficient of storage by use of gravity measurements. Ph.D. thesis, University of Arizona, Tucson
65. Leirião S, He X, Christiansen L, Andersen OB, Bauer-Gottwein P (2009) Calculation of the temporal gravity variation from spatially variable water storage change in soils and aquifers. J Hydrol 365(3–4):302–309. https://doi.org/10.1016/j.jhydrol.2008.11.040
66. Lambert A, Beaumont C (1977) Nano variation in gravity due to seasonal groundwater movements: implications for the gravitational detection of tectonic movements. J Geophys Res 82(2):297–306. https://doi.org/10.1029/JB082i002p00297
67. Goodkind JM (1986) Continuous Measurement of Nontidal Variations of Gravity. J Geophys Res 91(B9):9125–9134
68. Mukai A, Higashi T, Takemoto S, Nakagawa I, Naito I (1995) Accurate estimation of atmospheric effects on gravity observations made with a superconducting gravity meter at Kyoto. Phys Earth Planet Interiors 91(1–3):149–159
69. Pool DR, Eychaner JH (1995) Measurements of aquifer-storage change and specific yield using gravity surveys, groundwater 33(3):425–432. https://doi.org/10.1111/j.1745-6584.1995.tb00299.x
70. Bower DR, Courtier N (1998) Precipitation effects on gravity measurements at the Canadian Absolute Gravity Site. Phys Earth Planet Interiors 106:353–369. https://doi.org/10.1016/S0031-9201(97)00101-5
71. Smith AB, Walker JP, Western AW, Ellett KM (2005) Using ground based measurements to monitor changes in terrestrial water storage. In: 29th Hydrology and water resources symposium (CD Rom), Institute of Engineers Australia
72. Damiata BN, Lee TC (2002) Gravitational attraction of solids of revolution—part 1: vertical circular cylinder with radial variation of density. J Appl Geophys 50(3):333–349. https://doi.org/10.1016/S0926-9851(02)00151-9
73. Rodell M, Famiglietti JS, Chen J, Seneviratne SI, Viterbo P, Holl S, Wilson CR (2004) Basin scale estimates of evapotranspiration using GRACE and other observations. Geophys Res Lett 31:L20504. https://doi.org/10.1029/2004GL020873
74. Ellett KM, Walker JP, Rodell M, Chen JL, Western AW (2005) GRACE gravity fields as a new measure for assessing large-scale hydrological models. In: Zerger A, Argent RM (Eds). MODSIM 2005 international congress on modelling and simulation. Modelling and simulation society of Australia and New Zealand, December 2005, pp 2911–2917, ISBN: 0-9758400-2-9

75. Rodell M, Chen J, Kato H, Famiglietti JS, Nigro J, Wilson CR (2006) Estimating groundwater storage changes in the Mississippi River basin (USA) using GRACE. Hydrogeol J 15(1):159–166. https://doi.org/10.1007/s10040-006-0103-7
76. Andersen OB, Seneviratne SI, Hinderer J, Viterbo P (2005) GRACE-derived terrestrial water storage depletion associated with the 2003 European heat wave. Geophys Res Lett 32(L18405):2–5. https://doi.org/10.1029/2005GL023574
77. Neumeyer J, Barthelmes F, Dierks O, Flechtner F, Harnisch M, Harnisch G, Hinderer J, Imanishi Y, Kroner C, Meurers B, Petrovic S, Reigber C, Schmidt R, Schwintzer P, Sun HP, Virtanen H (2006) Combination of temporal gravity variations resulting from superconducting gravimeter (SG) recordings, GRACE satellite observations and global hydrology models. J Geod 79(10–11):573–585. https://doi.org/10.1007/s00190-005-0014-8
78. Yan JP, Hinderer M, Einsele G (2002) Geochemical evolution of closed-basin lakes: general model and application to Lakes Qinghai and Turkana. Sediment Geol 148(1–2):105–122. https://doi.org/10.1016/S0037-0738(01)00212-3
79. Syed T, Famiglietti J, Rodell M, Chen J, Wilson C (2008) Total basin discharge for the Amazon and Mississippi River basins from GRACE and a land-atmosphere water balance. Water Res Res 44:W02433. https://doi.org/10.1029/2006WR005779
80. Crowley JW, Mitrovica JX, Bailey RC, Tamisiea ME, Davis JL (2006) Land water storage within the Congo Basin inferred from GRACE satellite gravity data. Geophys Res Lett 33:L19402. https://doi.org/10.1029/2006GL027070
81. Schmidt R, Schwintzer P, Flechtner F, Reigber C, Güntner A, Döll P, Ramillien G, Cazenave A, Petrovic S, Jochmann H, Wünsch J (2006b) GRACE observations of changes in continental water storage. Global Planet Change 50(1–2):112–126. https://doi.org/10.1016/j.gloplacha.2004.11.018
82. Swenson S, Yeh PJ-F, Wahr J, Famiglietti J (2006) A comparison of terrestrial water storage variations from GRACE with in situ measurements from Illinois. Geophys Res Lett 33:L16401. https://doi.org/10.1029/2006GL026962
83. Winsemius HC, Savenije HHG, van de Giesen NC, van den Hurk B, Zapreeva EA, Klees R (2006) Assessment of gravity recovery and climate experiment (GRACE) temporal signature over the upper Zambezi. Water Res Res 42:W12201. https://doi.org/10.1029/2006WR005192
84. Awange JL, Anyah R, Agola N, Forootan E, Omondi P (2013) Potential impacts of climate and environmental change on the stored water of Lake Victoria Basin and economic implications. Water Res Res 49:8160–8173
85. Awange JL, Ogallo L, Kwang-Ho B, Were P, Omondi P, Omute P, Omulo M (2008) Falling Lake Victoria water levels: is climate a contribution factor? J Clim Change 89:287–297. https://doi.org/10.1007/s10584-008-9409-x
86. Awange JL (2012) Environmental monitoring using GNSS global navigations satellite systems. Springer, Heidelberg, Berlin
87. Awange JL, Forootan E, Kuhn M, Kusche J, Heck B (2014) Water storage changes and climate variability within the Nile Basin between 2002 and 2011. Adv Water Res 73:1–25. https://doi.org/10.1016/j.advwatres.2014.06.010
88. Ndehendehe C, Awange J, Agutu N, Kuhn M, Heck B (2016) Understanding changes in terrestrial water storage over West Africa between 2002 and 2014. Adv Water Res 88:211–230. https://doi.org/10.1016/j.advwatres.2015.12.009
89. Agutu N, Awange JL, Zerihun A, Ndehedehe C, Kuhn M, Fukuda Y (2017) Assessing multi-satellite remote sensing, reanalysis, and land surface models' products in characterizing agricultural drought in East Africa. Remote Sens Environ 194:287–302. https://doi.org/10.1016/j.rse.2017.03.041
90. Agutu NO, Awange JL, Ndehedehe C, Kirimi F, Kuhn M (2019) GRACE-derived Groundwater Changes over Greater Horn of Africa: temporal variability and the potential for irrigated agriculture. Sci Total Environ. https://doi.org/10.1016/j.scitotenv.2019.07.273
91. Bettadpur S (2007) UTCSR Level-2 processing standards document for level-2 product release 0004. Gravity recovery and climate experiment (GRACE) Rev 3.1, GRACE 327-742 (CSR-GR-03-03). Center for Space Research, The University of Texas at Austin

92. Crétaux J-F, Jelinski W, Calmant S, Kouraev A, Vuglinski V, Bergé-Nguyen M, Gennero M-C, Nino F, Abarca Del Rio R, Cazenave A, Maisongrande P (2011) SOLS: A lake database to monitor in the Near real-time water level and storage variations from remote sensing data. Adv Space Res 47:1497–1507. https://doi.org/10.1016/j.asr.2011.01.004
93. Pugh D (2004) Changing sea levels. Effect of tides, weather and climate. Cambridge University Press, Cambridge
94. Abdalati W, Zwally HJ, Bindschadler B, Csatho B, Farrell SL, Fricker HA, Harding D, Kwok R, Lefsky M, Markus T, Marshak A, Neumann T, Palm S, Schutz B, Smith B, Spinhirne J, Webb C (2010) The ICESat-2 laser altimetry mission. Proc IEEE 98(5):735–751. https://doi.org/10.1109/JPROC.2009.2034765
95. Becker M, Llowel W, Cazenave A, Güntner A, Crétaux J-F (2010) Recent hydrological behaviour of the East African Great Lakes region inferred from GRACE, satellite altimetry and rainfall observations. Comptes Rendus Geosci 342(3):223–233. https://doi.org/10.1016/j.crte.2009.12.010
96. Khandu (2016) Assessing climate change impacts on water resources in the Ganges–Brahmaputra–Meghna River Basin. PhD thesis, Curtin University
97. Ebert EE, Janowiak JE, Kidd C (2007) Comparison of Near-Real-Time Precipitation Estimates from Satellite Observations and Numerical Models. Bull. Am. Meteorol. Soc. 88(1):47–64. https://doi.org/10.1175/BAMS-88-1-47

Chapter 6
GNSS Reflectometry and Applications

Increasing sophisticated uses of Global Navigation Satellite Systems (GNSS) observables have led to a new era in remote sensing. Geodesists, geophysicists, and surveyors have all established large GNSS networks. Nearly all of them have open data policies and encourage broad usage of their data. The vast majority of GNSS data users focus on positioning, although the timing and atmospheric communities also value data from GNSS networks. Here, we have shown how to further extend the value of ground GNSS networks by describing how to routinely measure soil moisture, snow depth, and vegetation growth. These data are valuable both to scientists and water managers and a cost effective use of existing infrastructure.

–K. M. Larson [10]

6.1 Remote Sensing Using GNSS Reflectometry

6.1.1 Background

When positioning with Global Navigation Satellite Systems (GNSS), multipath signal is a reflected GNSS signal that is a nuisance and as such needs to be eliminated. Whereas this reflected signal on the one hand is a nuisance for positioning, for environmental monitoring purposes, it could be useful in monitoring sea-wind retrieval, seawater salinity detection, ice-layer density measurements and other remote sensing applications (e.g., topography, soil moisture and vegetation), see, e.g., [17].

In this approach, also known as the GNSS-reflectometry (GNSS-R) remote sensing, which works as a bi-static radar (i.e., where the transmitter and receiver are separated by a significant distance, [7], the microwave signals reflected from various surfaces are received and processed to extract useful environmental information about those surfaces. As can be seen in Fig. 6.1, GNSS satellites (GPS, Galileo,

J. Awange, *Lake Victoria Monitored from Space*,
https://doi.org/10.1007/978-3-030-60551-3_6

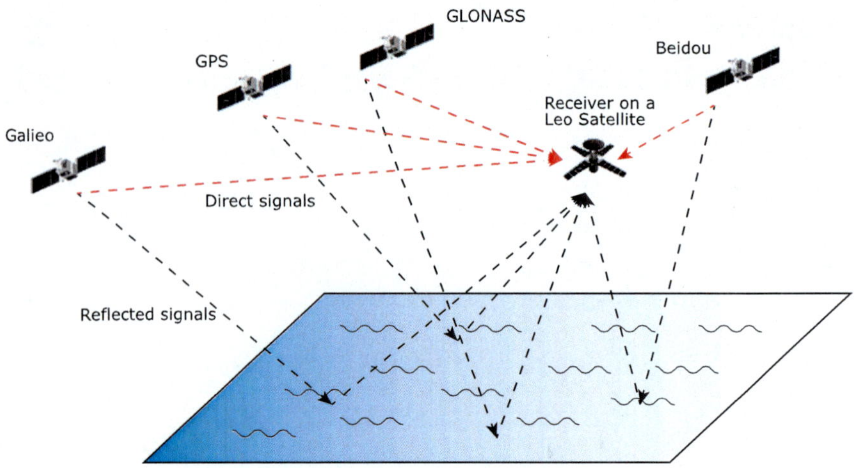

Fig. 6.1 Remote sensing using Global Navigation Satellite Systems (GNSS) reflected signals

GLONASS and Beidou) transmit signals to the receiver onboard low earth orbiting (LEO) satellites, but some signals are reflected by nearby surfaces. In this example, the reflected signals are received by the receivers, placed on LEO satellites such as GRACE (see Sect. 5.3.3). These reflected microwave signals, which could also be received by receivers situated on land, are processed in GNSS-R remote sensing to provide geophysical characteristics of the reflecting/scattering surface and in so doing give environmental monitoring parameters as we shall see in Sect. 6.1.2.

The possibility of using GNSS reflected signals for remote sensing sea surface heights was proposed by Martín-Neira [14], who used fixed-platform experiments to demonstrate that GNSS-reflection altimetry performed to an accuracy of ~20 m over the ocean, 450 m above Crater Lake, and 10 m over a pond, see e.g., [9, and the references therein]. In this pioneering work, Martín-Neira [14] suggested the use of delayed signals between the direct and reflected signals (e.g., Fig. 6.1) in what is known as passive reflectometry and interferometry system (PARIS), [7]. According to Lowe et al. [9], such GNSS altimetry would involve an orbiting receiver that obtains position and timing information from the GNSS constellation as usual, but measures ocean height using the arrival time of GNSS signals reflected from the surface. The advantage over mono-static radar altimeters discussed in 5.5 is that the receiver could produce about 25 simultaneous measurements (~ 55 GNSS satellites are fully operational, [7]), distributed over an area thousands of km across-track [9]. Such high number of independent observations obtained from GNSS over the same scene has the advantage of increasing the coverage of the area and providing more reliability of the estimated environmental monitoring parameter. Egido [3] points to two reasons why GNSS-R has gained increasing interest, i.e., (i) global availability and stability of GNSS signals, and (ii), the use of L-band radiation that makes it highly interactive with the natural scattering medium. Studies of GNSS-reflections from space include, e.g., [9, 13]. The advantages of GNSS-R remote sensing over traditional satellite scatterometry and radar altimetry are given, e.g., by [7, 17] as follows:

- provision of long term stable signals that are free without any need for additional transmitter,
- attractive of GNSS-R receivers that are small in size, light-weight, with low power consumption,
- plenty of signal sources, which in include GPS, Galileo, GLONASS, and Beidou/Compass, all which will contribute to improved spatial and temporal resolution,
- unlike radiometers, GNSS-R signals are not affected by background temperature.
- works at L-band range that is suitable for soil moisture monitoring
- use of spread-spectrum communication technology to enable the receiver to receive weak signals, and
- wide range of uses for such things as sea-wind retrieval, seawater salinity detection, ice-layer density measurement, humidity measurement of land, and the detection of moving objects.

6.1.2 Geometry and Observations

Jin et al. [7] provides good reading on GNSS-R geometry and observables. In this section, a brief outlook is provided to simplify the understanding of the GNSS-R concept. When waves originating from a single direction are reflected towards a given (single) direction, i.e., assuming a flat surface where the incident and the reflected angles are equal (Fig. 6.2), specular scattering is said to have occurred. If on the other hand the waves from a single direction are reflected into different directions, diffuse scattering is said to have occurred. Looking at Fig. 6.2, the *specular point* is the point with minimum (shortest) distance from the reflecting surface to the receiver. Away from the specular point, as the surface becomes rough, other points will also redirect the reflected signals towards the receiver, thus contributing towards the final reflected signals. The area occupied by these reflecting points is known as the *glistening zone* [3, 7] .

Assuming the specular point to be the center of origin of a cartesian coordinate system with the YZ plane being the reflection plane (Fig. 6.2), the distance β of any point to the specular point within the glistening zone can be calculated from the transmitted vector ρ_t and receiver vector ρ_r, see Edigo [3]. This distance to the specular point form the delay distance or iso-delay (blue lines in Fig. 6.2). Points with the same delay can be joined by lines forming ellipse with semi-major and semi-minor axes. Besides the iso-delay, points in the glistening zone have different doppler shift due to the transmitter—surface—receiver geometry, which can be computed from respective velocities, μ_t, μ_r [3]. Similarly, points with equal doppler shift can be joined by a line to form iso-Doppler (red lines in Fig. 6.2). The intersection of the

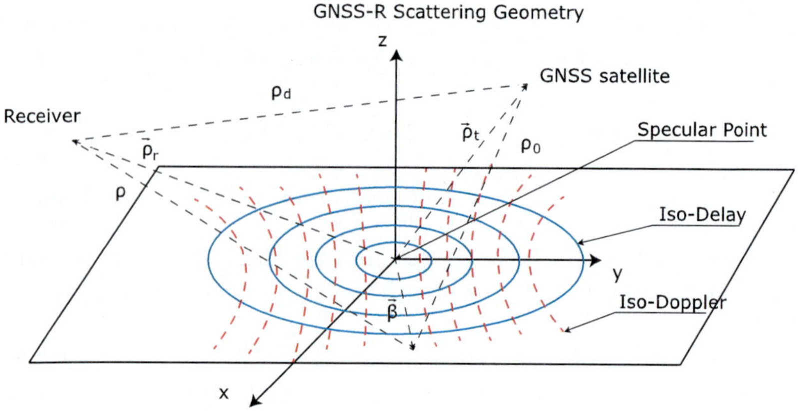

Fig. 6.2 GNSS-R geometry and observations

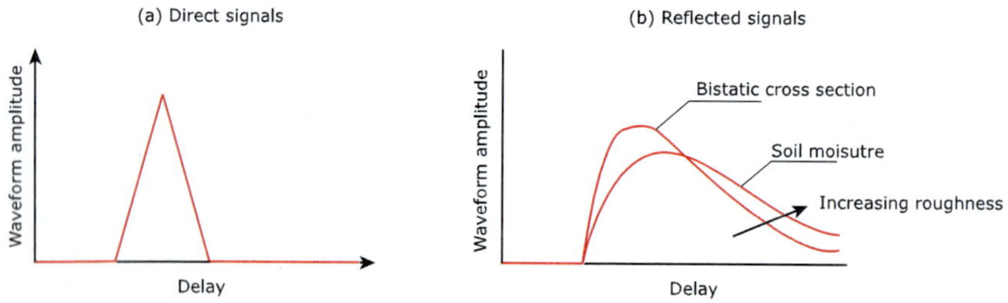

Fig. 6.3 GNSS reflectometry: Direct and reflected waveforms. Modified after Egido [3]

iso-delay and iso-doppler form the *delay doppler map* (DDM), which forms the first observable of GNSS-R signal that accounts for the average GNSS scattered power on the surface as a function of delay and frequency.

The second type of GNSS-R observations is obtained from performing cross correlation between the direct and the reflected signals with a pseudo random noise PRN replica code (for delay) and frequency shift with carrier frequency to obtain a *complex waveform*. By selecting the doppler shift of the specular point, the reflected cross correlation waveform can be obtained [4] and are depicted, e.g., in Fig. 6.3.

In general, there are two ways in which GNSS-R technique could be used to sense changes in environmental features such as soil moisture, snow depth and vegetation:

(a) using a second receiver to measure the reflected signals as performed by the GNSS reflectometry group that uses two receivers, or
(b) using a single receiver such as that used by geodesists, surveyors and geophysicists, see Fig. 6.1.

In (b), Awange [18, 19] opin that the geodesists view the reflected signal from multipath as a nuisance and attempt to model it during data processing. Using the vendor software such as Trimble Business Center (TBC), this is achieved through the scrutinization of the least squares generated residuals. For environmentalists, however, this reflected signal is exactly what they need to decipher changes in the environmental surfaces such as snow/ice and vegetation. Larson et al. [10] postulate that if one thinks on how best to measure multipath reflection rather that to model multipath corrections for carrier phase data, then one is better off using signal power, which is similar to the data used by the GNSS-reflectometry community in (a). GNSS-R receivers will slightly differ from those used by geodesist, surveyors and geophysicist to obtain positions. Geodesist GNSS receivers are tuned to measure the direct signals and suppress the reflected signals, while those of GNSS-R measure both direct and reflected signals. in addition to the direct signals, however, the geodesists receiver measures the interference of the direct and reflected signals through the signal power.

Geodetic GNSS receivers generate carrier-to-noise density data stored in Receiver Independent Exchange (RINEX) data format as signal-to-noise ratio (SNR). SNR are functions of satellite elevation angles and manifest themselves in lower elevation satellites (i.e., those setting or rising, 25° and below). GNSS-R on the other hand will receive both right hand circularly polarized (RHCP) and left hand circularly polarised (LHCP) components of the reflected signal. RHCP include the direct signal while LHCP measures the indirect (reflected) signal, see e.g., Fig. 6.3. Larson et al. [10] used both RHCP and LHCP to generate signal-to-noise ratio (SNR) as a function of the reflecting surface and the elevation angle. For a planar horizontal reflection (e.g., Fig. 6.1), the frequency of the interference of the direct and reflected signal observed in SNR data is constant as a function of the sine of the elevation angle [10]. Larson et al. [10] called this dominant frequency that can be estimated from the spectral density of the signal or extracted using periodogram as the *effective reflector height* and analysed its changes to derive the ice-depth and soil moisture changes. In essence, change in the effective reflector height implies change in the surface around the antenna [10]. Also, by analysing the amplitude of the multipath reflected signals at a GNSS site, soil moisture can be estimated [7]. In general, the measuring principle from remote sensing satellites is largely based on the dielectric properties of the reflecting surface (e.g.,wet and dry soils have different dielectric constants). Larson et al. [10] employ the amplitude of the reflection observed in the SNR data, which depend on the dielectric constant of the surface material (e.g., vegetation with high water content has much smaller SNR than vegetation with very low water content) to monitor changes in snow, soil moisture and vegetation.

6.2 Environmental Applications

Such GNSS remote sensing using reflected signals find use, e.g., in the provision of altimetric precision and spatial resolution necessary to map mesoscale eddies, which has been the most prominent limitation of conventional radar altimeters [9]. Other applications of GNSS-R remote sensing include water reservoir level and ocean monitoring [4, 5]. In addition, over the last few years, there has been increasing interest in this technique for applications such as soil moisture monitoring, where the observations relating to the flux of water to- and from- the land surface can be gleaned from GNSS multipath measurements of, e.g., snow depth and soil moisture [11, 12]. These measurements derive changes in the properties of a site's environment from changes in the amplitude and frequency of the multipath interference (relating, respectively, to attenuation properties and position of reflective surfaces) [6]. These developments have led to the establishment of new research themes targeting the measurement of land bio-geophysical parameters [4].

6.2.1 Sensing Changes in Soil Moisture

Soil moisture plays a crucial role in several fields that include agriculture [1], hydrology [2, 15], engineering (i.e., flood prediction, modelling surface runoffs and soil erosion) [8], drought monitoring [1], surface runoff after rainfall events [3], among others. In agriculture, it is well known that vegetation (crop) water content originates from soil moisture, which plays a double role of providing the essential water needed to support plant growth on the one hand while on the other hand, it is used to monitor changes in vegetation water contents. Monitoring of vegetation water content is vital for informing the impacts of climate varaibility/change on crops. Soil moisture also finds use in regulating the energy balance during land-atmosphere interaction through evapotranspiration. Its accurate monitoring is therefor essential in informing, e.g., the impacts of climate varaibility/change besides playing an active role in food production.

Traditionally, it has been monitored using passive radiometers that are less sensitive to surface characteristics but influenced more by background temperature. In use are also the active sensors such as radars that are less sensitive to soil moisture but are affected by surface characteristics, see e.g., [7]. Both the passive and the active methods are, however, expensive. Its monitoring can also be through in-situ sensors, e.g., portable sensors that can be pushed directly into the ground or buried sensors that are connected either to a fixed meter or a central monitoring station [3]. In-situ methods, however, suffer from their limited coverage that does not encompass regional or global scales. Owing to this spatial inadequacy of the in-situ sensors, remote sensing techniques (i.e., passive, e.g., optical spectrometers and microwave radiometers) and active (e.g., microwave radars)) have been employed to measure the scattering characteristics of the soil. These scattering characteristics are in turn used to provide information of the soil moisture content.

Radiometers, which measures surface temperature (i.e., from the sun or natural bodies) are endowed with high temporal resolution but poor spatial resolution. Instruments include optical that sense the top soil layer due to their short wavelength to microwave that have the advantage of being all-weather, day and night operational capability, and cloud penetration. For sensing soil moisture, the P-band (50 cm) and L-band (20 cm) are good [7]. GNSS reflectometry (GNSS-R) works within the L-band, i.e., GNSS' band [18, 19]. As opposed to the passive radiometers that measures the sun's radiation scattered off the Earth's surface or the natural radiation from bodies, active radars generate their own source of energy to illuminate and measure reflected signals back to the instrument (i.e., monostatic) or a separate receiver (i.e., bi-static, to which GNSS-R belong), see e.g., [3]. Radars can be classified as imaging (e.g., real and synthetic aperture radars (SAR)) and non-imaging (e.g., satellite altimetry discussed in Sect. 5.5.1 and scatterometers used to measure cryosphere), e.g., [3]. To maximize on the advantages of both passive and active instruments, a combination of both has been used, e.g., in the soil moisture active and passive (SMAP) mission launched on 31st January 2015, and whose mission ended with data at kilometer spatial resolution valuable for scientific study. Another soil moisture satellite mission with kilometer spatial resolution is the soil moisture and ocean salinity (SMOS) launched on the 2nd of November 2009 with a three year life span, which has been exceeded.[1] GNSS-R could be useful in calibrating and validating soil moisture products from these two missions as it can provide soil moisture (similar to SMAP and SMOS) to 0–5 cm depth [7, 12].

At a GNSS station at San Jose, California, Larson et al. [10] obtained GNSS-R measurements of volumetric soil moisture using the amplitude of SNR discussed in Sect. 6.1.2 and compared them with daily precipitation and found a strong correlation between the two products at 0–5 cm depth (Fig. 6.4d). They caution, however, that the use of GNSS-R is challenged by the presence of snow on top of the soil and also when the soil is covered by vegetation with very high water content.

6.2.2 *Sensing Changes in Vegetation*

Vegetation, like soil moisture, plays a crucial role in agriculture, hydrology, drought monitoring through, e.g., normalised difference vegetation index (NDVI) and carbon cycle, among others, see e.g., [1, 3, 7, 16, 20, 21]. For climate change, for instance, carbon dioxide (CO_2) is known to be a component of the greenhouse gas that contribute to global warming. Vegetation contributes to regulating the carbon dioxide in the atmosphere through photosynthesis process and as such, is vital for regulating global change. In-situ methods adopted for monitoring of vegetation changes measures vegetation parameters such as above ground biomass, vegetation water content and plant heights, see e.g., [3]. Similar to the case of soil moisture, the spatial inadequacy of the in-situ vegetation measuring methods necessitate the use of remote sensing techniques with regional and global coverage.

Satellite methods employed to measure vegetation changes use the scattering and attenuation of the electromagnetic waves. Sensors include optical (e.g., light detection and ranging—LIDAR) and microwave (e.g., synthetic aperture radar—SAR), both which are active sensors. Passive sensors of vegetation include multispectral and hyper-spectral optical sensors such as MODIS (moderate resolution imaging spectroradiometer) and NDVI (normalized difference vegetation index).

[1] https://earth.esa.int/web/guest/missions/esa-operational-eo-missions/smos.

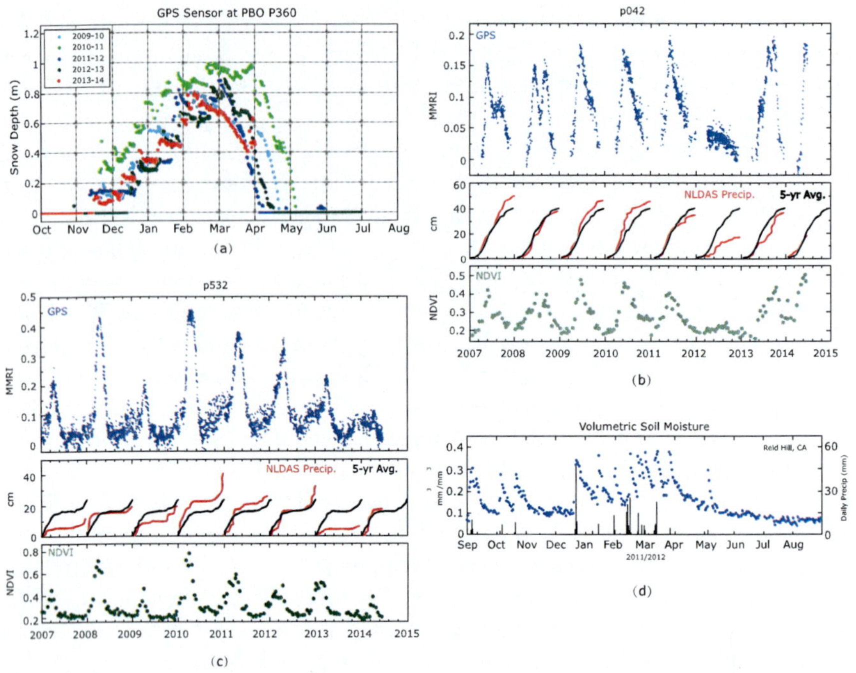

Fig. 6.4 GNSS reflectometry test results of Larson [10]; **a** Five years of snow depth time series of GPS site at P360 in southern Idaho **b** GPS site at Wheatland, Wyoming (top), GPS vegetation measurement (also called normalised microwave reflection index—NMRI) compared with normalised difference vegetation index (middle) and the cumulative precipitation from the North America Land Data Assimilation System (NLDAS) **c** GPS vegetation growth index compared with NDVI and accumulated NLDAS at site P532 located northwest of Santa Barbara, California, and **d** daily measurement of volumetric soil moisture measured with GNSS (Blue) and daily precipitation from NLDAS. *Source* Larson [10]

Using signal to noise ratio (SNR), see Sect. 6.1.2, Larson et al. [10] defined the GNSS-R derived vegetation changes based on the changes in the reflection amplitude as normalised microwave reflection index (NMRI), where a value of zero is assigned to vegetation with the lowest water content. Using two test sites (eastern Wyoming and California), they compared the NMRI vegetation water content generated from GNSS-R and those obtained from NDVI data generated at 16 days interval using MODIS sensor with a spatial resolution of 250 m. Their study revealed a close correlation between the GNSS-R derived NMRI and NDVI (i.e., a correlation of 0.86) in Wyoming, with both products capturing the 2012 drought that impacted on the vegetation (Fig. 6.4b). Just like in Wyoming, both NMRI and NDVI captured the 2014 drought in California (Fig. 6.4c). Larson et al. [10] point out that the GNSS-R sensing of vegetation has a shorter seasonal length compared to NDVI that has consistently longer growth season. They conclude that given that GNSS-R is sensitive to vegetation water content while NDVI is correlated to chlorophyl production, combining both products could provide better constraints to penologists studying the influence of climate variations on periodic plant life cycle. Further applications of GNSS satellites to sense vegetation are discussed in [18, 19].

6.2.3 Sensing Changes in Cryosphere

In Sect. 5.5.2, ICESat was presented as an altimetry satellite that is useful for monitoring changes in ice sheet. GNSS-R provide an alternative to sensing snow and ice, two components of climate system and total water storage (surface water, groundwater, soil moisture, biomass, ice and snow, see Sect. 5.3.3). Monitoring them is therefore essential in understanding how ice respond to climate change, predicting the size of ice in the polar regions, and for hydrological studies. In-situ methods for monitoring snow and ice are sparse on the one hand while the remote sensing methods are imprecise with poor spatial resolution, e.g., [7].

Jin et al. [7] present GNSS-R sensing of snow and ice in two-fold (i) relating the thickness of the amplitude of the reflected signal as a function of the incident angle or relative amplitude between polarization, and (ii), the use of multipath modulated signal. Larson et al. [10] made their first snow depth measurements in 2009 at a flat mesa site south of Boulder (USA) and reported the success of the method in retrieving the snow depths. They repeated the experiment over 5 year time period in two sites; Niwot Ridge Colorado (due to its topographical variability, extreme cold and very high wind) and Island Park, Idaho (USA). From the time series of the 5 year period, they concluded that the method was robust with very few data outages and when compared to in-situ measurements taken every two weeks, the method proved to be very accurate (Fig. 6.4a).

6.2.4 Sensing Changes in Lakes and Oceans

For sensing lakes or oceans level changes, satellite altimetry techniques discussed in Sect. 5.5 would be preferable to GNSS-R method. However, the cross-track distance of satellite altimetry method is usually large with low spatial resolution making the method unsuitable for monitoring changes in smaller lakes. It is in such instances that the GNSS-R method becomes attractive, see e.g., [8].

For oceans, the GNSS-R method offers the capability of sensing surface roughness (wind) besides the sea surface altimetry. Due to its temporal and spatial resolution, it could therefore complement the traditional satellite altimetry approach and assist in detecting, e.g., tsunami among others, see [7]. It should be mentioned that GNSS satellites can be used to obtain sea surface heights directly using buoys fitted with GNSS receivers [18, 19]. In contrast, GNSS-R uses a second receiver that receives the reflected signals from the sea surface and then computes the time delay that is multiplied by the speed of light to give the equivalent range delay. Knowing the position of the GNSS satellites accurately as well as that of the receiver measuring time delay (i.e., the reflected signal), the obtained range delay is exploited to give the instantaneous sea surface heights after the atmospheric, instrumental, and other errors have been accounted for, see e.g., Jin et al. [7] for more elaboration on the techniques and the applications of GNSS-R to measure ocean surface roughness. These application of GNSS-R discussed above could be vital tools for future monitoring of lake victoria.

6.3 Concluding Remarks

This chapter has presented in a nutshell the concepts of GNSS reflectometry (GNSS-R) and their applications to sensing the environment (e.g., soil moisture, snow/ice and hydrology). Details of the techniques are not elaborately covered here since they have been treated in other works such as [4, 7, and the references therein]. Here, the focus was to present the basics of GNSS-R method and showcase its potentials for environmental sensing within lake Victoria basin (LVB). With the proliferation of GNSS continuous operating reference stations discussed in [18, 19], there exists the potential of using their signal power through the signal-to-noise ratio (SNR) to remote sense the environment as demonstrated, e.g., by Larson et al. [10]. The development of the new L2C GPS signal and the freely available Galileo signals [18, 19] further adds weight to the potentials of GNSS-R for sensing the environment. However, even with such praise of the GNSS-R method, limitations do exist. Two such limitations are presented by Larson et al. [10] as (i) changes in environmental features of interest such as soil moisture, snow depth and vegetation cannot be measured in all GNSS sites since some are situated on top of buildings while others are near car parks where the reflected signals will be contaminated, and (ii), not all the receiver independent exchange (RINEX) data contain the desired observable, i.e., SNR or where they are available, they could have been degraded. This chapter, therefore, motivates scholars to explore the possibility of GNSS-R as a tool for sensing within LVB. This chapter complements Sect. 5.5 that measures lake surface height using satellite altimetry, albeit from a different perspective that could be advantageous. Having said that, it should be pointed out that GNSS-R technology is still at its infancy stage and possibly expensive and out of reach for users in the region. Nonetheless, it remains one of the techniques that could be exploited in future to complement studies such as that of [20].

References

1. Agutu N, Awange JL, Zerihun A, Ndehedehe C, Kuhn M, Fukuda Y (2017) Assessing multi-satellite remote sensing, reanalysis, and land surface models' products in characterizing agricultural drought in East Africa. Remote Sens Environ 194:287–302. https://doi.org/10.1016/j.rse.2017.03.041

2. Awange LJ, Gebremichael M, Forootan E, Wakbulcho G, Anyah R, Ferreira CG, Alemayehu T (2014) Characterization of Ethiopian mega hydrogeological regimes using GRACE, TRMM and GLDAS datasets. Adv Water Resour 74:64–78. https://doi.org/10.1016/j.advwatres.2014.07.012
3. Egido AE (2013) GNSS reflectometry for land remote sensing applications. PhD thesis dissertation, Starlabs, Barcelona. https://www.researchgate.net/publication/280732466_GNSS_Reflectometry_for_Land_Remote_Sensing_Applications. Accessed 25 Jan 2017
4. Egido AE, Delas M, Garcia M, Caparrini M (2009) Non-space applications of GNSS-R: from research to operational services. Examples of water and land monitoring systems. In: IEEE international geoscience and remote sensing symposium, IGARSS, Cape Town, pp II-170–II-173
5. Gleason S, Hodgart S, Sun Y, Gommenginger C, Mackin S, Adjrad M, Unwin M (2005) Detection and processing of bistatically reflected GPS signals from low Earth orbit for the purpose of ocean remote sensing. IEEE Trans Geosci Remote Sens 43(6):1229–1241. https://doi.org/10.1109/TGRS.2005.845643
6. Hammond WC, Brooks BA, Bürgmann R, Heaton T, Jackson M, Lowry AR, and Anandakrishnan S (2010) The scientific value of high-rate, low-latency GPS data, a white paper. http://www.unavco.org/community_science/science_highlights/2010/realtimeGPSWhitePaper2010.pdf. Accessed 06 Jun 2011
7. Jin S, Cardellach E, Xie F (2014) GNSS remote sensing. Theory, methods, and applications. Springer, Dordrecht
8. Jin S, Komjathy A (2010) GNSS reflectometry and remote sensing: a new objectives and results. Adv Space Res 46:111–117
9. Lowe ST, Zuffada C, Chao Y, Kroger P, Young LE, LaBrecque JL (2002) 5-cm-Precision aircraft ocean altimetry using GPS reflections. Geophys Res Lett 29(10):1375. https://doi.org/10.1029/2002GL014759
10. Larson KM, Small EE, Braun JJ, Zavorotny VU (2014) Environmental sensing. A revolution in GNSS applications. Inside GNSS July/August 36-46
11. Larson KM, Gutmann ED, Zavorotny VU, Braun JJ, Williams MW, Nievinski FG (2009)
12. Larson KM, Small EE, Gutmann ED, Bilich AL, Braun JJ, Zavorotny VU (2008) Use of GPS receivers as a soil moisture network for water cycle studies. Geophys Res Lett 35:L24405. https://doi.org/10.1029/2008GL036013
13. Lowe ST, LaBrecque JL, Zuffada C, Romans LJ, Young L, Hajj GA (2002) First spaceborne observation of an earth-reflected GPS signal
14. Martín-Neira M (1993) A passive reflectometry and interferometry system (PARIS): application to ocean altimetry. ESA J 17(4):331–335
15. Ndehedehe CE, Awange JL, Corner RJ, Kuhn M, Okwuashi O (2016) On the potentials of multiple climate variables in assessing the spatio-temporal characteristics of hydrological droughts over the Volta Basin. Sci Total Environ 557?58: 819–837. https://doi.org/10.1016/j.scitotenv.2016.03.004
16. Omute P, Corner R, Awange J (submitted) NDVI monitoring of Lake Victoria water level and drought. Water Resource Management
17. Yang D, Zhou Y, Wang Y (2009) Remote Sensing with reflected signals. GNSS-R data processing software and test analysis. Inside GNSS, Sept/Oct., pp 40–45
18. Awange JL (2012) Environmental monitoring using GNSS. Global navigation satellite system. Springer, Berlin
19. Awange JL (2018) GNSS environmental sensing. Revolutionizing environmental monitoring. Springer, Berlin
20. Morgan B, Awange JL, Saleem A, Kexiang H (2020) Understanding vegetation variability and their "hotspots" within Lake Victoria basin for the 2003–2018 period. Applied Geography
21. Omute P, Corner R, Awange JL (2012) The use of NDVI and its derivatives for monitoring Lake Victoria's water level and drought conditions. Water Resour Manage 26(6):1591–1613

Chapter 7
Improved Remotely Sensed Satellite Products

"Remotely sensed products provide a valuable alternative but come with a penalty of being mostly incoherent with each other as they originate from different sources, have different underlying assumptions and models. Improved remotely sensed satellite products for studying Lake Victoria's water storage changes is thus essential."—M. Khaki [34][1]

7.1 Summary

Lake Victoria (LV), the world's second largest freshwater lake, supports a livelihood of more than 42 million people and modulates the regional climate. Studying its changes resulting from impacts of climate variation/change and anthropogenic is, therefore, vital for its sustainable use. Owing to its sheer size, however, it is a daunting task to undertake such study relying solely on in-situ measurements, which are sparse, either missing, inconsistent or restricted by governmental red tapes. Remotely sensed products provide a valuable alternative but come with a penalty of being mostly incoherent with each other as they originate from different sources, have different underlying assumptions and models. This Chapter based on the work of [34] pioneers a procedure that uses a Simple Weighting approach to merge LV's multi-mission satellite precipitation and evaporation data from various sources and then improves them through a Postprocessing Filtering (PF) scheme to provide coherent datasets of precipitation (**p**), evaporation (**e**), water storage changes (Δ**s**), and discharge (**q**) that accounts for its water budget closure (see Fig. 7.1). Principal component analysis (PCA) is then applied to the merged-improved products to analyze LV's spatio-temporal changes resulting from impacts of climate variation/change. Compared to the original unmerged data (0.62 and 0.37 average correlation for two samples), the merged-improved products are largely in agreement (0.91 average correlation). Furthermore, smaller imbalances between the merged-improved products are obtained with precipitation (37%) and water storage changes (35%) being the

[1]This is an invited Chapter from Dr. Mehdi Khaki, School of Engineering, University of Newcastle, Callaghan, New South Wales, Australia.

J. Awange, *Lake Victoria Monitored from Space*,
https://doi.org/10.1007/978-3-030-60551-3_7

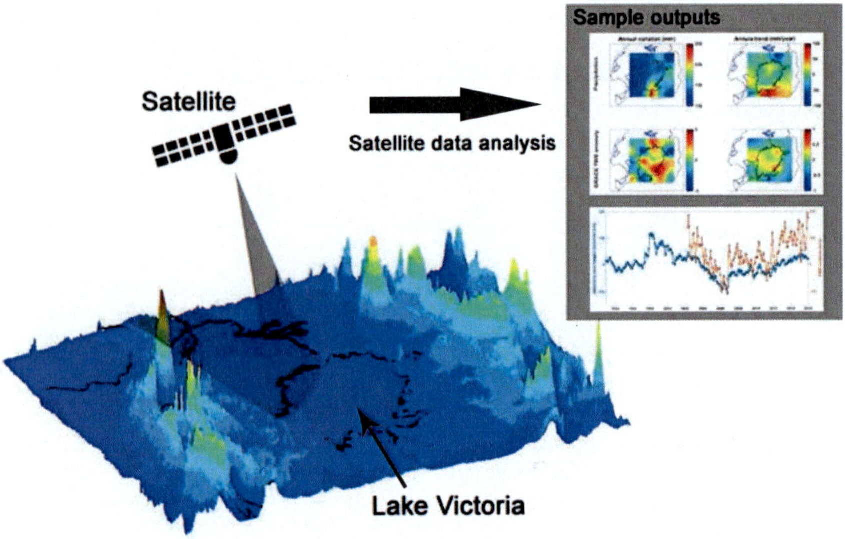

Fig. 7.1 Graphic abstract summarizing the chapter's work. Source [34]

largest contributors to LV's water budget. This data improvement scheme could be applicable to any inland lake of a size similar to LV.

7.2 Need for Merged-Improve Remote Sensing Products

Lake Victoria, spanning an area of 69,295 km^2 with a basin size of more than 280,000 km^2, e.g., [53, 76], is the largest lake in the developing world, and the world's second largest freshwater lake after Lake Superior in the US. The lake, which is shared by Kenya, Tanzania, and Uganda directly supports the livelihood of more than 42 million people, with the population projected to triple by 2050, see e.g., [12, 50]. Furthermore, being the source of the White Nile, i.e., one of the main streams of the Nile river, the lake supports the livelihood of Egypt, Sudan and South Sudan, e.g., [6]. Moreover, it is known to modulate the regional climate, e.g.,[4, 48]. Any significant change in the lake water storage, triggered e.g., by climatic impacts, e.g., [19, 52, 75] or anthropogenic factors such as dam expansion, e.g., [2, 18], therefore, is likely to affect millions of people who directly depend on it for livelihood plus others the world over who indirectly depend on it. Therefore, it is essential to continuously monitor its behaviour in terms of water storage changes and effective climate parameters as undertaken, e.g., by [3, 68, 76] among others.

Its monitoring has often taken on various forms, e.g., use of ground-based in-situ measurements, e.g., [45, 49], land surface models (LSM), e.g., [15, 35], and satellite remote sensing products, see, e.g., [20, 58, 64, 66, 74]. Due to its wide area and limited number of in-situ stations (e.g., rain and water level gauges), monitoring of the lake solely based on ground-based measurements, i.e.,"boots on the ground", becomes practically impossible. Moreover, there is no reliable regional land hydrological model in the area for undertaking such monitoring. Satellite remote sensing, on the other hand, due to their vast coverage, high spatio-temporal resolutions, and easier access provide better tools for analyzing the hydroclimate variations within Lake Victoria Basin (LVB).

A number of literatures have studied Lake Victoria using various satellite remotely sensed products, e.g., satellite radar altimetry to observe the lake's water level variations, e.g., [5, 64, 71] and their importance for flood monitoring [9], the Gravity Recovery and Climate Experiment (GRACE) for studying the lake's total water storage (TWS) changes, e.g., [3, 6, 25], satellite precipitation data for studying the lake's rainfall, e.g., [5, 36], a combination of both ground-based and remotely sensed observations for studying the lake's water balance, e.g., [68, 75]. Despite these plethora of studies, a precise study of the hydrological processes of Lake Victoria using merged and improved coherent datasets from multiple sources, what would also benefit other inland lake waters the world over, is still missing. For example, although [32] used a multi-mission satellite data to study various water storage including surface and subsurface water components over the Nile basin, their study does not account for the discrepancy between different datasets from various sources. It is also important to further study the water storage changes within the water balance equation to analyze the interrelationship between the different water components.

Due to the fast emerging satellite platforms, especially in the last two decades, there are different data sources for various data types, making the extraction of the most reliable datasets from the available products, e.g., a merged and improved rainfall data from various precipitation sources, a necessity for providing improved datasets. Moreover, the balance between different data types (i.e., precipitation, evaporation, water storage changes, and water discharge) that is normally addressed using the water balance equation stands to benefit from using such merged and improved datasets. Traditionally, hydrological model and data assimilation are used to establish the balance between different components, e.g., [31, 55, 56, 59]. Here, however, in the absence of an accurate model over the LVB, use is made of a data combination strategy to obtain a coherent data set of four water cycle components, i.e., precipitation ($\mathbf{p}$), evaporation ($\mathbf{e}$), water storage changes ($\Delta\mathbf{s}$), and discharge ($\mathbf{q}$). This could enable one to accurately analyze the lake's hydrology and the associated climatic variation/change impacts. The main objectives of this chapter, therefore, are (i) generate improved coherent water cycle components ($\mathbf{p}$, $\mathbf{e}$, $\Delta\mathbf{s}$ and $\mathbf{q}$) from different sources over lake Victoria, (ii) explore the changes in the lake's water storage and its water level using these improved coherent datasets, and (iii), investigate climatic impacts on the lake's water storage changes based on the improved datasets in (i). To achieve these goals, use is made of a proposed two-step filtering step by [1]. The filter applies a Simple Weighted (SW) approach to merge remotely sensed products over Lake Victoria from multi-mission satellites and filter them employing a Postprocessing Filter (PF) to generate improved coherent products of precipitation ($\mathbf{p}$), evaporation ($\mathbf{e}$), water storage changes ($\Delta\mathbf{s}$), and discharge ($\mathbf{q}$) by accounting for balance between them. These improved products are useful in not only analysing changes of the lake as a consequence of climate and anthropogenic impacts but also in correcting for the imbalance in the components of the water balance model ($\Delta \mathbf{s} = \mathbf{p} - \mathbf{e} - \mathbf{q}$). This procedure for improving remotely sensed data could potentially be applied to any inland lake of basin scale around the world.

Multi-mission satellite and ground-based products used include three precipitation products of Tropical Rainfall Measuring Mission Project (TRMM) [69], Global Precipitation Climatology Centre (GPCC) [61], and Climate Prediction Center (CPC) unified gauge dataset [16]. For evaporation, three data products of MODIS (Moderate Resolution Imaging Spectroradiometer) Global Evapotranspiration Project (MOD16) [43], Global Land Evaporation Amsterdam Model (GLEAM) [41], and ERA-interim [65] are employed. Water storage changes from GRACE and water discharge time series from two ground stations (Jinja and Entebbe) are also used. In addition, satellite altimetry data from TOPEX/Poseidon (T/P) and its follow-on missions Jason-1 and -2, as well as ENVISAT (Environmental Satellite) are applied for the analysis of surface water variations. The altimetry data are used to build virtual stations covering the period from 1992 to 2016 over Lake Victoria (Fig. 7.2). Lake level variations at virtual stations, and associated precipitation and TWS time series are used to analyze Lake Victoria's behavior during the study period. To improve on the satellite altimetry range estimations, which are erroneous when used over an inland body of waters and rivers [9, 14, 28], the Extrema Retracking (ExtR) algorithm of [27] is employed to retrack satellite waveform data. Furthermore, principal component analysis (PCA) [38, 57] is used to better investigate spatio-temporal variations of Lake Victoria water storage and its relationship to climatic impacts.

The remainder of the chapter is organised as follows. In Sect. 7.3 datasets are presented while Sect. 7.4 provides the method. The results and discussion are presented in Sect. 7.5, and the chapter concluded in Sect. 7.6.

7.3 Satellite Datasets

The datasets employed comprise (i) precipitation products (Tropical Rainfall Measuring Mission Project (TRMM) [69], Global Precipitation Climatology Centre, (ii) evaporation products (MODIS Global Evapotranspiration Project, Global Land Evaporation Amsterdam Model, and ERA-interim, (iii) water storage changes from GRACE, (iv) water discharge time series from two ground stations, and (v), satellite altimetry data from TOPEX/Poseidon (T/P) and its follow-on missions Jason-1 and -2, and also ENVISAT.

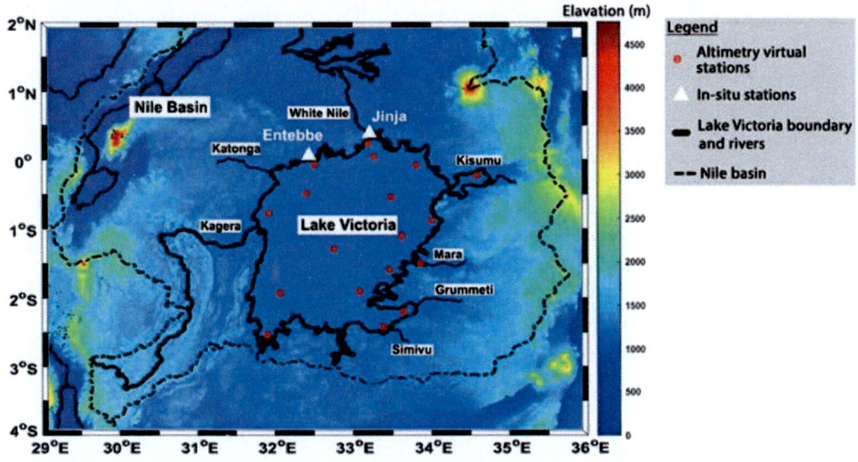

Fig. 7.2 Location of virtual altimetry stations (red) and the two water level gauge stations in Uganda (white). Source [34]

7.3.1 GRACE Products

Monthly GRACE level 2 (L2) potential coefficients products up to degree and order (d/o) 90 are obtained for the period 2002–2016 from the ITSG-Grace 2014 gravity field model [40] to estimate TWS changes, see also [29, 30]. Lower spherical harmonic degrees components are replaced with more accurate estimates of [67] (degree 1 coefficients) and [17] (Degree 2 and order 0 coefficients). The L2 gravity fields are then converted into $1^\circ \times 1^\circ$ TWS fields, see [72]. Colored/correlated noises in the products are reduced using the Kernel Fourier Integration (KeFIn) filter proposed by [33], which also accounts for signal attenuations and leakage effects caused by smoothing. The KeFIn filter works through a two-step post-processing algorithm. The first step mitigates the measurement noise and the aliasing of unmodelled high-frequency mass variations, while the second step contains an efficient kernel to decrease the leakage errors. Details of this filter can be found in [33].

7.3.2 Precipitation, Evaporation, and Discharge

Precipitation datasets from the Tropical Rainfall Measuring Mission Project (TRMM; 3B43 version 7) products [69], Global Precipitation Climatology Centre (GPCC) [61], and CPC unified gauge dataset [16] covering the period from 1998 to 2016 at monthly $1^\circ \times 1^\circ$ spatial resolution are employed.

Evaporation datasets are acquired from MODIS Global Evapotranspiration Project (MOD16) [44], Global Land Evaporation Amsterdam Model (GLEAM) [41], and

ERA-interim [65] for the same temporal period. These evaporation products are then converted to monthly $1^\circ \times 1^\circ$ spatial resolution similar to those of precipitation.

Water discharge time series are obtained from (i) the Jinja and Entebbe stations in Uganda, and (ii), from different sources including the Global Runoff Data Centre (GRDC; http://www.bafg.de/), a report produced by Power Planning Associates (PPA, 2007), and River Watch project (http://floodobservatory.colorado.edu/). Similar to precipitation and evaporation datasets, all these discharge products are resampled to monthly average height variations. Figure 7.2 shows the locations of discharge stations.

7.3.3 Satellite Radar Altimetry

TOPEX/Poseidon (T/P), Jason-1, and Jason-2 data ($\sim$9.915 day temporal resolution) of the Sensor Geographic Data Records (SGDR), which contains 20-Hz waveform data as well as ENVISAT 18-Hz SGDR product (35-day temporal resolution) from RA-2/MWR are used. This includes 360 cycles of T/P covering 1992–2002, 260 cycles of Jason-1 from 2002 to 2008, 277 cycles of Jason-2 covering 2008 to 2016, and 112 cycles of ENVISAT. T/P and Jason-1 data are both derived from the Physical Oceanography Distributed Active Archive Center (PO.DAAC), Jason-2 data is provided by AVISO, and ENVISAT data is obtained from European Space Agency (ESA). Before using these dataset, geophysical corrections that include solid earth tide, pole tide, and dry tropospheric [8] are applied. The data sets are then converted to a monthly scale and used to build virtual time series over different points (see Fig. 7.2) located on the satellite ground tracks over Lake Victoria. At each virtual point, several points belonging to the same satellite cycle are considered, and the median value of the retracked altimetry-based water levels computed to address the hooking effects [23]. This effect is derived from off-nadir measurements when a satellite locks over a water body before or after passing above it [11, 63]. A summary of the datasets used is presented in Table 7.1 while their detailed discussions are provided in Chap 5.

7.4 Simple Weighting (SW) and Postprocessing Filtering (PF) Scheme

7.4.1 Data Merging and Filtering

A two-step data combination approach proposed by [1] is applied, where first, a Simple Weighting (SW) approach is employed to merge different precipitation and evaporation data sets leading to new merged (precipitation and evaporation) products. These merged precipitation and evaporation products together with those of GRACE

Table 7.1 A summary of the datasets used in this chapter

Product	Platform	Reference	Source
Terrestrial water storage (TWS)	GRACE	[40]	Satellite remote sensing
Precipitation	TRMM-3B43	[69]	Satellite remote sensing
Precipitation	GPCC	[61]	Based on in-situ data
Precipitation	CPC	[16]	Gauge-based
Evapotranspiration	MOD16	[43, 44]	Satellite remote sensing
Evapotranspiration	GLEAM	[41]	Model and satellite remote sensing
Evapotranspiration	ERA-interim	[65]	Reanalysis dataset
Altimetry water height	T/P, Jason-1	http://podaac.jpl.nasa.gov	Satellite remote sensing
Altimetry water height	Jason-2	http://avisoftp.cnes.fr/	Satellite remote sensing
Water discharge	Jinja and Entebbe stations	Ministry of Energy and Mineral Development Kampala (Uganda)	In-situ
Water discharge	GRDC	http://www.bafg.de/GRDC/EN/Home/homepage_node.html	In-situ
Water discharge	PPA	Power Planning Associates (PPA, 2007)	In-situ
Water discharge	River Watch	http://floodobservatory.colorado.edu/	In-situ

TWS and discharge are passed through a Postprocessing Filtering (PF) procedure to generate improved precipitation ($\mathbf{p}$), evaporation ($\mathbf{e}$), water storage changes ($\Delta\mathbf{s}$), and discharge ($\mathbf{q}$) that accounts for water budget closure. Compared to other techniques, the SW approach has been shown to perform better e.g., [1]. During the SW step, the filter assigns a weight to each water component (e.g., for each precipitation and evaporation product). The PF step that is based on the water balance equation then checks the water budget closure (Eqs. 7.1 and 7.2) using a Kalman-based scheme.

$$\begin{aligned} \mathbf{p} - \Delta\mathbf{s} - \mathbf{e} - \mathbf{q} &= \mathbf{0}, \\ \mathbf{X^T}.\mathbf{G} &= \mathbf{0}, \end{aligned} \tag{7.1}$$

with $\mathbf{X}^{\mathrm{T}} = (\mathbf{p}, \Delta\mathbf{s}, \mathbf{e}, \mathbf{q})$ (T indicates matrix transpose), calculated from the first step (SW), and $\mathbf{G}^{\mathrm{T}} = (\mathbf{1}, -\mathbf{1}, -\mathbf{1}, -\mathbf{1})$. A Kalman-like solution following [55] is then applied by,

$$\begin{aligned} \mathbf{X_a} &= \mathbf{K}.\mathbf{X}, \\ \mathbf{K} &= (\mathbf{I} - \mathbf{BG}(\mathbf{G}^{\mathrm{T}}\mathbf{BG})^{-1}\mathbf{G}^{\mathrm{T}}), \end{aligned} \tag{7.2}$$

where $\mathbf{I}$ is the identity matrix, and $\mathbf{B}$ is the error covariance matrix of $\mathbf{X}$ from SW. The application of PF guarantees that the estimated flux nets in $\mathbf{X_a}$ are balanced, see details in [1, 46].

7.4.2 *Extrema Retracking (ExtR)*

Satellite radar altimetry, originally designed to monitor sea level changes, nowadays are also used to monitor inland water bodies, see e.g., [8] and rivers, see e.g., [7, 10, 70]. Nevertheless, the waveform retracking, which refers to the re-analysis of the waveforms, a time-series of returned power in the satellite antenna [21, 24], is required to improve the accuracy of measured ranges (or sea surface height; SSH) over inland waters [13]. Here, to retrack satellite radar altimetry data, a developed Extrema Retracking (ExtR) algorithm proposed by [27] is applied to generate refined virtual lake level heights that are used in the water storage analysis step. It should be pointed out that this water analysis step uses improved water storage changes obtained from the SW and PF steps (see Fig. 7.4). Our motivation for selecting the ExtR algorithm is due to its processing speed and its promising results that were obtained over the Caspian Sea when compared to the Off Center of Gravity (OCOG) [73], the NASA β-Parameter Retracking [39], and Threshold Retracking [22]. The ExtR is applied to the altimetry-derived waveforms to retrack datasets, which is necessary for inland applications of satellite radar altimetry. The algorithm includes three steps [27]; (i) applying a moving average filter to reduce the random noise of the waveforms, (ii) identifying extremum points of the filtered waveforms, and (iii), exploring the leading edges among all detected extremum points. Range corrections are applied using the offset between the position of the leading edges and their on-board values.

Two gauge stations around Lake Victoria (see Fig. 7.2) located at Jinja (1992–1995) and Entebbe (1992–2009) from the Ministry of Energy and Mineral Development Kampala (Uganda), Old Aswan (1996–2009), Esna Barrage (1996–2009), and Naga Hammadi Barrage (1996–2007), and Assiut Barrage (1996–2009) from [26], and Nubaria (1997–2007) from [60] are used to examine the performance of the ExtR filter. Retracked time series of two closest virtual stations to the in-situ stations are compared to the in-situ water level measurements. The average bias and standard deviation (STD) of average errors for both stations are presented in Fig. 7.3. Significant decreases in both bias and STD are found after applying the filtering

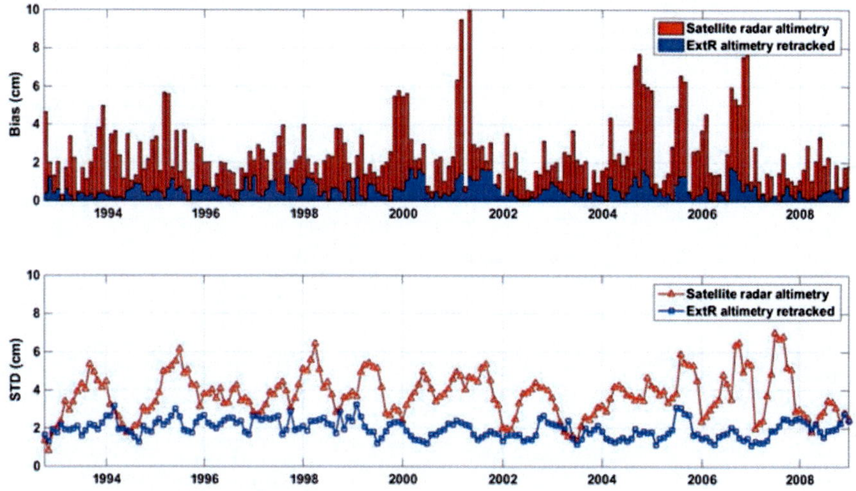

Fig. 7.3 Average bias and STD of the error time series, i.e., the difference between altimetry data and in-situ measurements. These graphs are generated using retracked time series of two closest virtual stations to the in-situ stations and corresponding in-situ water level measurements. The figure shows that the ExtR filter (blue graphs) reduces retracked errors and thus improves the satellite altimetry products employed in this chapter. Source [34]

process (cf. Fig. 7.3). The figure shows the capability of the ExtR filter for reducing errors and justifying its usage in this study.

7.4.2.1 Climate Variability Impacts

In order to investigate the impacts of climate variability/change on LV's water storage changes, correlation analysis is used. Hereafter, reported correlation values between any two variables are calculated as the average of correlation between their time series at all grid points. Furthermore, principal component analysis (PCA) [38, 57] is applied on the improved precipitation, evaporation, and water storage time series (after filtering in Sect. 7.4.1) to better analyze the spatio-temporal changes of water storages and climatic indicators. This is done to examine the climate patterns within the LV area and to investigate their connections to water storage changes. Since precipitation and evaporation are the major effective parameters on water storage recharge, the process helps to study the role of climate variability on water storage variations. A schematic illustration of the applied processing steps in this chapter, i.e., data integration procedure, retracking, and climatic impacts exploration, is provided in Fig. 7.4.

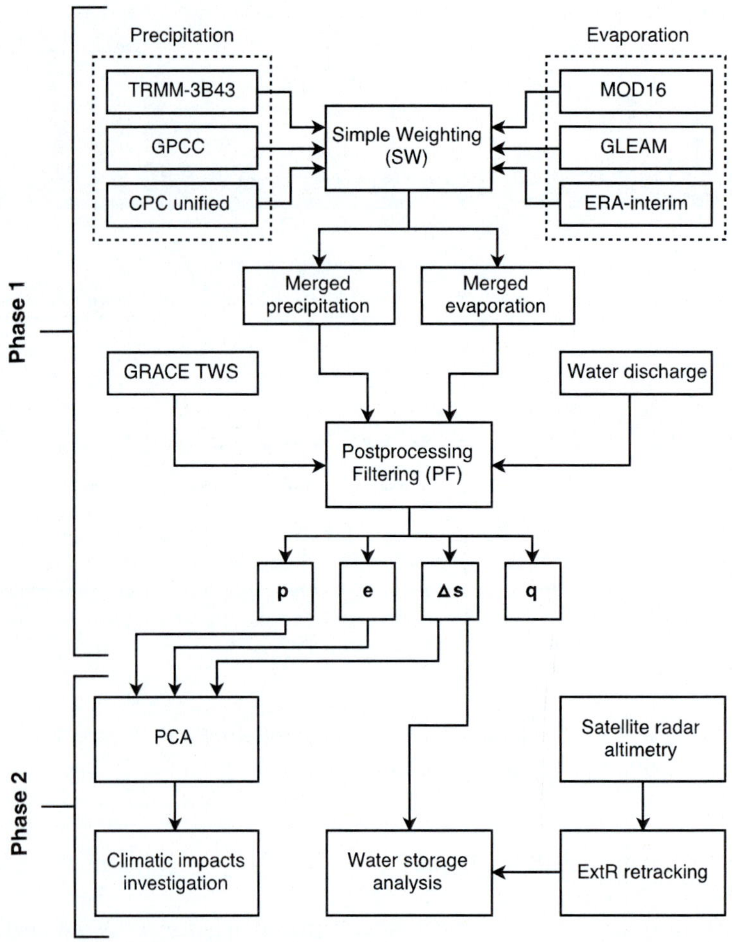

Fig. 7.4 A schematic illustration of the applied methodology. The algorithm operates in two steps. First, the SW approach is employed to merge the precipitation and evaporation data sets. The merged products together with those of GRACE TWS and discharge are then subjected to the PF filter in the second step to produce improved water budget parameters. Source [34]

7.5 Merged-Improved Products and Applications

First, we present the two-step (SW+PF) filtering results and the impact of the process on individual data type, as well as on the balance between them. Afterwards, spatio-temporal variations of precipitation and TWS and their interactions as major components of water fluxes are investigated.

7.5.1 Parameters: Coherent Filtered Products

To demonstrate the usefulness of the SW+PF for filtering remotely sensed datasets for Lake Victoria, correlations between the original and filtered time series of each water cycle component and other filtered components are calculated in order to assess how the filtering process increases the agreement between different water cycle components. Table 7.2 shows the average improvements in the correlations between every two products, i.e., how each estimated water flux is correlated with other flux observations. Note that this is done to assess the effect of the applied method to produce a more coherent data, which does not necessarily lead to a higher accuracy. Correlation values between the original and filtered water fluxes, e.g., original and filtered **p** and all data products of **e**, Δ**s**, and **q** are calculated to allow for estimation of achieved improvement in the filtered data. It is evident that in all the cases, improvements are achieved between any two filtered datasets. For example, between the original GPCC products and filtered precipitation time series, the later is 13.48% more correlated to filtered water storage changes. It can also be seen that the obtained improvements are different for various products. In general, for precipitation, higher increase in correlation is achieved from GPCC while less improvements are found in CPC, which is gauge-based. A similar correlation improvement can also be seen for evaporation, where different products (e.g., MOD16, GLEAM, and ERA-interim) receive various weights in the process, which correspondingly lead to various levels of improvements. Based on the results in Table 7.2, it can be concluded that the filtered **p**, **e**, Δ**s**, and **q** products are largely in agreement.

Table 7.2 Average correlation improvements (%) between different variables. Note that **p**, **e**, Δ**s**, and **q** refer to the filtered water cycle components of precipitation, evaporation, water storage changes, and discharge. Improvements in the correlation (r) values are calculated as $[(r_{filtered\ results} - r_{original\ data})/r_{original\ data}] \times 100(\%)$

	p	e	Δs	q
p (compared to TRMM-3B43)	0	8.51	12.93	9.81
p (compared to GPCC)	0	13.48	16.75	11.32
p (compared to CPC)	0	2.73	6.14	5.73
e (compared to MOD16)	9.50	0	11.65	6.69
e (compared to GLEAM)	8.36	0	8.01	7.17
e (compared to ERA-interim)	18.18	0	14.44	9.20
Δ**s** (compared to GRACE TWS)	16.56	12.94	0	15.58
q (compared to initial discharge)	11.11	07.08	9.52	0

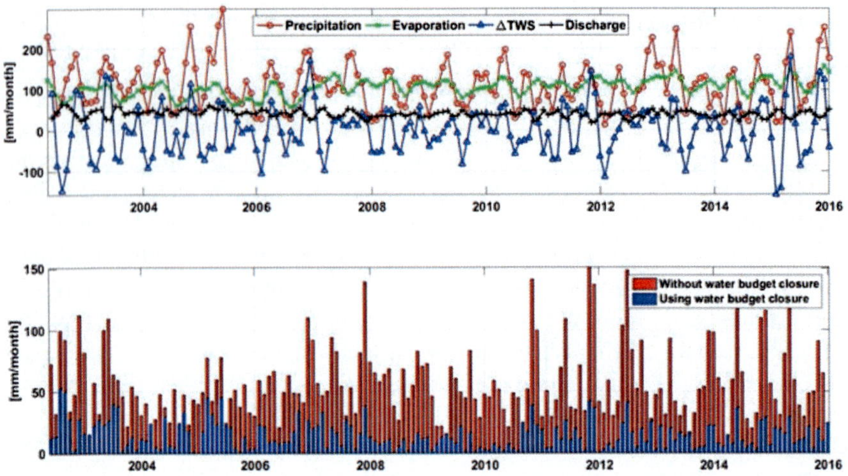

Fig. 7.5 Water components variations after filtering using SW+PF (**a**). The imbalance errors before and after applying the filter are shown in (**b**). Source [34]

In addition to correlation improvements above, based on water balance equation, the SW+PF filtering algorithms also corrects for imbalance between the water cycle components. The post-processed water fluxes from the application of the SW+PF filter is displayed in Fig. 7.5. Four filtered water budget components of precipitation, evaporation, TWS changes and discharge show different performance in the water balance equation (Fig. 7.5a). Precipitation and water storage changes are seen to have the largest contributions, i.e., 37% and 35%, respectively. Evaporation shows 19% contribution while runoff depicts the least contribution of 9%. The large value of the evaporation contribution, corroborated also by the findings of [42], indicates that a big part of precipitation over the lake area cannot recharge the outflow river (e.g., White Nile).

Figure 7.5b illustrates the imbalance error before and after using the SW+PF filtering, showcasing the capability of the filters to reduce the imbalance between water cycle components in order to provide a more coherent data sets. This is also evident from the correlations between altimetry level variations and the fluxes before and after filtering. The average correlation improvement of 12% is obtained between lake height variations and all the four components after applying the SW+PF filtering algorithm. In spite of this improvement after applying the filter, the imbalance between component can still be seen. This can be attributed to various factors such as observation errors, the contribution of groundwater in- and out-flows and its interaction with surface storage, and also the impact of extreme climate, which can be underestimated in reanalysis and remote sensing observations contrary to the in-situ discharge measurements. In what follows, these improved (merged and filtered) products are used to analyze trends in Lake Victoria's water in the face of climate variation/change impacts.

7.5.2 Analysis of Climate Impact on Lake Victoria

To investigate the impacts of climatic variation/change on the Lake Victoria water storage, first a comparison of the average variations of precipitation, evaporation, and water storage changes within the lake is made. Figure 7.6 shows the average time series of three data types from various sources including filtered and unfiltered products over the entire LVB. Note that 6 months running mean is applied to filter out high-frequency variations leading to better representations of variations and trends. As can be clearly seen, the filtered results are in a large agreement (0.91 average correlation between each two time series) compared to the two other samples from original (unfiltered) datasets, thus indicating the capability of the applied filtering method for achieving coherent data. Water storage changes largely follow the precipitation and evaporation patterns in Fig. 7.6 top panel. This shows that climate is the most effective factor in the lake's water storage changes. As expected, there is also a large agreement between precipitation and water discharge, especially after the filtering process. This agreement is better discussed by comparing Fig. 7.6 top and bottom panels, in which the original datasets are plotted. In addition to the time series' patterns, it can also be seen that the filtering approach affects the time series' magnitudes. For example, the magnitude of precipitation after the filtering is different from the original data in Fig. 7.6 middle and bottom panels. From Fig. 7.6, several significant positive and negative variations can be seen. Increases in rainfall in 2005, 2007, 2013, and 2016 cause similar rise in water storages. On the contrary, declines are observed in all time series during 2006 and 2014. Furthermore, it can be seen in Fig. 7.6 (bottom and middle panels) that a larger discrepancy exists between precipitation and two other flux observations. Due to the large evaporation rate over Lake Victoria, this larger interaction with water storage is expected. Such a connection, which can be absent on other water bodies depend on their characteristics, and can affect the water flux covariance matrix and violate the assumption of independence observation made on water budget closure. Here, however, the impact of this partial dependency between water storage and evaporation is neglected mainly due to the fact that no information is available in this regard.

To better understand water storage changes and the associated climatic impacts over the lake, PCA method is employed to the merged filtered precipitation, evaporation, and water storage changes. Figure 7.7 shows the spatial variations of these datasets within LVB corresponding to the first three empirical orthogonal functions (EOFs) of the PCA analysis. The spatial variability of water storage changes matches those of precipitation and evaporation in most of the areas proving that they are the dominant indicators of the impact of climate on LVB water storage. It can be seen that the main rainfall pattern exists in the central (EOF1) parts of the lake corresponding to a similar pattern in evaporation and water storage. The main water storage patterns, as expected, are observed in the central parts of the Lake Victoria (EOF1) as a result of recharge from rainfall in this area. EOF2 shows considerable positive signals in the western parts. This could be attributed to the contribution of Kagera river to the lake's water changes. Kagera river, which originates from Burundi, is the

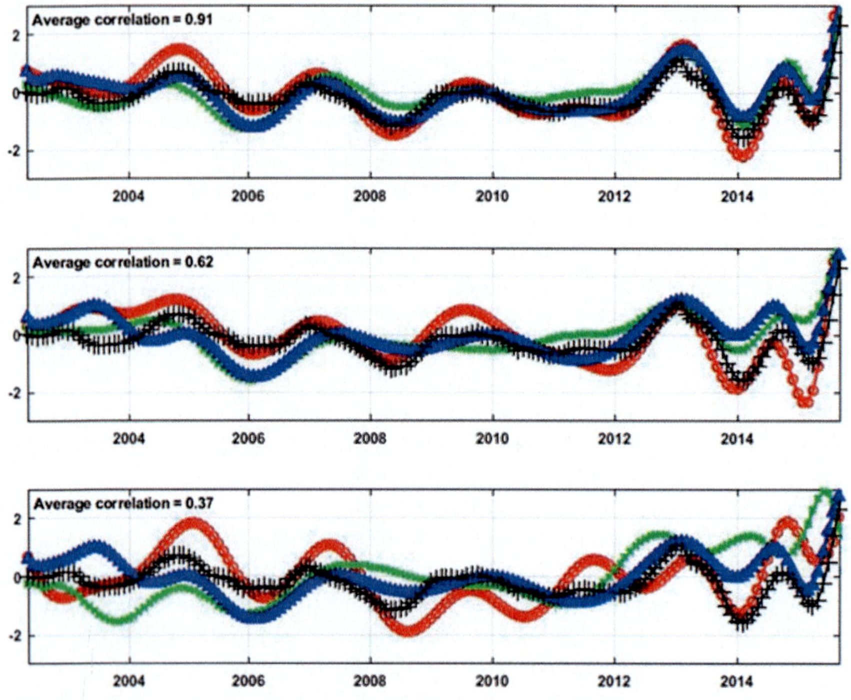

Fig. 7.6 Average (6 months running mean) precipitation (red), evaporation (green), water discharge (black), and water storage changes (blue) from different sources including filtered products (top), original TRMM, MOD16, and GRACE (middle), and original GPCC, GLEAM, and GRACE (bottom). Note that normalized values (based on the time series STD) are presented for a better visual comparison. Note that precipitation and evaporation data in the middle and bottom panels are selected randomly to show how different they can act. Source [34]

largest inflow into Lake Victoria. To a lesser degree, larger rainfall, evaporation and water storage can be observed in the eastern parts (EOF3) of the lake resulting from the effect of the south east monsoon trade winds, e.g., [3, 4], and possible recharge from the Grummeti, Simivu, and Mara rivers. Major rainfall spatial variabilities in the central (EOF1) and western (EOF2) parts corroborate the findings in [3, 4], which shows that these parts are responsible for most of the rainfalls occurring over Lake Victoria and its water recharge thereby resulting in a larger evaporation and water storage changes.

Figure 7.8 shows the corresponding first three principal components (PCs) time series. The dominant seasonal (PC1) and annual (PC2) rainfall patterns can be seen, which are in agreements with those of evaporation and water storage changes. Some significant anomalies can also be observed in precipitation, e.g., the large negative anomalies observed in 2006 (PC2) and 2014 (PC1 and PC3), and significant positive variations observed, e.g., in 2005 and 2007 (PC1, PC2, and PC3), 2010 (PC1), 2013 (PC1 and PC2), and 2016 (PC1). These variations can also be seen in evaporation

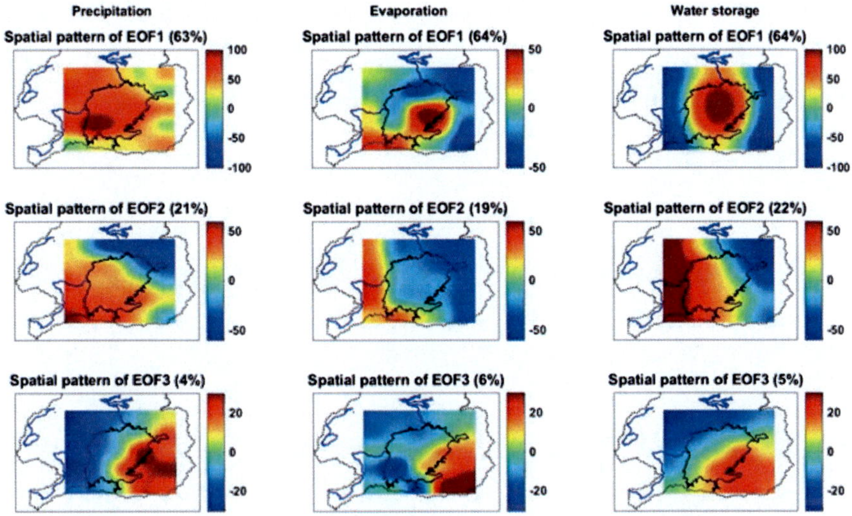

Fig. 7.7 Spatial variations of precipitation, evaporation, and water storage changes from the first three empirical orthogonal functions (EOFs) of PCA (units are mm). The central (EOF1), western (EOF2), and the eastern parts (EOF3) show large amount of variations. Source [34]

and water storage changes' time series, especially the rises in 2005, 2007, and 2013. The 2007 ENSO rainfall effect, e.g., [6, 51] is evident in PC1 for all three datasets. A negative trend is found between 2003 and 2005 for water storage time series similar to those of evaporation (PC1). This could be attributed to excessive water usages reported in the works of [3, 68] while such a negative trend is absent in rainfall time series. Large rainfalls, generally after 2013 result in positive water storage and evaporation trends. A decrease in rainfall is captured after 2011 due to the drought that affected the region, see, e.g., [5], which caused a decline in water storage variations. These similar patterns suggest a close tie between water storage variabilities and climatic impacts.

These similar patterns in climatic indicators and water storage changes, in terms of spatial (cf. Fig. 7.7) and temporal (cf. Fig. 7.8) variabilities, suggest a close tie between water storage variabilities and climatic impacts. This can also be seen in Figs. 7.9 and 7.10, which show annual spatial variations and trends, and temporal variations of precipitation and water storage over the area, respectively.

It can be seen in Fig. 7.9 that the major variations exist in the south-eastern parts, for both precipitation and water storage. A similar pattern is found for trends (right panels in Fig. 7.9). This means that precipitation plays the main role in the lake's water storage variations as already reported in other studies, e.g., [48]. One can see the same effects from the time series in Fig. 7.10, where the agreement between rainfall changes (top panel) and the lake's water storage changes (bottom panel) emphasizes the impact of rainfall on Lake Victoria. Above-average rainfalls during 2007 El'niño significantly affected water storage variations corresponding to the large anomalies.

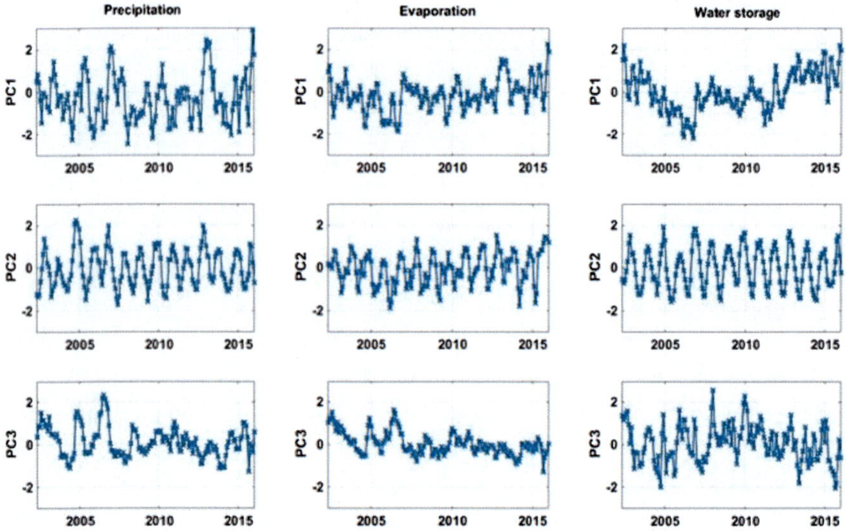

Fig. 7.8 The first three principal components (PCs) time series of precipitation, evaporation, and water storage changes from PCA. While the seasonal pattern (PC1) and annual (PC2) are dominant, several considerable positive and negative variations (e.g., in 2005 and 2007 (PC1 and PC2), and 2014 (PC3)) can also be seen in PCs. Source [34]

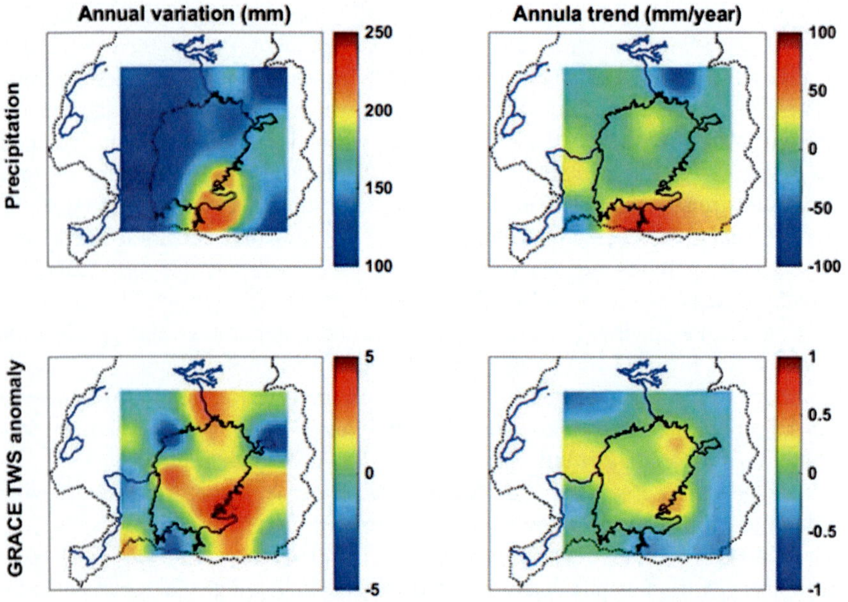

Fig. 7.9 Average annual variations and trends of precipitation and the filtered water storage time series at each grid point. Source [34]

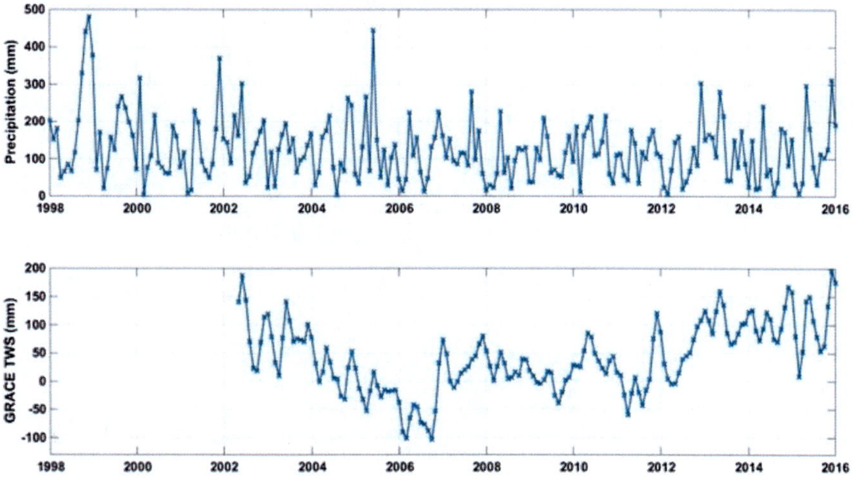

Fig. 7.10 Spatially averaged time series of precipitation and the filtered water storage changes within the Lake Victoria. Source [34]

Nevertheless, as previously mentioned, the large contribution of evaporation does not allow water storage changes to perfectly match precipitation time series variations, e.g., in 2012 and after 2014. The impact of climate variability influences the balance between precipitation and evaporation and subsequently impacts the lake's depth and arguably its areal extent and correspondingly water storage changes, e.g., [54]. Another effective factor on water storage changes is groundwater within the area, which has been under larger influences by the growing population in recent years. Nevertheless, there is not much strong evidence of interaction between groundwater and surface water mainly due to the lack of ground-based groundwater measurements.

To better study Lake Victoria's stored water changes, analyzing its surface water variations is essential. To this end, altimetry-derived surface water changes are plotted in Fig. 7.11 and compared with TWS changes from GRACE. The water level height variations in Fig. 7.11 depicts a large agreement to water storage variations. A large negative trend is found between 1998 and 2006 before a remarkable positive anomaly due to the effects of 2007 ENSO. The lake's water level variation also closely follows rainfall pattern. Larger rainfalls before 1998 result in a water level increase in the same period, which ends with a remarkable positive anomaly in 1998 due to an excessive ENSO rainfall in 1997. Decreases in water level are also observed for the period of 2002 to 2004 and after 2007 similar to water storage variations. This large agreement between GRACE-derived TWS variations and altimetry-derived surface water changes suggests that the impact of groundwater and its interaction with surface storage is minimal. This justifies the common assumption that has been made by a number of previous studies, see e.g., [37, 47, 62], in which they ignore the groundwater contribution to the Lake Victoria.

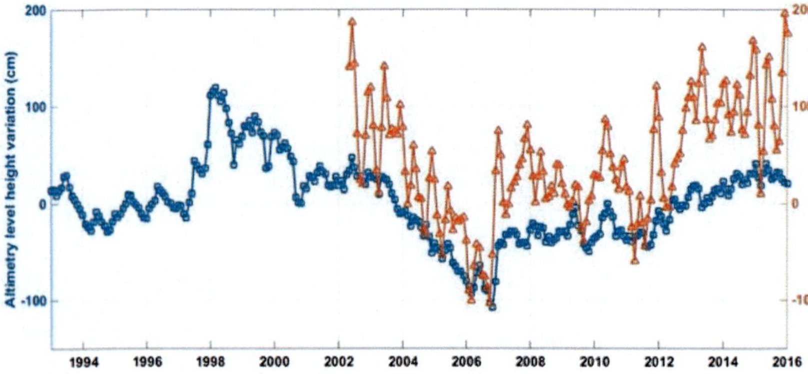

Fig. 7.11 Comparison between average time series of altimetry-derived lake level variation (after applying ExtR retracking) and the filtered water storage variations. Source [34]

7.6 Concluding Remarks

The present chapter (i) investigated the capability of the two-step data-driven approach of Simple Weighting (SW) and Postprocessing Filtering (PF) to improve remotely sensed datasets through merging and filtering to obtain four water cycle key components; precipitation, evaporation, discharge, and water storage variations, and (ii), explored the impacts of climate variabilities on water storage changes using the improved datasets in (i). The application of this approach, for the first time for LVB, results in a more efficient analysis of water fluxes that preserve water balance. The filtered water fluxes were largely in agreement compared to the original unfiltered datasets. This shows that SW+PF merging and filtering approach can effectively reduce imbalances between different observations over a limited scale inland water bodies. It was also found that there is a remarkably smaller imbalance between the post-processed time series with various rates of contribution for each water component, e.g., 37% and 35% for precipitation and water storage changes, respectively, as the largest contributions. The achieved coherent datasets allowed for a better analysis of the lake's water changes. Based on these, major rainfall spatial variabilities were observed in the central and western parts of the lake corresponding to the similar pattern in water storage changes. In addition, various strong anomalies were found in the filtered time series, e.g., in 2006 and 2014 (being negative), and 2005, 2007, and 2010 (being positive). The chapter showed that the climatic variation/change through the precipitation and evaporation (as indicators) are the main sources of the water storage changes within the lake. Moreover, an average correlation of 0.93 was found between water storage changes and the lake's water level variations, which suggests that the main part of water storage changes within the lake refers to the variation of surface storages. These findings suggest the possible application of the applied algorithm to any inland lake that permits the use of satellite remote sensing, especially GRACE for studying water storage changes.

References

1. Aires F (2014) Combining datasets of satellite retrieved products. Part I: methodology and water budget closure. J Hydrometeorol 15(4):1677–1691
2. Aman A (2014) Egypt seeks to halt Ethiopian dam, Al-Monitor: the Pulse of the Middle East. AlMonitor Retrieved April 23, 2014. http://www.almonitor.com/pulse/originals/2014/02/egypt-lobby-renaissance-dam-ethiopia.html
3. Awange JL, Sharifi MA, Ogonda G et al (2008) The falling lake victoria water level: GRACE, TRIMM and CHAMP satellite analysis of the lake basin. Water Resour Manag 22:775. https://doi.org/10.1007/s11269-007-9191-y
4. Awange JL, Anyah R, Agola N, Forootan E, Omondi P (2013) Potential impacts of climate and environmental change on the stored water of Lake Victoria Basin and economic implications. Water Resour Res 49(12):8160–8173
5. Awange JL, Forootan E, Kusche J, Kiema JBK, Omondi PA, Heck B, Fleming K, Ohanya SO, Goncalves RM (2013) Understanding the decline of water storage across the Ramser-Lake Naivasha using satellite-based methods. In: Advances in water resources, vol 60, pp 7–23, ISSN 0309-1708. http://dx.doi.org/10.1016/j.advwatres.2013.07.002
6. Awange JL, Gebremichael M, Forootan E, Wakbulcho G, Anyah R, Ferreira VG, Alemayehu T (2014) Characterization of Ethiopian mega hydrogeological regimes using GRACE, TRMM and GLDAS datasets. In: Advances in water resources, vol 74, December 2014, pp 64–78, ISSN 0309-1708. http://dx.doi.org/10.1016/j.advwatres.2014.07.012
7. Berry PAM, Garlick JD, Freeman JA, Mathers EL (2005) Global inland water monitoring from multi-mission altimetery. Geophys Res Lett 32:L16401. http://dx.doi.org/10.1029/2005GL022814
8. Birkett CM (1995) The contribution of TOPEX/POSEIDON to the global monitoring of climatically sensitive lakes. J Geophys Res 25:179–25
9. Birkett C, Murtugudde R, Allan T (1999) Indian Ocean climate event brings floods to East Africa's lakes and the Sudd Marsh. GRL 26(8):1031–1034
10. Birkett CM, Mertes LAK, Dunne T, Costa MH, Jasinski MJ (2002) Surface water dynamics in the amazon basin: application of satellite radar altimetry. J Geophys Res 107. http://dx.doi.org/10.1029/2001JD000609
11. Boergens E, Dettmering D, Schwatke C, Seitz F (2016) Treating the hooking effect in satellite altimetry data: a case study along the mekong river and its tributaries. Remote Sens 8:91. http://dx.doi.org/10.3390/rs8020091
12. Bremner J, Lopez-Carr D, Zvoleff A (2013) Using new methods and data to assess and address population, fertility, and environment links in the Lake Victoria Basin, population and the environment, session 247. https://www.iussp.org/en/event/17/programme/paper/6136
13. Brown GS (1977) The average impulse response of a rough surface and its applications. IEEE Trans Anten Propag 25:67–74
14. Calmant S, Seyler F, Cretaux JF (2008) Monitoring continental surface waters by satellite altimetry. Surv Geophys 29. http://dx.doi.org/10.1007/s10712-008-9051-1
15. Chamberlain JM, Bain CL, Boyd DFA, McCourt K, Butcher T, Palmer S (2014) Forecasting storms over Lake Victoria using a high resolution model. Met Apps 21:419–430. http://dx.doi.org/10.1002/met.1403
16. Chen M, Xie P, Janowiak JE, Arkin PA (2002) Global land precipitation: a 50-yr monthly analysis based on gauge observations. J Hydrometeorol 3:249–266
17. Cheng MK, Tapley BD (2004) Variations in the Earth's oblateness during the past 28 years. J Geophys Res, Solid Earth 109:B09402. http://dx.doi.org/10.1029/2004JB003028
18. Consulate General of Ethiopia (2012) Grand Ethiopian Renaissance Dam. Los Angeles, CA
19. Conway D (2002) Extreme rainfall events and lake level changes in East Africa: recent events and historical precedents. In: Odada EO, Olago DO (eds) The east african great lakes: limnology, palaeolimnology and biodiversity. Advances in global change research, vol 12. Kluwer, Dordrecht, pp 63–92

20. Crétaux J-F, Abarca-del-Ríoo R, Bergé-Nguyen M, Arsen A, Drolon V, Clos G, Maisongrande P (2016) Lake volume monitoring from space surveys. Geophysics 37:269–305. https://doi.org/10.1007/s10712-016-9362-6
21. Davis CH (1995) Growth of the greenland ice sheet: a performance assessment of altimeter retracking algorithms. IEEE Trans Geo-Sci Remote Sens 33(5):1108–1116. https://doi.org/10.1109/36.469474
22. Davis CH (1997) A robust threshold retracking algorithm for measuring ice-sheet surface elevation change from satellite radar altimeters. IEEE Trans Geosci Remote Sens 35(4). http://dx.doi.org/10.1109/36.602540
23. Frappart F, Calmant S, Cauhope M, Seyler F, Cazenave A (2006) Preliminary results of envisat ra-2-derived water levels validation over the amazon basin. Remote Sens Environ 100(2):252–264
24. Gomez-Enri J, Vignudelli S, Quartly G, Gommenginger C, Benveniste J (2009) Bringing satellite radar altimetry closer to shore. Remote Sens, SPIE Newsroom. http://dx.doi.org/10.1117/2.1200908.1797
25. Hassan AA, Jin S (2014) Lake level change and total water discharge in East Africa Rift Valley from satellite-based observations. In: Global and planetary change, vol 117, pp 79–90. ISSN 0921-8181. https://doi.org/10.1016/j.gloplacha.2014.03.005
26. Ismail SS, Samuel MG (2011) Response of river Nile dredging on water levels. In: Fifteenth international water technology conference, IWTC-15 2011, Alexandria, Egypt
27. Khaki M, Forootan E, Sharifi MA (2014) Satellite radar altimetry waveform retracking over the Caspian Sea. Int J Remote Sens 35(17):6329–6356. https://doi.org/10.1080/01431161.2014.951741
28. Khaki M, Forootan E, Sharifi MA, Awange J, Kuhn M (2015) Improved gravity anomaly fields from retracked multimission satellite radar altimetry observations over the Persian Gulf and the Caspian Sea. Geophys J Int 202(3):1522–1534. https://doi.org/10.1093/gji/ggv240
29. Khaki M, Hoteit I, Kuhn M, Awange J, Forootan E, van Dijk AIJM, Schumacher M, Pattiaratchi C (2017) Assessing sequential data assimilation techniques for integrating GRACE data into a hydrological model. Adv Water Resour 107:301–316, ISSN 0309-1708. http://dx.doi.org/10.1016/j.advwatres.2017.07.001
30. Khaki M, Schumacher MJ, Forootan E, Kuhn, M, Awange, E, van Dijk, AIJM (2017) Accounting for spatial correlation errors in the assimilation of GRACE into hydrological models through localization. Adv Water Resour, Available online 1 August 2017, ISSN 0309-1708. https://doi.org/10.1016/j.advwatres.2017.07.024
31. Khaki M, Ait-El-Fquih B, Hoteit I, Forootan E, Awange J, Kuhn M (2017) A two-update ensemble kalman filter for land hydrological data assimilation with an uncertain constraint. J Hydrol, ISSN 0022–1694: https://doi.org/10.1016/j.jhydrol.2017.10.032
32. Khaki M, Awange J, Forootan E, Kuhn M (2018) Understanding the association between climate variability and the Nile's water level fluctuations and water storage changes during 1992–2016. In: Science of the total environment, vol 645, pp 1509–1521, ISSN 0048-9697. https://doi.org/10.1016/j.scitotenv.2018.07.212
33. Khaki M, Forootan E, Kuhn M, Awange J, Longuevergne L, Wada W (2018) Efficient basin scale filtering of GRACE satellite products. In: Remote sensing of environment 204: 2018, pp 76–93, ISSN 0034-4257. https://doi.org/10.1016/j.rse.2017.10.040
34. Khaki M, Awange JL (2019) Improved satellite remotely sensed products for studying climate-induced changes over lake victoria. Sci Total Environ 652:915–926. https://doi.org/10.1016/j.scitotenv.2018.10.279
35. Khan SI, Adhikari P, Hong Y, Vergara H, Adler RF, Policelli F, Irwin D, Korme T, Okello L (2011) Hydroclimatology of lake victoria region using hydrologic model and satellite remote sensing data. Hydrol Earth Syst Sci 15:107–117. https://doi.org/10.5194/hess-15-107-2011
36. Kizza M, Rodhe A, Xu CY, Natle KH, Halldian S (2009) Temporal rainfall variability in the Lake Victoria Basin in East Africa during the twentieth century. Theor Appl Climatol 98:119–135
37. Kite GW (1982) Analysis of lake victoria levels. Hydrol Sci J 27(2, 6):99–110

38. Lorenz, E (1956) Empirical orthogonal function and statistical weather prediction. Technical report science report no 1, Statistical forecasting project. MIT, Cambridge
39. Martin TV, Zwally H, Brenner AC, Bindschadler RA (1983) Analysis and retracking of continental ice sheet radar altimeter waveforms. J Geophys Res 88(C3):1608. https://doi.org/10.1029/JC088iC03p01608
40. Mayer-Gürr T, Zehentner N, Klinger B, Kvas A (2014) ITSG-Grace2014: a new GRACE gravity field release computed in Graz. In: GRACE science team meeting (GSTM), Potsdam am: 29.09.2014
41. Miralles DG, Holmes TRH, de Jeu RAM, Gash JH, Meesters AGCA, Dolman AJ (2011) Global land-surface evaporation estimated from satellite-based observations. Hydrol Earth Syst Sci 15:453–469
42. Mohamed YA, van den Hurk BJJM, Savenije HHG, Bastiaanssen WGM (2005) Impact of the Sudd wetland on the Nile hydroclimatology. Water Resour Res 41:W08420. http://dx.doi.org/10.1029/2004WR003792
43. Mu Q, Heinsch FA, Zhao M, Running SW (2007) Development of a global evapotranspiration algorithm based on MODIS and global meteorology data. Remote Sens Environ 111:519–536. https://doi.org/10.1016/j.rse.2007.04.015
44. Mu Q, Zhao M, Running SW (2011) Improvements to a MODIS global terrestrial evapotranspiration algorithm. Remote Sens Environ 115:1781–1800
45. Mugidde R, Hecky RE, Hendzel LL, Taylor WD (2003) Pelagic nitrogen fixation in lake victoria (East Africa). J Great Lakes Res 29(2):76–88
46. Munier S, Aires F, Schlaffer S, Prigent C, Papa F et al (2015) Combining datasets of satellite retrieved products for basin-scale water balance study. Part II: Evaluation on the Mississippi Basin and closure correction model. J Geophys Res: Atmosp, Am Geophys Union, 2014, 119:100–116
47. Nicholson SE, Yin X (2001) Rainfall conditions in equatorial East Africa during the nineteenth century as inferred from the record of Lake Victoria. Clim Change 48(2–3):387–398
48. Nicholson SE, Some B, Mccollum J, Nelkin E, Klotter D, Berte Y, Diallo BB, Gaye I, Kpabeba G, Ndiaye O, Noukpozounkou JN, Tanu MM, Thiam A, Toure AA, Traore AK (2003) Validation of TRMM and other rainfall estimates with a high-density gauge dataset for West Africa. Part II: validation of TRMM rainfall products. J Appl Meteor 42:1355–1368
49. Njuru PG (2014) An overview of the present status of Water Quality of Lake Victoria, Kenya: a limnological perspective. In: Ministry of environment and natural resources, LVEMP water quality component kisumu, Kenya. http://hdl.handle.net/1834/6970
50. Okungu JO, Njoka S, Abuodha JOZ, Hecky RE (2005) An introduction to Lake Victoria catchment, water quality, physical limnology and ecosystem status (Kenyan sector). Lake victoria environment management project (LVEMP) Kisumu, Kenya. http://hdl.handle.net/1834/7140
51. Omondi P, Awange JL, Ogallo LA, Ininda J, Forootan E (2013) The influence of low frequency sea surface temperature modes on delineated decadal rainfall zones in Eastern Africa region. Adv Water Resour 54:161–180
52. Omondi PA, Awange JL, Forootan E, Ogallo LA, Barakiza R, Girmaw GB, Fesseha I, Kululetera V, Kilembe C, Mbati MM, Kilavi M, King'uyu SM, Omeny PA, Njogu A, Badr EM, Musa TA, Muchiri P, Bamanya D, Komutunga E (2014) Changes in temperature and precipitation extremes over the Greater Horn of Africa region from 1961 to 2010. Int J Climatol 34:1262–1277. https://doi.org/10.1002/joc.3763
53. Onganga O, Awange JL (2005) Lake victoria. In: Ecology, resources, environment, world lakes conference, vol 11. 31 Oct.-4 Nov. 2005. Nairobi. KE
54. Owor M, Taylor R, Mukwaya C et al (2011) Groundwater - surface water interactions on deeply weathered surfaces of low relief in the Upper Nile Basin of Uganda. Hydrogeol J 19:1403. https://doi.org/10.1007/s10040-011-0779-1
55. Pan M, Wood EF (2006) Data assimilation for estimating the terrestrial water budget using a constrained ensemble Kalman filter. J Hydrometeorol 7(3):534–547

56. Pan M, Sahoo AK, Troy TJ, Vinukollu RK, Sheffield J, Wood EF (2012) Multisource estimation of long-term terrestrial water budget for major global river basins. J Clim 25(9):3191–3206
57. Preisendorfer RW (1988) Principal component analysis in meteorology and oceanography. Elsevier, New York, p 425
58. Piper BS, Plinston DT, Sutcliffe JV (1986) The water balance of Lake Victoria. Hydrol Sci J 31(1):25–37
59. Sahoo AK, Pan M, Troy TJ, Vinukollu RK, Sheffield J, Wood EF (2011) Reconciling the global terrestrial water budget using satellite remote sensing. Remote Sens Environ 115(8):1850–1865
60. Samuel MG (2014) Limitations of navigation through Nubaria canal. Egypt J Adv Res 5:147–155. https://doi.org/10.1016/j.jare.2013.01.006
61. Schneider U, Fuchs T, Meyer-Christoffer A, Rudolf B (2008) In: Centre GPC (ed), Internet publication
62. Sene KJ, Plintson DT (1994) A review and update of the hydrology of Lake Victoria in East Africa. Hydrol Sci J 50(1–2):177–208
63. Seyler F, Calmant S, Santos da Silva J, Filizola N, Roux E, Cochonneau G, Vauchel P, Bonnet M-P (2008) Monitoring water level in large trans-boundary ungauged basins with altimetry: the example of ENVISAT over the Amazon basin. J Appl Remote Sens 7150:715017. http://dx.doi.org/10.1117/12.813258
64. Sichangi AW, Makokha GO (2017) Monitoring water depth, surface area and volume changes in Lake Victoria: integrating the bathymetry map and remote sensing data during 1993–2016. Model Earth Syst Environ 3:533–538. https://doi.org/10.1007/s40808-017-0311-2
65. Simmons AJ, Uppala S, Dee D, Kobayashi S (2007) ERA-interim: new ECMWF reanalysis products from 1989 onwards, ECMWF Newsletter No 110 – Winter 2006/07
66. Song C, Huang B, Ke L (2015) Heterogeneous change patterns of water level for inland lakes in High Mountain Asia derived from multi-mission satellite altimetry. Hydrol Process 29:2769–2781. https://doi.org/10.1002/hyp.10399
67. Swenson S, Chambers D, Wahr J (2008) Estimating geocentervariations from a combination of GRACE and ocean model output. J Geophys Res 113:B08410. http://dx.doi.org/10.1029/2007JB005338
68. Swenson S, Wahr J (2009) Monitoring the water balance of Lake Victoria, East Africa, from space. J Hydrol 370:163–176. https://doi.org/10.1016/j.jhydrol.2009.03.008
69. Tropical Rainfall Measuring Mission (TRMM) (2011) TRMM (TMPA/3B43) Rainfall Estimate L3 1 month 0.25 degree x 0.25 degree V7, Greenbelt, MD, Goddard earth sciences data and information services center (GES DISC), Accessed [Data Access Date]. https://disc.gsfc.nasa.gov/datacollection/TRMM_3B43_7.html
70. Tseng KH, Shum CK, Yi Y, Fok HS, Kuo CY, Lee H, Cheng X, Wang X (2013) Envisat altimetry radar waveform retracking of quasi-specular echoes over the ice-covered qinghai lake. Terrest Atmosp Ocean Sci 24:615–627. https://doi.org/10.3319/TAO.2012.12.03.01(TibXS)
71. Uebbing B, Kusche J, Forootan E (2015) Waveform retracking for improving level estimations from Topex/Poseidon, Jason-1 and -2 altimetry observations over African lakes. IEEE Trans Geosci Remote Sens 53(4):2211–2224
72. Wahr JM, Molenaar M, Bryan F (1998) Time variability of the earth's gravity field: hydrological and oceanic effects and their possible detection using GRACE. J Geophys Res 108(B12):30205–30229. https://doi.org/10.1029/98JB02844
73. Wingham DJ, Rapley CG, Griffiths H (1986) New techniques in satellite altimeter tracking systems. In: ESA proceedings of the 1986 international geoscience and remote sensing symposium (IGARSS 86) on remote sensing. Todays solutions for tomorrows information needs, vol 3, pp 1339–1344
74. Woodward G, Warren PH (2007) Body size and predatory interactions in freshwaters: scaling from individuals to communities. In: Hildrew AG, Raffaelli D, Edmonds-Brown R (eds) Body size: the structure and function of aquatic ecosystems. Cambridge University Press, Cambridge, pp 98–117

75. Yin X, Nicholson SE (1998) The water balance of Lake Victoria. Hydrol Sci J Sci Hydrol 43(5):789–811
76. Awange JL, Saleem A, Sukhadiya RM, Ouma YO, Kexiang H (2019a) Physical dynamics of Lake Victoria over the past 34 years (1984–2018): is the lake dying? Sci Total Environ 658:199–218. https://doi.org/10.1016/j.scitotenv.2018.12.051

Part III
Sensing the Lake and Its Basin

Chapter 8
Physical Dynamics of the Lake: Is It Dying?

For Lake Victoria, the second largest freshwater lake in the world, changes in its physical dynamics that have occurred due to climatic change and anthropogenic impacts have not been thoroughly studied. For example, articles written on Lake Victoria referenced various figures for the dimensions of the lake (66,400–69,485 km^2 for its area; 300–412 km for its maximum length; 240–355 km for its maximum width; and 3300–4828 km for its shorelines). These discrepancies are largely due to the difficulties of obtaining accurate data because of both the sheer size of the lake and the lack of resources that have been committed for exploratory research by regional governments.

–J.L. Awange [14]

8.1 Summary

Understanding changes in the physical dynamics of lakes (e.g., shorelines and area) is important for their management as well as for strategic development before, during, and after climate extremes (e.g., floods and droughts) in order to inform policy formulations, planning and mitigation measures. For Lake Victoria, the second largest freshwater lake in the world, changes in its physical dynamics that have occurred due to climatic change and anthropogenic impacts have not been thoroughly studied. For example, articles written on Lake Victoria referenced various figures for the dimensions of the lake (66,400–69,485 km^2 for its area; 300–412 km for its maximum length; 240–355 km for its maximum width; and 3300–4828 km for its shorelines). These discrepancies are largely due to the difficulties of obtaining accurate data because of both the sheer size of the lake and the lack of resources that have been

J. Awange, *Lake Victoria Monitored from Space*,
https://doi.org/10.1007/978-3-030-60551-3_8

committed for exploratory research by regional governments. This chapter presents the work of Awange et al. [14] who uses a suite of remotely sensed Landsat, Sentinel-2, Moderate Resolution Imaging Spectroradiometer (MODIS), Google Earth Pro imagery, Climate Hazards Group InfraRed Precipitation with Station (CHIRPS) rainfall, Multivariate El' Niño-Southern Oscillation (ENSO) Index (MEI) and altimetry data for the past 34 years (1984–2018) together with exterior sources (37 published literature; 1969–2018) of the lake's physical parameters to (i) study changes in the physical parameters of the lake; surface area, shoreline, length, and width, and establish its current (2018) state, (ii) identify and analyse the lake's hotspots (i.e., regions with significantly noticeable areal changes), and (iii), assess the impacts of climate signature (seasonal, annual, and global ENSO teleconnection) and anthropogenic activities on the lake and its hotspots (see Fig. 8.1). To achieve these aims, manual digitization, Modified Normalized Difference Water Index (MNDWI), Normalized Difference Vegetation Index (NDVI), and Principal Component Analysis (PCA) are employed for the analysis. The results indicate the mean area of the lake from the 1984, 2002, 2017 and 2018 images to be 69,295 km^2 (i.e., 812 km ^{2} or 1.2% more

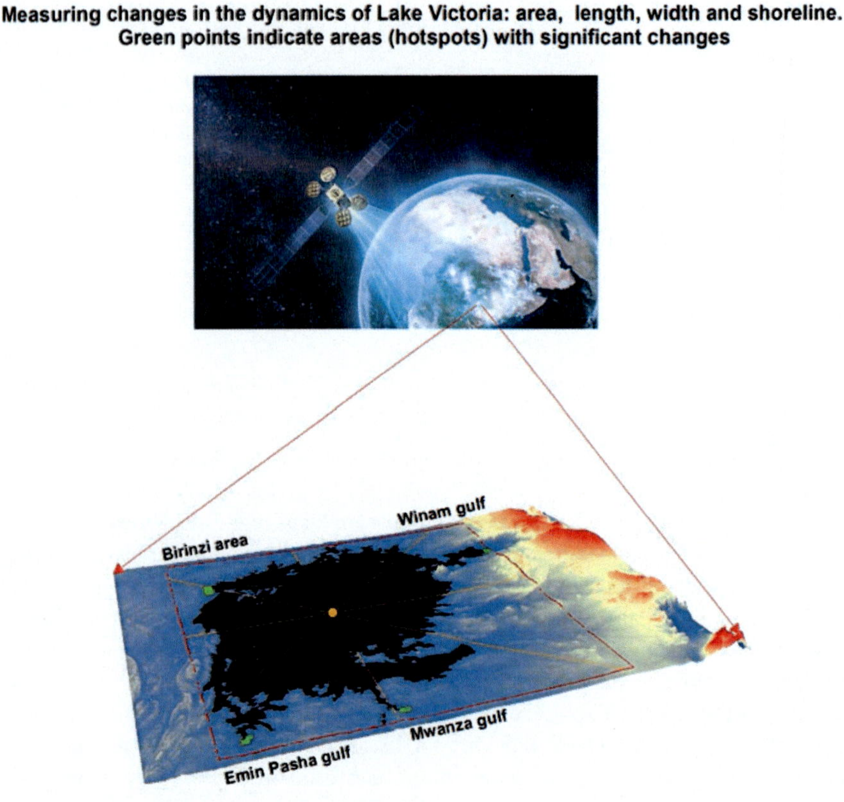

Fig. 8.1 Graphic abstract summarizing the chapter's work. *Source* [14]

than the mean value of the 37 sampled literatures), and the 2018 value to be 69,216 km^2 (i.e., differs by $\sim$733 km^2 (1.1%) from the mean value quoted in 37 previous studies sampled). As to whether the lake is actually dying, the study found that the largest tropical lake has shrunk compared to its 1984 areal value by 203 km^2 (0.3%) due to both climate change and anthropogenic factors. This areal decrease primarily occurred in four regions identified as hotspots: Birinzi area (40%), Winam Gulf (20%), Emin Pasha Gulf (38%) and Mwanza Gulf (55%). Furthermore, these hotspots were affected by the expansion of Nalubaale dam during 2002–2006, with areal decreases of 31, 10, 21 and 44% in Birinzi, Winam Gulf, Emin Pasha Gulf and Mwanza Gulf areas, respectively. Seasonal analysis show an increase of 9 km^2 in the lake's area during the heavy rainy seasons March-May (MAM) on the one hand, while on the other hand, the ENSO events enlarged the lake's area by 0.23 and 0.45% in 2007 and 2010, respectively.

8.2 Lake's Dynamics: Background

Studying the physical characteristics and dimensions of lakes and river drainage systems is important to better understand how they are impacted by natural and anthropogenic activities, see e.g., [36, 42, 64, 85, 107]. Lake Victoria, the largest tropical and second largest freshwater lake in the world, supports socio-economic sectors in the East African region [93] and thereby the livelihood of more than 42 million people, whose population is projected to triple by 2050, see e.g., [20, 57, 74]. It is thus the primary source of water for domestic consumption for all the major cities around it [93]. However, its physical parameters (length, width and surface area) have not been adequately studied and the associated impacts of climate variability/change and anthropogenic activities are not well understood, see e.g., [9, 10, 56, 58]. Information on the physical dimensions of Lake Victoria varies significantly by source. For example, according to [8], the lake has an irregular boundary with an area of about 68,635 km^2. References [55, 63] estimate the surface area to be around 68,800 km^2, [22, 69, 86] cite 68,000 km^2, while [71] on the other hand quotes a value of 69,000 km^2. The length and width of the lake also vary, depending on authors. For example, values of length and width of 400 km and 320 km, respectively, are quoted by [8], while [48] record 412 km and 355 km, respectively. Outlying values of width of 240 km is recorded by [74] and 250 km by [93]. Reference [55] gives 300 km as the length from the North to the South and 280 km width from the East to the West. These variations in the cited physical parameters of the lake are a result of its sheer size, which imposes the difficulties in in-situ data collection, thus, necessitating the use of remotely sensed data.

Remotely sensed data have been widely used to study the lakes' geomorphology globally. For example, [35] used 30-m-resolution Landsat data to produced a global inland surface water dataset while in Africa, the areal changes of Lake Manyara have been studied using Modified Normalized Difference Water Index (MNDWI) on Moderate Resolution Imaging Spectroradiometer (MODIS) data [28]. Reference [90] monitored Lake Chad's boundaries, vegetation growth, and surface temperature patterns using Landsat and National Oceanic and Atmospheric Administration (NOAA) satellite data, while monitoring of the fluctuation in spatial extent and the water levels of Lake Naivasha have been undertaken using Gravity Recovery and Climate Experiment (GRACE), Tropical Rainfall Measuring Mission (TRMM), and Landsat satellite products by [11]. Apart from the individual lakes' studies, the African water bodies (e.g., Lakes Turkana, Chew Bahir, Volta, Chad, River Niger, River Congo, etc.) have collectively been studied using MODIS products by [27].

Although remotely sensed data have been broadly applied on the African lakes as discussed above, for the greatest lake in the continent, most previous studies have focused largely on monitoring changes in its water levels/storage, e.g., [3, 9, 10, 13, 18, 56, 65, 77, 97] and related impacts of climate variability/change, e.g., [9, 12, 51, 68, 72, 82] rather than changes in its physical dynamics; information that would be useful in informing policy formulations, planning, and mitigation of climate extremes (e.g., floods and droughts) whose impacts, and those of human activities, have been observed, e.g., by [9, 10, 73, 91].

Of the studies that address the physical dynamics of Lake Victoria and the related changes, e.g, by [21, 81], a general approach has been to perform such analysis within a wider global context that does not permit an in-depth analysis of the localised changes of the lake's physical dynamics. For instance, [21] studies volume changes of 135 lakes globally for the period 1984 to 2015 using altimetry and Landsat surface data where Lake Victoria is grouped under constant area lakes with a minimum area of 66,123 km^2 and maximum area of 66,765 km^2. [81] on the other hand uses 3 million Landsat images to quantify changes in global surface water (inland and coastal) over the past 32 years (1984–2015) and provides an online interactive data (e.g., https://global-surface-water.appspot.com/) where one can see changes in these inland water bodies over the period. The only documented work that specifically addresses the physical dynamics of Lake Victoria at a local rather than global level is by [98] who uses coarse 500 m resolution remotely sensed MODIS imagery to study changes in the lake's water level and volume for the past 22 years (1991–2012). However, [98] does not address the changes in the lake's physical dimensions (i.e., changes in surface area and shoreline length with time), thereby opening room for the use of high-resolution remote sensing data to study changes in the lake's physical dynamics associated with climate variability/change and anthropogenic activities. This will be an important step towards realizing a sustainable management of this world's largest tropical lake.

Adding to the contributions above to better the understanding of changes in Lake Victoria's physical dynamics over the past 34 years (1984–2018), therefore, this chapter employs, for the first time, the high spatio-temporal resolution Sentinel-2 data and manual digitization method not used in the previous studies to (i) accurately determine the lake's current physical parameters (surface area, shoreline and extent) and evaluate their changes since 1984, (ii) identify hotspot regions (i.e., areas with significant areal changes), and (iii), assess the impacts of climate (seasonal, annual and ENSO global teleconnection) and anthropogenic activities on the lake's physical characteristics and that of its hotspots. To this end, manual digitization and Modified

Normalized Difference Water Index (MNDWI) are used to analyse Landsat (1984, 2002 and 2017) and Sentinel-2 (2018) imagery that cover the entire lake. Digital images from Google Earth Pro are then used to visually observe the trend of changes in surface area within each identified hotspot region. The impacts of climate variation/change and anthropogenic activities on the lake and its hotspot regions are evaluated by analyzing changes in rainfall using PCA (principal component analysis), variations in water levels and changes in vegetation cover using Normalised Difference Vegetation Index (NDVI).

The remainder of the chapter is organized as follows. Section 8.3 introduces Lake Victoria and details the data and methods used in the chapter, while the results and discussion are presented in Sect. 8.4 before concluding in Sect. 8.5.

8.3 Data and Methods

8.3.1 Lake Victoria

Lake Victoria (Fig. 8.2a, located at latitude 31°39′E–34°53′E and at longitude 0°20′N–3°S, 1135 m above mean sea level) in East Africa is bordered by Kenya (17%) to the East, Uganda (33%) to the North-West and Tanzania (50%) to the South [10], with a catchment area of 184,000 km^2 [56, 63]. The lake significantly influences the economic and social development of East Africa [8, 54], although it

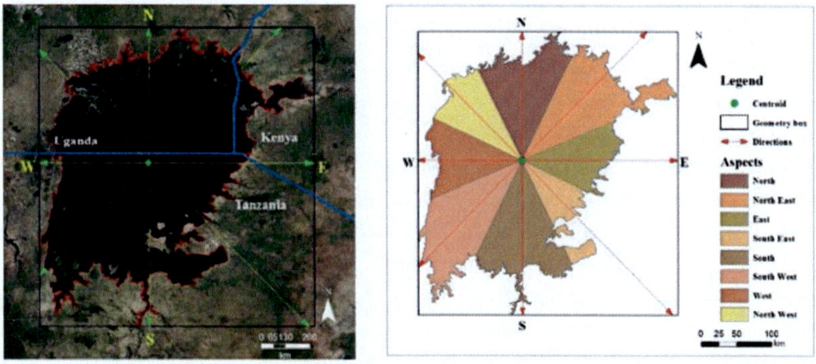

(a) Lake Victoria (b) Aspect zones for identification of the dynamic hotspots

Fig. 8.2 Lake Victoria **a** with the lake's boundary indicated in red, its center by the green point, the geometry box covering maximum extents of the lake by the black rectangle, and the directions by the green arrows where the N-S and E-W are the maximum length and width, respectively, and **b** the aspect zones from the digitised Landsat and Sentinel-2 images are merged from which the calculated surface areas are divided into increased and decreased areas used to identify the hotspots. *Source* [14] has considerably less volume than smaller lakes in the region [8]. It is also a source of the second largest river in the world, the White Nile [10].

The seasonal climate of the lake is mainly controlled by the inter-tropical convergence zone (ITCZ), which crosses the lake twice a year giving it its bimodal rainfall pattern; the long rains of March-May (MAM) and short rains of September-November (SON) [8, 12, 68, 93]. Rainfall is the crucial source of Lake Victoria's water balance and provides 80% of its recharge [9, 56]. The northern and western shores of the lake receives heavy rainfall contributed by the South-East trade winds, which take moisture from Lake Victoria while passing and condenses to give a heavy rain [10].

8.3.2 Data

Multi-spectral satellite images and hydrological data are employed to study the lake's physical dynamics (changes) and the associated impacts of climate variability/change and anthropogenic activities. The dataset are presented in Sects. 8.3.2.1–8.3.2.7, and summarized in Table 8.1.

8.3.2.1 Landsat

Level-1 Landsat satellite imagery captured during the short rainy season between September to January of the respective years are used to observe spatio-temporal changes in the lake's parameters (length, width, shoreline length and surface area). This data provide valuable remotely sensed imagery that highlight changes in land cover and terrestrial ecosystems over the lake's basin caused by increase in population and climate change that occurred over the past four decades [105]. Surface data that was available from 1972 to the present widened the scope of the spatio-temporal analysis. However, the sheer size of the lake and the presence of cloudy areas reduced the amount of usable data. For example, a lot of the imagery captured over the study area are either cloudy (>50%) or for the 1990–2000 period, not large enough to cover the entire lake, thus limiting the usable Landsat scenes needed to cover the extent of the expansive lake area to six distinct images for 1984 and 2002, and seven for 2017 (i.e., December 2016 and January 2017 images are combined). The paths and rows of the images are 170–171 and 60–62, respectively. 171–62 path and row are also used for Landsat-8 imagery from 2017 during the mosaicking process.

8.3.2.2 Sentinel-2

To validate the results of the Landsat data discussed above, and also to identify the lake's current parameters (2018), remotely sensed imagery captured by the high spatio-temporal Sentinel-2 sensor, recently employed by [31] for the study of the

Table 8.1 Summary of the datasets used. (Note the asterick* indicate spatial resolution (in meters for the images). *Source* [14])

Description	Source	Study period	Data resolution		Cloud coverage	Usage	Data access
			Spatial*	Temporal (days)	(%)		
Multi-spectral images	Landsat 5	Sep-Nov 1984	30	16	0–40	Lake's dynamics	https://earthexplorer.usgs.gov/
	Landsat 7	Sep-Nov 2002	30	16	0–46	Lake's dynamics	https://earthexplorer.usgs.gov/
	Landsat 8	Dec-Jan 2016–2017	30	16	0–10	Lake's dynamics	https://earthexplorer.usgs.gov/
	Landsat 8	Jan 2017	30	16	0–10	Lake's dynamics	https://earthexplorer.usgs.gov/
	Sentinel-2	Jan 2018	10	5	0–10	Lake's dynamics	https://earthexplorer.usgs.gov/
	MODIS	May 2007, 2010	250	8	N/A	Lake's dynamics	https://lpdaac.usgs.gov/
	(MOD09Q1)	Dec 2006, 2009	250	8	N/A	Lake's dynamics	https://lpdaac.usgs.gov/
	Google Earth	2002–2006	N/A	N/A	N/A	Anthropogenic impacts	
Precipitation	CHIRPS	1984-Feb 2018	0.05°	30	N/A	Impacts of climate change	http://chg.geog.ucsb.edu/data/chirps/
Water level height	Altimetry	1992-Jan 2018	N/A	10–35	N/A	Impacts of climate change	http://hydroweb.theia-land.fr/
ENSO index	MEI	1996-Feb 1998	N/A	N/A	N/A	Impacts of climate change	https://www.esrl.noaa.gov/psd/enso/mei/
ENSO index	MEI	2006–2007	N/A	N/A	N/A	Impacts of climate change	https://www.esrl.noaa.gov/psd/enso/mei/
ENSO index	MEI	2009-2010	N/A	N/A	N/A	Impacts of climate change	https://www.esrl.noaa.gov/psd/enso/mei/
Evaporation	Merra-2	1984–2017	0.5° × 0.625°	monthly	N/A	Compare with area change	https://disc.gsfc.nasa.gov/
Air temperature	Merra-2	1984–2017	0.5° × 0.625°	monthly	N/A	Compare with area change	https://disc.gsfc.nasa.gov/

Liberian coast, is used. The European Space Agency (ESA) launched Sentinel-2 in June 2015 to monitor the environment, vegetation, land cover and natural hazards [30, 32]. Sentinel-2 provides superior imagery with consistent land coverage and relatively high spatial resolution (10 m compared to 30 m of Landsat) and 5 days' scene revisit (compared to 16 days for Landsat, see Table 8.1) [30]. In addition, Sentinel-2 satellites have 13 multi-spectral bands, which have various spatial resolutions (10, 20, and 60 m). Four of these bands (blue, green, red and NIR) have 10 m spatial resolution, which are effective for monitoring water bodies [32]. The study relied on these four bands to generate high resolution imagery needed for accurately calculating the 2018 parameters of the lake. Fourteen Sentinel-2 images were obtained from USGS Earth Explorer to cover the entire lake (see the link in Table 8.1).

8.3.2.3 MODIS

Terra MODIS (MOD09Q1) products are employed in this chapter to observe the areal changes of the lake caused by global ENSO teleconnection. MODIS images have 250 m spatial resolution in red and NIR bands, and high temporal resolution (8 days) [47]. The 2007 and 2010 ENSO years are analysed for changes in the lake's surface area whereby the images' 2 scenes before and after the ENSO event of each year are downloaded from LP DAAC website (https://lpdaac.usgs.gov/) to cover the entire study area.

8.3.2.4 Google Earth Pro images

The freely available interactive interface Google Earth Pro images that provide digital images of the Earth's surface [99] are used for the dynamic hotpsots' analysis for the 2002–2006 period to assess anthropogenic impacts on the lake's dynamics. Images from December of each year are applied for the areal change analysis for the selected hotspots.

8.3.2.5 CHIRPS (Rainfall Data)

Climate Hazard Group Infrared Precipitation with Station (CHIRPS) data was developed by the U.S Geological Survey (USGS) and Climate Hazards Group (CHG) to study climate extremes and environmental change. CHIRPS delivers daily, pentadal and monthly high resolution (0.05°) gridded, unbiased data with regular updates since 1981 [37]. References [37, 61] have successfully studied rainfall analysis in the African continent using CHIRPS datasets. This chapter quantified CHIRPS monthly rain-fall data for the study period (1984–Feb 2018) to evaluate the impacts of climate variation/change on surface area changes.

8.3.2.6 Satellite Altimetry Data

Envisat, Jason-1, and Geosat Follow-On (GFO) satellite altimetry data spanning from 1992 to February 2018 obtained from Hydroweb website (http://hydroweb.theia-land.fr/) are here used to compare the behaviour of the surface area of the lake with variations in water levels. This dataset has already been corrected for atmospheric related errors. The surface water levels of lakes and rivers are calculated in meters above mean sea level and are available from 1992 to the present.

8.3.2.7 Multivariate ENSO Index (MEI)

Six variables (sea level pressure, surface zonal, meridional wind components, sea surface temperature, sea air temperature and cloudiness) are used to create the MEI [104] data spanning the period between 1950 and 2018, which are available at https://www.esrl.noaa.gov/psd/enso/mei/. Here, only data that covered 2006–2007 and 2009–2010 are used to study the impacts of the 2007 and 2010 ENSO events on the lake'e dynamics.

8.3.2.8 Merra-2 Reanalysis

The second Modern-Era Retrospective analysis for Research and Applications (MERRA-2) is an atmospheric reanalysis dataset that is derived from Goddard Earth Observing System Model and Atmospheric Data Assimilation System [19, 40]. The MERRA-2 tavgM_2d_flx_Nx: Surface Flux Diagnostics, is selected to estimate evaporation and air temperature above Lake Victoria, to capture their relationships with surface area. The data are available from 1980 to present, with 0.5° × 0.625° spatial and monthly resolution (downloaded at https://disc.gsfc.nasa.gov/datasets/M2TMNXFLX_V5.12.4/summary?keywords=merra-2).

8.3.3 Methods

Manual digitization, Modified Normalised Difference Water Index (MNDWI) and Normalised Difference Vegetation Index (NDVI) are employed to analyze the satellite imagery while Principal Component Analysis (PCA) is implemented on rainfall data. This section discusses the image processing methods used, including; atmospheric correction, removing cloud data, mosaicking and image registration. A summary of the image processing approach and the employed methods is presented in Fig. 8.3.

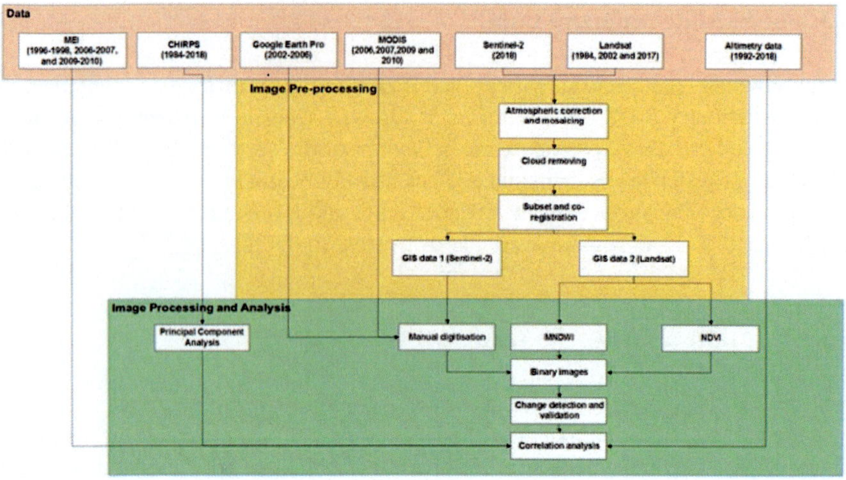

Fig. 8.3 A summary of the data (see Table 8.1) and the processing/analysis methods used in this chapter. *Source* [14]

8.3.3.1 Image Pre-processing

Atmospheric correction (e.g., radiometric correction and dark object substraction) is implemented on 33 satellite images to reduce the atmospheric disturbance and enhance the quality of data, see e.g., [39, 87]. To cover the lake, the study mosaiced 6 Landsat images for each of 1984 and 2002, 7 images for 2017 and 14 Sentinel-2 scenes for 2018. The percentage of cloud cover was high in most Landsat images and a bulk of this imagery is used to remove cloudy areas from mosaiced images by clipping, masking and mosaicking technique. Landsat images from 1984 and 2002 are registered and clipped with Landsat 2017 imagery, which is selected as a reference. The snap raster environment option is used to align pixels in the images (1984 and 2002) with the adjusted raster layer (Landsat 2017). The obtained Root Mean Square Error (RMSE) during image registration is < 0.5 pixel.

8.3.3.2 Manual Digitisation

Although manual digitisation technique, e.g., [29, 102, 108] is a time consuming approach, it is a more precise and accurate method compared to the automated MNDWI approach [26]. A person can easily identify land features from a digitised product, which can be verified using other data [26]. Because of this, and to ensure accurate and consistent results that identify and quantify changes in Lake Victoria's parameters, it is carried out on four Landsat images (1984, 2002, and 2017 at a uniform scale of 1:3000) and for Sentinel-2 image (2018 at 1:2000). Moreover,

Landsat images have cloud cover within the boundaries of the lake that impact on the accuracy of the automated procedure.

Whereas [29] used a scale of 1:5000 to delineate river channels, large scales of 1:3000 for Landsat and 1:2000 for Sentinel-2 are selected for a better and accurate boundary delineation of Lake Victoria. During the digitization of each image, bands 2, 3 and 5 of Landsat images are combined to distinguish the boundaries between water and land. The NIR band of Sentinel-2 imagery is only used to clearly discriminate water and land boundary during digitisation. To handle cloud cover while delineating boundaries, additional imagery are used as references. For example, clouds and shadows are observed while digitizing the image of 1984, and as such, either the 2002 or 2017 image is used as a reference to identify the 1984 boundary of each region. Additionally, 4 MODIS images are also digitised to quantify areal changes of the Lake during the 2007 and 2010 ENSO events.

To provide accurate values of lengths and widths, a centroid of the lake together with a geometrical boundary box (Fig. 8.2a) is created using ArcGIS. Using this centroid and the generated boundary box, lengths and widths are measured from eight directions as shown in the figure, resulting in the maximum length and width given by the N-S and E-W measurements, respectively.

The lengths of the shoreline are then computed using the polyline shapefile, and the surface area calculated using the polygon shapefile. Digitized boundaries of the lake for the 1984, 2002 and 2018 years are then merged and the centroid of the lake calculated and used to divide digitised boundaries into eight aspects based on the measured directions (Fig. 8.2b). Changes in surface area are divided into areas that decreased, increased and remained unchanged. Maximum decreased aspects are visually chosen and based on the aspect ratio, i.e., N-S, E-W, NE-SW, and NW-SE, areas with observed significant changes in surface area are designated as hotspots. Dynamic changes in these hotspots are then rigorously analyzed.

8.3.3.3 Modified Normalised Difference Water Index (MNDWI)

In addition to the more accurate manual digitization approach discussed in Sect. 8.3.3.2 above, and in order to assess whether same behavioral pattern is captured by different approaches, i.e., manual digitization and automated MNDWI, Modified Normalised Difference Water Index (MNDWI) [106] is employed on Landsat imagery because it more accurately identifies water bodies than Normalised Difference Water Index (NDWI) [38, 43, 106]. Noise from the vegetation and built-up regions can be reduced using green and mid-infrared (MIR) bands [106], which also enable water boundaries to be precisely discriminated from land surfaces. MNDWI, with values ranging between -1 (vegetation, soil and build-up areas) and $+1$ (clear water bodies), provide water features with higher positive values due to the absorption of MIR light than near-infrared (NIR) from NDWI [106]. Bands 3 and 6 are green and MIR in Landsat 8 imagery while bands 2 and 5 are green and MIR in Landsat 5 and 7. The green band reflects maximum water surface while MIR highlights non-water features with negative values in MNDWI results [28]. There are instances where binary images from MNDWI inappropriately classifies boundaries in areas where clouds and shadows are present. In such cases, manual editing and merging are used to obtain the original boundary of the lake. MNDWI equations for Landsat 5 and 7 (Eq. 8.1) and Landsat 8 (Eq. 8.2) show green as G and mid infrared as MIR. The results of MNDWI are compared with those of manual digitization as mentioned above for pattern behavior assessment. For Landsat 5 (TM) and 7 (ETM+):

$$MNDWI = \frac{\text{band 2 (G)} - \text{band 5 (MIR)}}{\text{band 2 (G)} + \text{band 5 (MIR)}} \tag{8.1}$$

For Landsat 8 (OLI),

$$MNDWI = \frac{\text{band 3 (G)} - \text{band 6 (MIR)}}{\text{band 3 (G)} + \text{band 6 (MIR)}} \tag{8.2}$$

8.3.3.4 Normalised Difference Vegetation Index (NDVI)

Whereas MNDWI discussed in Sect. 8.3.3.3 is useful in identifying the boundary between land and water, Normalised Difference Vegetation Index (NDVI) is employed to depict the vegetation cover along the lake's shoreline for further anthropogenetic impact analysis. Although a number of vegetation indices have been developed, NDVI (Eq. 8.3) is the most common and widely used index to calculate the vegetation cover. For example, [11, 28, 77] employed it to evaluate the impact of land use/land cover on areal changes of Lake Victoria. NDVI compares red and near infrared (NIR) bands of Landsat 5 (TM) and 7 (ETM+) to produce a ratio between -1 and $+1$ based on the extent of surface reflectance. The high reflectance of NIR characterises green vegetation while red wavelength absorbs chlorophyll, with the positive values corresponding to green vegetation (i.e., values > 0.2 define vegetation cover) [84]. The calculated NDVI from the images are classified into areas of bare land, dense and moderate vegetation and water bodies using manual thresholding in ArcGIS environment. Vegetation areas are calculated from binary images in which one class comprises of dense and moderate vegetation and bare land, while the other class comprise of the lake's water. The NDVI in proximity of hotspots are calculated for 1984 (Landsat TM), 2002 (Landsat ETM+) and 2017 (Landsat OLI) to analyse vegetation cover changes. Bands 3 and 4 represent red and Near infrared (NIR) for Landsat 5 and 7 while bands 4 and 5 represent red and Near infrared (NIR) in Landsat 8 in Eq. 8.3.

$$NDVI = \frac{\text{NIR} - \text{Red}}{\text{NIR} + \text{Red}} \tag{8.3}$$

8.3.3.5 Principal Component Analysis (PCA)

Principal Component Analysis (PCA [15]), widely used in climate and water storage change studies (e.g., [10, 12, 46]), is a well-established approach to identify variance in physical fields related to meteorology [83], and is a useful approach for reducing the size of large datasets without losing much information on the one hand, e.g., [16, 44], while allowing appropriate examinations by simplifying a complex sets of inter-relationships into two or more new variables on the other hand, e.g., [94]. Its outcomes are empirical orthogonal functions (EOFs), which is a linear combination of basis and mean fields, and the principal components (PCs), i.e., a linear combination of time-dependent coefficients [103]. In this chapter, PCA is applied to rainfall data to assess the impacts of climate variability/change on the lake's dynamics. Assuming a data matrix $\mathbf{X}$ contains rows representing the observations over time n over p locations, PCA decomposes the data matrix as:

$$\mathbf{X} \simeq \mathbf{P}_j \mathbf{E}_j^T, \tag{8.4}$$

where $\mathbf{E}_j^T$ contains the unit length eigenvectors (i.e., empirical orthogonal functions EOFs or spatial components) in its columns arranged with respect to the magnitude of eigenvalues and $\mathbf{P}_j$ is their corresponding temporal components (i.e., principle components PCs). The detail can of PCA can be found, e.g., in [46, 83].

8.3.3.6 Water Level Changes

The average of altimetry data from 1992 to 2018 are calculated to analyse the annual water level changes, which are correlated with areal changes in hotspot regions to understand anthropogenic impacts on Lake Victoria's physical dynamics. Furthermore, averages are calculated for the two main rainy seasons; March-May (MAM) and September-November (SON) in order to analyse seasonal water level changes and the resulting changes in the physical dynamics of the lake.

8.3.3.7 Change Detection and Validation

The physical parameters of the lake are calculated from both manual digitisation and MNDWI procedures and the results discussed in Sect. 8.4.1. The areal changes of the Lake are identified by intersecting the lake's boundaries from Landsat images (1984, 2002, 2017) and Sentinel-2 (2018). The lake's areas with significant areal changes are then referred to as hotspots whose surface areal changes are sought. For validation purpose, a simple shoreline positional accuracy approach [41] is adopted, where extreme lengths are used to validate the shoreline's positions from Landsat data with the Sentinel-2 (2018) derived shoreline used as a reference. The shoreline from Landsat 2017 imagery is segmented into points with intervals of 100 m and intersected with a 30 m buffer layer from the 2018 Sentinel-2 imagery. Following the delineation of the shorelines for each year, an accuracy assessment is performed using error matrix [25]. Overall accuracy and Kappa coefficients are calculated for the unchanged areas of the shorelines within 1 km buffer distance to include both water and land.

8.3.3.8 Evaporation and Temperature Analysis

The variables of 'evaporation from turbulence' and 'surface air temperature' in MERRA-2 Surface Flux Diagnostics are extracted in order to assess significant trends and their relations with the surface area changes of Lake Victoria. Annual total volumes are calculated by accumulating annual values to obtain evaporation parameters while the annual temperature mean are calculated by averaging monthly values. Finally, the standardized anomalies are calculated from these annual values separately for comparative purposes.

8.3.3.9 Correlation Analysis

Correlation analysis is performed between annual water level changes and annual hotspot surface areal changes for the 2002–2006 period, and the corresponding *p*-values calculated to test the significance of the obtained correlations coefficients. The graphical and spatial illustrations of correlation analysis are presented and discussed in Sect. 8.4.4.

8.4 Results and Discussion

8.4.1 Changes in the Dimensions of the Lake

The dimensions of Lake Victoria are calculated for 1984 and used as a baseline upon which values of the other years up to 2018 are compared. The mean values of parameters obtained from this study using Landsat and Sentinel-2 imagery are compared to those from 37 previous studies sampled in Table 8.2. Worth noting from the results is that the present (2018) surface area of the lake is 69,216 km^2 with a shoreline length of 4572 km. These values are higher than the mean values of those studies in Table 8.2 (i.e., average area of 68,483 km^2 and shoreline length of 3999 km, for the duration between 1969 and 2018, when these studies were carried out). These corresponds to 1.1% increase in the surface area and 17.3% increase in shoreline length compared to the mean values of those studies. Pegged against the baseline value of 1984, the present surface area of the lake shrunk by 0.3% whereas the shoreline length increased by 1.6% (see Table 8.2). The contradictory decrease in surface area of Lake Victoria on the one hand and increase in shoreline on the other hand could be attributed to the uncertainty in shoreline values quoted by the previous studies where a majority of them are based on coarse resolution images as opposed to the present study that uses higher resolution, i.e., Sentinel-2 and Landsat products. Moreover, the extraction of the lake's shoreline in this study was undertaken using manual digitisation where human eyes can detect and map boundaries of water bodies very well with high level of accuracy

Figure 8.4a, b highlighted the changes in surface area identified using manual digitisation and the automated MNDWI method on Landsat and Sentinel-2 images. The results indicate that the largest surface area of Lake Victoria obtained from manual digitisation is the 1984 value of 69,419 km^2. It shrunk to 69,321 km^2 in 2002, which is considerably higher than the 66,916 km^2 value for 2002 computed by [98] who observed noticeable drop in surface area from 2002 to 2006, and an upward trend between 2007 and 2012. Further reductions are seen in 2017 and 2018 with surface areas of 69,223 and 69,216 km^2, respectively (Fig. 8.4a). The surface areas quoted by most studies listed in Table 8.2 are below 69,000 km^2, with a majority citing a value of 68,800 km^2. This is possibly because most of those studies quoted the surface area originally from [55]. Incidentally, [55] does not state the data nor method used to obtain this area but instead refers to earlier sources. The closest value of the lake's area cited above the 69,000 km^2 mark is that of [93], i.e., 69,463 km^2, which is close to the mean value of 69,295 km^2 obtained by this study for the period (1984–2018).

The reason for the large variation in the mean value of the lake's surface area obtained in this chapter compared to the mean value from the 37 published literature is due to the higher resolution (e.g., 10 m for Sentinel-2) imagery used in this chapter, which is by far higher than the coarse resolution (e.g., 500 m for MODIS) used in the previous studies see e.g., [98]. Between 2017 and 2018, the surface area declined by 7 km^2, which was close to the annual decline rate of 5.97 km^2 computed for the values over the 34 years, considered in this work. A similar trend is seen from the results of the automated MNDWI method in Fig. 8.4b, which indicate a 174 km^2 reduction in the total surface area of the lake between 1984 to 2017. Although both methods have a near identical annual surface area decline rate, the lake's surface area from both methods differed in 1984 by 92 km^2 and in 2017 by 70 km^2 on the one hand, but on the other hand, was near identical (i.e., 69,319 km^2 for MNDWI and 69,321 km^2 for manual digitisation approach) in 2002. This difference between the two methods could be attributed to the presence of clouds and shadows that lead to misclassification of water boundaries when using the automated MNDWI method. Also, manual digitisation is more accurate method for water bodies delineation as human eyes can detect the boundary of the lake in greater details. This difference notwithstanding, the results of the MNDWI indicate a similar trend to those of the manual digitisation, thereby, lending credence to the observation that the lake's surface area has been shrinking since 1984. There was also variation in shoreline length between 1984–2018 (Fig. 8.4d) where it dramatically decreased from 4498 km in 1984 to 4312 km in 2002 (186 km decrease). The shoreline increased by nearly 9 km between 2002 and 2017 before it increased significantly by 251 km in 2018.

Table 8.2 Parameters of Lake Victoria sampled from 37 previous studies covering the period 1969–2018. Note M_1 = Mean of sources values. *Source* [14]

Sources	Area (km^2)	Shoreline (km)	Length (km)	Width (km)
[1]	66,400			
[2]	68,800			
[4]	69,000			
[5]	68,800			
[6]	68,800			
[7]	68,000			
[8]	68,635	3,300	400	320
[17]	68,500	3,440	400	240
[22]	68,000			
[23]	68,800		412	355
[24]	68,800			
[33]	68,000			
[34]	66,400			
[45]	68,800			
[48]	68,890		412	355
[50]	69,485			
[52]	68,000			
[53]	68,800		412	355
[55]	68,800		300	280
[59]	68,900			
[62]	68,800			
[63]	68,800	3,500		
[66]	68,000			
[69]	68,000			
[71]	69,000			
[74]	68,800	3,500	400	240
[76]	68,800			
[78]	68,800	4,828	412	355
[80]	68,800	4,828		
[86]	68,000			
[88]	67,000			
[92]	67,000			
[93]	69,463		400	250
[96]	68,800			
[100]	68,800			
[101]	68,800			
[110]	68,800			
Mean (M1) all studies	**68,483**	**3,899**	**394**	**306**
Mapped 1984 lake's physical parameters	**69,419**	**4,498**	**390**	**365**
Mapped 2018 lake's physical parameters	**69, 216**	**4, 572**	**386**	**362**
Mean values (1984, 2002, 2017 and 2018)	**69,295**	**4,426**	**388**	**364**
M1-2018 difference %	**1.1%**	**17.3%**	**−2.0%**	**18.5%**
1984–2018 % changes	**−0.3%**	**1.6%**	**−1.0%**	**−0.8%**

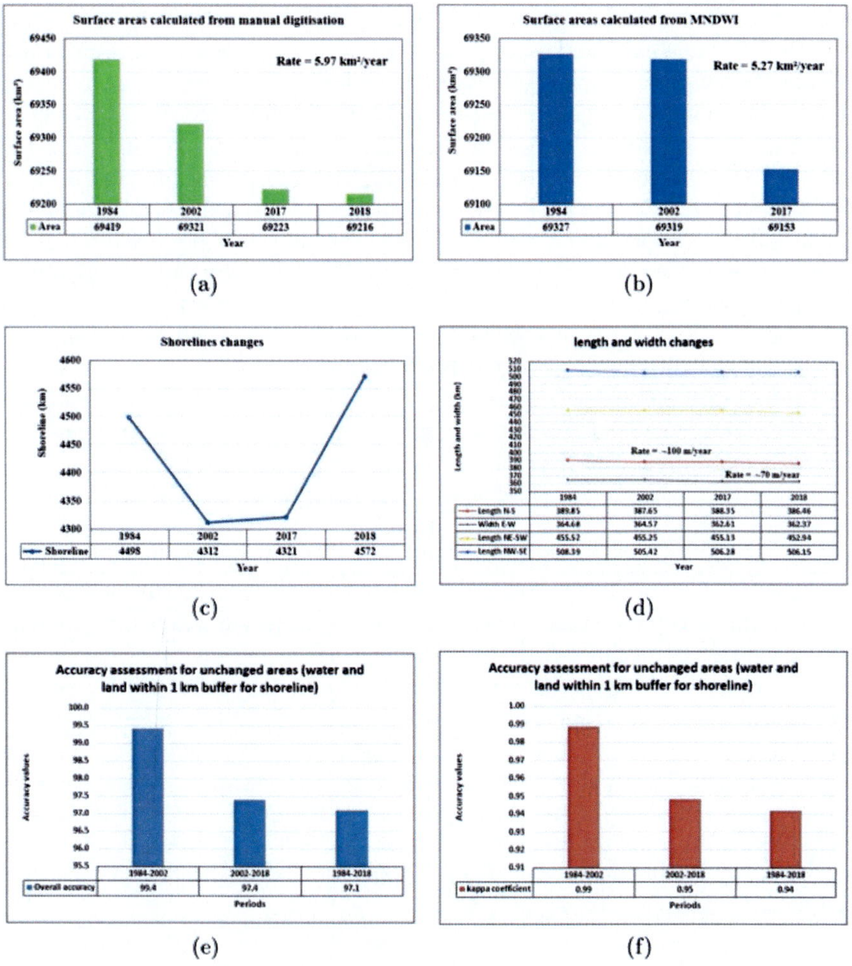

Fig. 8.4 Changes in the lake's parameters; **a** and **b** show the downward trends in the surface area for selected years obtained from the manual digitization and automated MNDWI approaches, **c** the variation in shoreline, **d** changes in the lengths (N-S) widths (E-W) and diagonal directions between 1984 and 2018, while **e** and **f** show the accuracy assessment results for the unchanged areas within 1 km buffer distance from the shoreline. Both manual and automated approaches are undertaken to assess the resulting trend in the Lake's surface area. *Source* [14]

Whereas the obtained 2017 maximum length is 388 km and maximum width is 362 km (see Fig. 8.4d), [8, 17, 23, 48, 53, 55, 74, 78, 93] cited higher lengths and smaller widths. The average maximum length and width from the studies listed in Table 8.2 are 394 km and 306 km, respectively, for the 1969–2018 duration. Figure 8.4d shows the lengths that were calculated from various directions based on a geometric box that covered the expanse of the lake and digitised boundaries from Landsat and Sentinel-2 data. Maximum lengths and widths are calculated from N-S and E-W, respectively. This calculations also include the lengths and width from NE-SW and NW-SE directions for the sole purpose of extracting the aspect zones needed to identify the hotspots. The results indicate that the lengths and widths of Lake Victoria did not dramatically change from 1984 to 2018. The length reduced from 389.85 km in 1984 to 387.65 km in 2002 while the width decreased by just 0.11 km during this period (Fig. 8.4d). From 2002 to 2017 the length of the lake increased by 0.7 km while the width decreased by 1.96 km. The length and width both decreased in 2018, and reached 386.46 km and 362.37 km, respectively. Compared to the base values of 1984, the current state of the lake in terms of length and width indicates a reduction of 3.4 km (-1.0%) and 2.3 km (-0.8%), respectively (Fig. 8.4d), with corresponding rate of changes in the length and width of $\sim$100 m/year and $\sim$70 m/year, respectively, for the period of 1984–2018.

The information on the original sources of the values quoted in literature in Table 8.2 nor the methods employed to get those values is unclear. It is possible, however, that different physical dimensions of the lake were measured (i.e., the quoted lengths and widths referred to different points of the lake) as well as the use of coarse resolution data or a less accurate methodological approach, see e.g., [98]. Our use of the more accurate manual digitisation method combined with high resolution Sentinel-2 imagery could suggest that previous studies either under-estimated or over estimated the physical parameters (maximum length and width, shoreline length and surface area of the lake.

The accuracy of the results for the lake boundary delineation can be affected by scattering from the complex water and ground surfaces [109]. Reference [29] stated that random errors in sensors and different spatial resolutions can produce errors during the detection of boundaries between land and water. Accordingly, the results considered a ± 1 pixel uncertainty in Landsat, Sentinel-2 and MODIS datasets. Furthermore, the result for the accuracy assessment indicated that the result for this analysis have an overall accuracy ranging from 97.1 to 99.4% (Fig. 8.4e) and Kappa coefficient from 0.94 to 0.99 (Fig. 8.4f).

8.4.2 Variations in Birinzi, Winam, Emin Pasha and Mwanza

Applying the aspects calculation approach to the lake (Fig. 8.2b), four zones that have large areal decline are identified as North-West (Birinzi in Uganda—33.9 km^2), South (Mwanza Gulf in Tanzania- 21.5 km^2), South-West (Emin Pasha Gulf in Tanzania—14.6 km^2), and North-East (Winam Gulf in Kenya—12.8 km^2) (Fig. 8.5a). These results corroborate the findings of [81], which shows significant surface areal decrease in these hotspots over the period 1984-2015 (see, e.g., https://global-surface-water.appspot.com/). Birinzi, Winam Gulf and Emin Pasha Gulf hotspots are further divided into regions 1 and 2 (see Fig. 8.6) for more in-depth analysis. The remaining aspect zones experienced < 15 km^2 reduction in surface area and are not considered further.

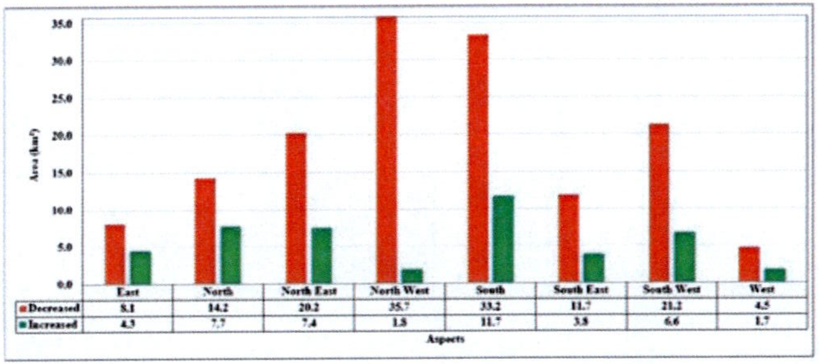

(a) Changes in the surface area based on the aspect calculation (see Fig. 8.2(b)) for 2002 and 2018. Major decline in area are noticed in North-West (Birinzi in Uganda), South (Mwanza Gulf in Tanzania), North-East (Winam Gulf in Kenya), and South-West (Emin Pasha Gulf in Tanzania)

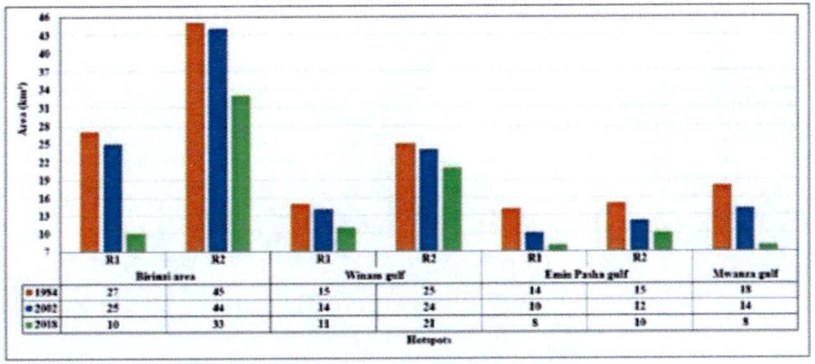

(b) Surface areas of the hotspots in 1984, 2002, and 2018

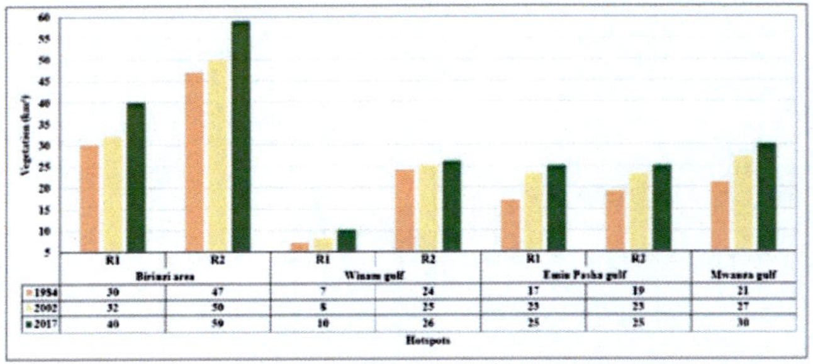

(c) Vegetation cover within the hotspots in 1984, 2002, and 2017

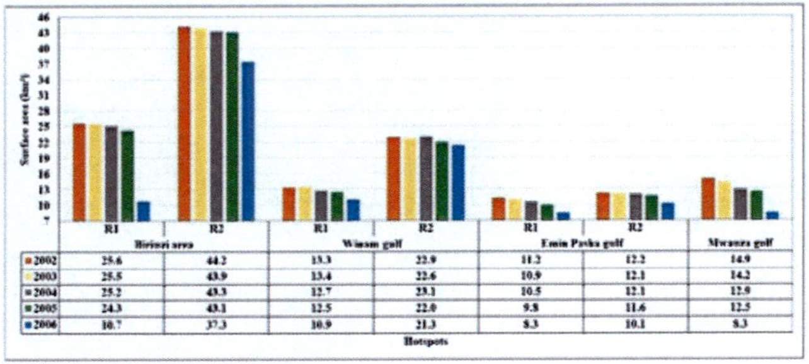

(d) Surface areas of the hotspots for the period 2002-2006

Fig. 8.5 Application of Aspects approach to the lake; **a** and **b** highlight four significant hotspots as the North-West (Birinzi), South (Mwanza Gulf), South-West (Emin Pasha Gulf) and North-East (Winam Gulf) of the lake. R1 and R2 refers to region 1 and 2, respectively for each hotspot (see Fig. 8.6 for regions R1 and R2). **c** the vegetation cover which has been increasing in all the selected regions whereas the surface areas of those regions have been decreasing, and **d** the maximum changes in the surface area between 2005 and 2006 in all the hotpsot regions. *Source* [14]

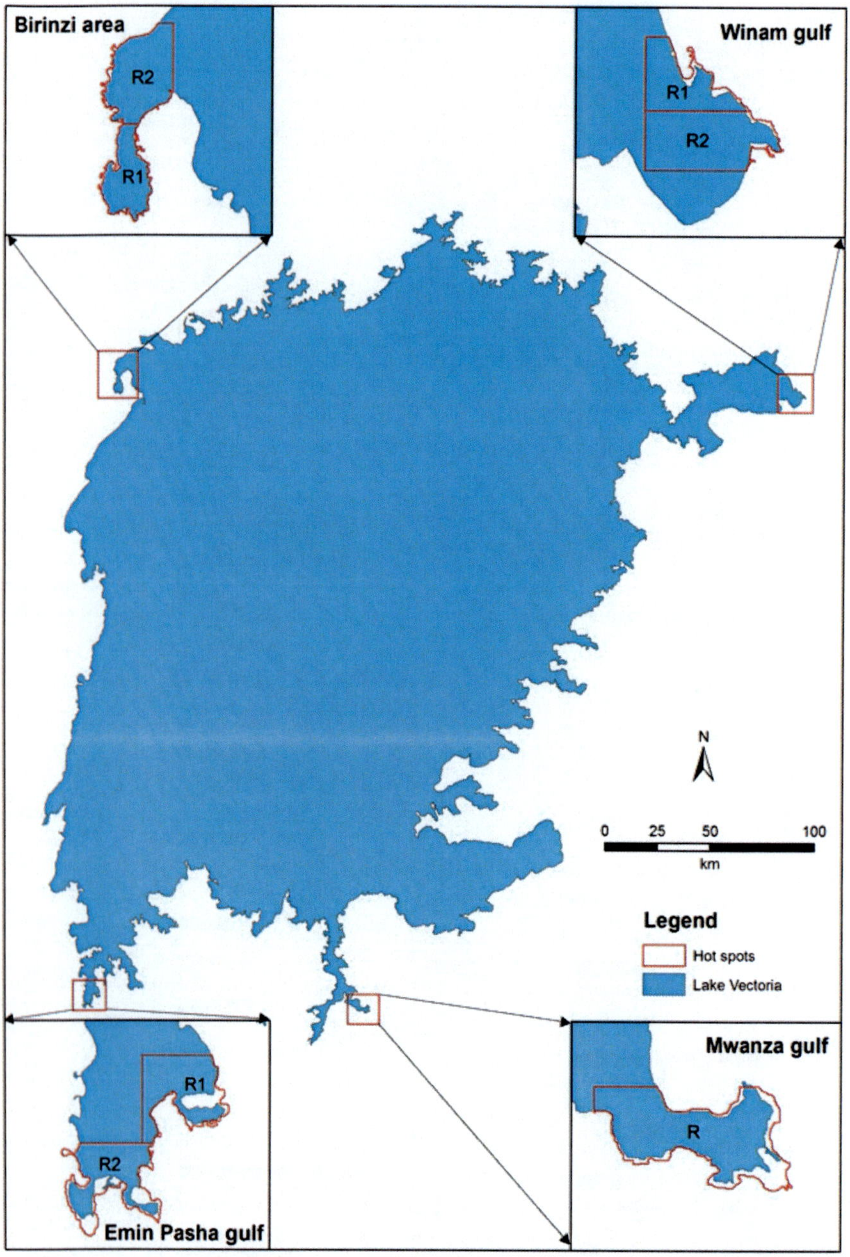

Fig. 8.6 Lake Victoria's hotspots shown by red boxes. These are places with significant areal changes observed from the most significant aspects shown in Fig. 8.2b (see also Fig. 8.5a). The areas are zoomed to show further divisions into region 1 (R1) and region 2 (R2) for in-depth analysis. *Source* [14]

Changes in surface area and vegetation cover (i.e., bare land, moderate and dense vegetation cover) within these hotspots are highlighted in Fig. 8.5b, c. The NDVI for each year, i.e., 1984 (Landsat TM), 2002 (Landsat ETM+) and 2017 (Landsat OLI) are calculated to understand the impacts of the vegetation cover on the lake's dynamic changes within these hotspots. Areal changes between 1984 and 2002 are nominal within Birinzi area where they decrease by 2 and 1 km^2 in regions R1 and R2, respectively (see Fig. 8.6). Compared to 2002, however, over the same region, significant changes in surface areas are observed in 2018, i.e., a decline of 15 km^2 (region R1) and 11 km^2 (region R2). Vegetation cover on the other hand has an upward trend where it increased by 8 and 9 km^2 between 2002 and 2017 in regions R1 and R2, respectively.

The area surrounding the region where Nyando River enters Lake Victoria in the Winam Gulf lost just 1 km^2 of surface area in each region in 2002. In 2018, the surface area of Winam Gulf (regions R1 and R2) had similar reductions (3 km^2). In addition, NDVI shows that vegetation cover in the Winam Gulf increased from 31 km^2 in 1984 to 36 km^2 in 2017 (Fig. 8.5b). The tail sections of Lake Victoria (Mwanza and Emin Pasha Gulfs) also shrank in surface areas and increased in vegetation cover. The surface areas of regions R1 and R2 of Emin Pasha Gulf are 14 and 15 km^2 in 1984, and receded to 10 and 12 km^2, respectively by 2002, and are under 10 km^2 in 2018. Mwanza Gulf has a similar pattern as surface area decreased from 18 km^2 in 1984 to 8 km^2 in 2018. Vegetation cover grew between 6 and 9 km^2 in Emin Pasha Gulf (regions R1 and R2) and Mwnaza Gulf, respectively. Spatial maps of areal and vegetation changes from Google Earth Pro imagery presented in Fig. 8.7 visually confirm the changes in surface areas and vegetation cover within the hotspots discussed above.

These results indicate that the surface areas of these hotposts shrank considerably over the past 34 years. A possible explanation of these shrinkages in areas could be attributed to population growth and land use/ land cover (LULC) changes. With an average population density of those living in close proximately to the lake being approximately 165 persons per km^2, and growing at a rapid rate of 3% annually [71], and with approximately 70% of the residents in the three riparian countries relying on agriculture for livelihood [54], there are bound to be significant use in fertile areas surrounding the lake through irrigation means to its detriment. For instance, deforestation and the expansion of agricultural land was noted by [75] to increase the volume and regularity of flood runoff in the Winam Gulf in Kenya, where approximately 7900 ha of forest land was converted to agricultural land between 1986 and 2000. Conversion of forests to agricultural lands on the one hand, and widespread irrigation necessitated by rising population [8] on the other hand, has seen topsoil erosion increase dramatically during the past several decades.

Olang and Fürst [75] and ICRAF [49] observed these land use/land cover changes in the Nyando River basin within Winam Gulf, an essential part of Lake Victoria's water system with a drainage basin that covers 3550 km^2. Increase in anthropogenic activities within Nyando river basin for instance played a significant role in depositing sediments and phosphorous in Lake Victoria [49]. For example, [49] found that there was between three to four-fold increase in sedimentation during the previous

100 years over Winam Gulf of the lake, possibly explaining the noticeable decline in surface area in the hotspot. This possibility is given credence by [70] who points out that the Kenyan rivers together with river Kagera in Tanzania are the major contributors of high sedimentation due to deforestation and poor agricultural practices. Further, land use studies in the Nyasho, Nyakato, Mwisenge and Bweri regions of Tanzania, which are close to the shores of Lake Victoria, indicate that approximately 10% of the surrounding land have been converted to settlements and farms during the period between 1994 and 2008 [67]. This LULC changes likely had a significant contribution to the observed decline in the surface areas of the two hotspots (Emin Pasha and Mwanza Gulfs) within Tanzania.

8.4.3 Climate Variability/Change Influence

Anthropogenic factors have been discussed in Sect. 8.4.2 above to contribute to the changes in the surface areas of the four hotspots. Yet, they are not the only culprits. Climate variability/change is also known to contribute to the changes in the observed lake dynamics. An analysis, therefore, of annual and seasonal rainfall on the one hand, and the impact of global ENSO teleconnection on Lake Victoria on the other hand is undertaken to explore the variation in water levels and the dimensions of the lake in relation to these climate variables.

8.4.3.1 Annual and Seasonal Rainfall Analysis

Rainfall is the primary source of Lake Victoria's water contributing about 80% of its total recharge [8, 10]. Changes in rainfall, therefore, directly impact on its water levels and surface area. Principal component analysis (PCA) of the CHIRPS rainfall identifies the first three spatial (EOFs) and temporal (PCs) patterns (Fig. 8.8) with cumulative variance over 99% (79.8, 16.0 and 4.0%).

First of all, the variability of the dominant seasonal rainfall pattern of Lake Victoria is captured on the western parts of Lake Victoria according to PC1 and EOF1, respectively. This is supported by [51, 56], who indicate the two main rainy seasons, March-May (MAM, the long rainy season) and September-November (SON, the short rainy season) contribute about 65% of annual rainfall in the Lake Victoria basin and 67% in East Africa. PC2 captures the annual rainfall pattern, where the northern parts and southern parts have opposite trend according to EOF2. EOF3 shows an east-west dipole structure. Comprehensively analysing all the EOFs and PCs (1 to 3), it is noticeable that most variable rainfall are locate in southwestern parts. The long rainy season (MAM) has increasing rainfall over the whole lake, while the short rainy season (SON) has only significant increase in rainfall over the southern parts (also indicated by [51]). Moreover, the linear trend of PC1 shows a slight decease, while PC2 and PC3 show increasing trend over the study period. This can be interpreted as (i) the rainfall over the whole lake is increasing, which agrees with the results of [11], that most of stations around the Lake Victoria recorded

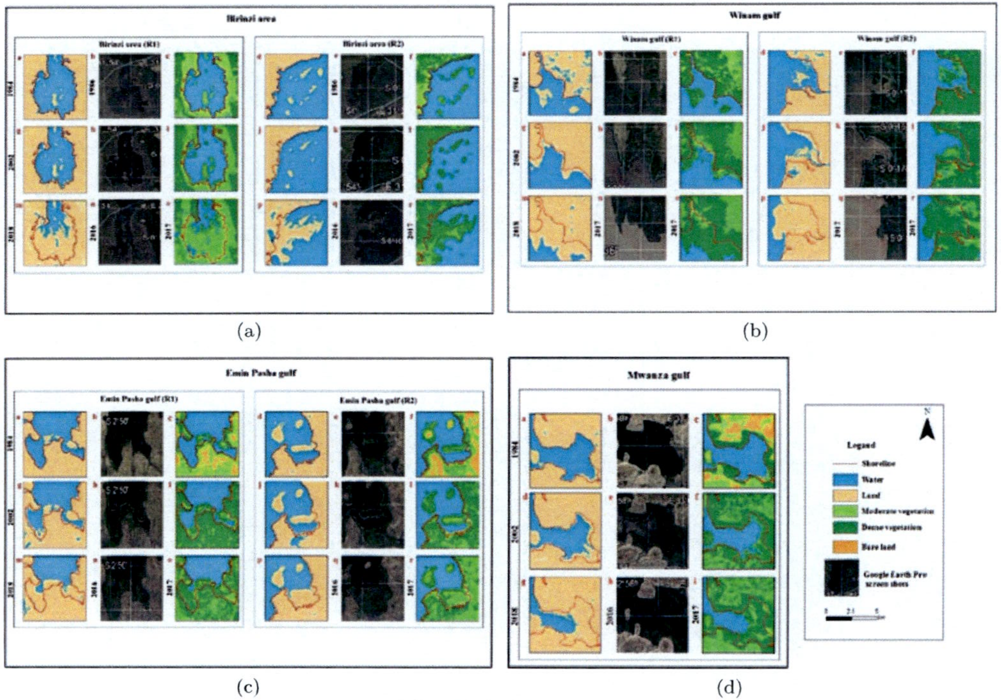

Fig. 8.7 Shrinkages in surface areas of the hotspots can be seen in the first, second, and third rows of each hotspot derived from the manual digitisation approach (first column), Google Earth Pro (second column), and NDVI (third column), respectively. R1 and R2 refers to regions 1 and 2, respectively for each hotspot (see Fig. 8.6). Regions 1 and 2 are the most affected parts of the lake for the period of 1984–2018. These regions occur in Birinzi part of Uganda, **a** while sedimentation plays a major role to diminish the lake's boundary near Winam Gulf located in the Kenya where the Nyando River enters the lake. **b** The regions of Emin Pasha Gulf and Mwanza Gulf shown in **c** and **d** are located in Tanzania. *Source* [14]

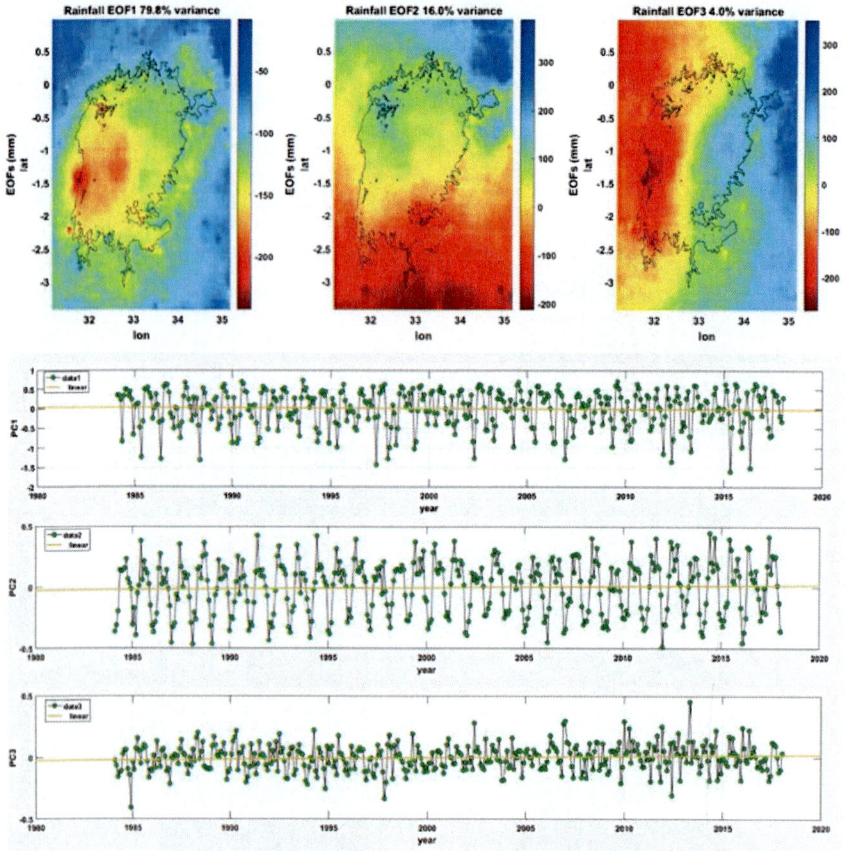

Fig. 8.8 The upper three figures shows the spatial patterns of rainfall over Lake Victoria. EOF1 shows variable seasonal rainfall pattern over the western parts, while EOF2 indicates the opposite trend of annual pattern between the northern and southern parts of the lake. EOF3 shows a dipole pattern of rainfall in the eastern parts of the lake, where extremes, e.g., heavy rainfall are noticed. In the figure, 'lat' denotes latitude and 'lon' longitude. The lower three figures shows the temporal pattern of rainfall over Lake Victoria. The seasonal rainfall is more important according to PC1, while PC2 shows the annual rainfall as nearly stable over Lake Victoria. PC3 shows the presence of climate extremes during the years 1985, 1997, and 2013. *Source* [14]

increasing rainfall during the 1921–2009 period, and (ii), the rainfall in the northern parts of the lake showed more increase compared to the southern parts, corroborating the finding of [68] as well as [56].

In terms of extreme climate, e.g., heavy rainfall captured by PCs in year of 1985–1986, 1997–1998, 2012–2013 and 2015–2016; and the dry period in year of 2000–2001, 2004–2005 and 2014–2015, are analysed with impacts of the global ENSO teleconnection in the next section, e.g., ENSO in 2016–2017 brought 39% of total annual rainfall in that year and increased the total area of the lake by around 9 km^2.

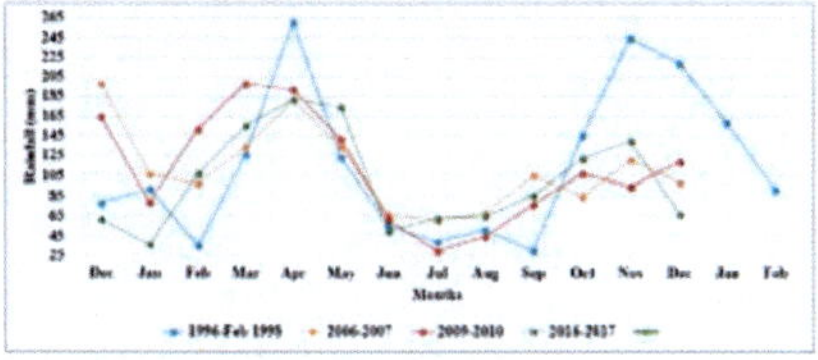

(a) Monthly rainfall

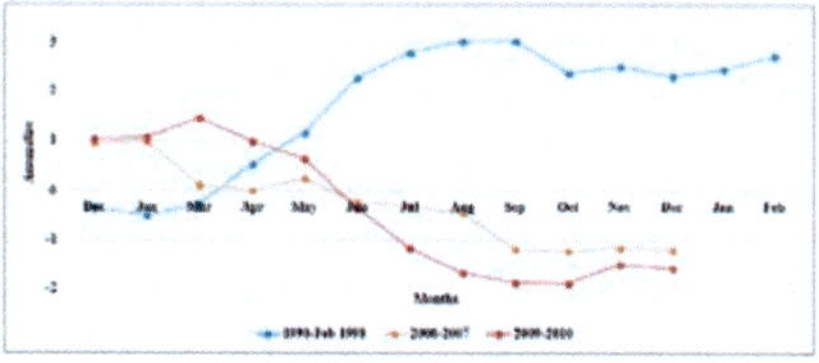

(b) Multivariate ENSO index

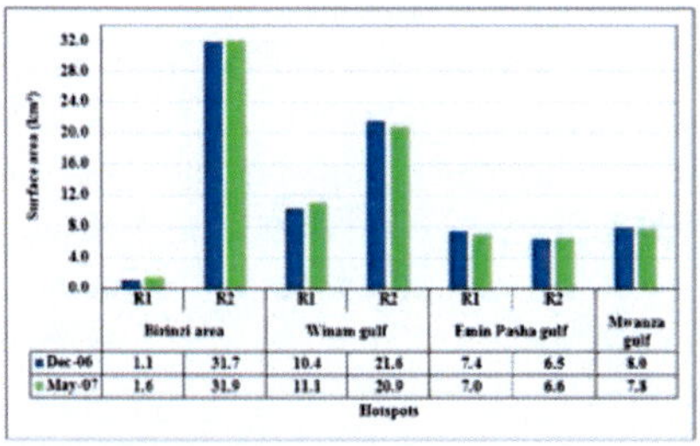

(c) Hotspots changes during the 2007 ENSO event

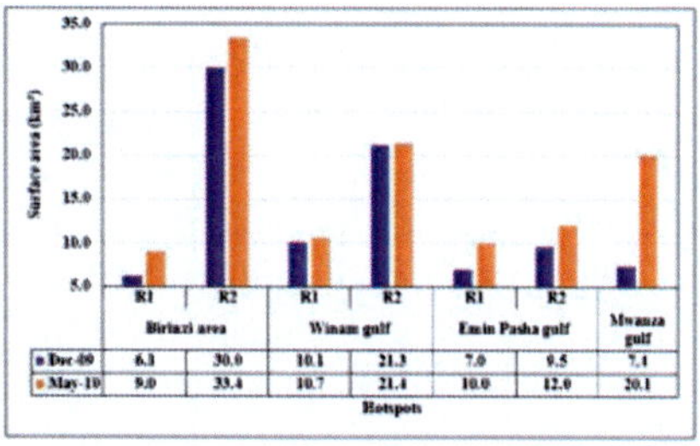

(d) Hotposts changes during the 2010 ENSO event

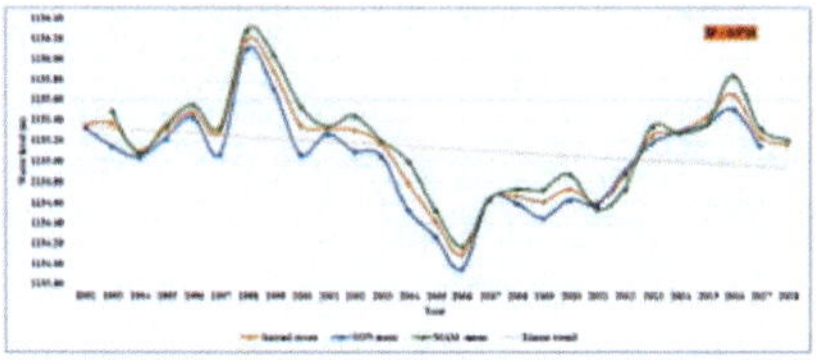

(e) Annual and seasonal mean trends of Lake Victoria's water level for the period 1992-2018

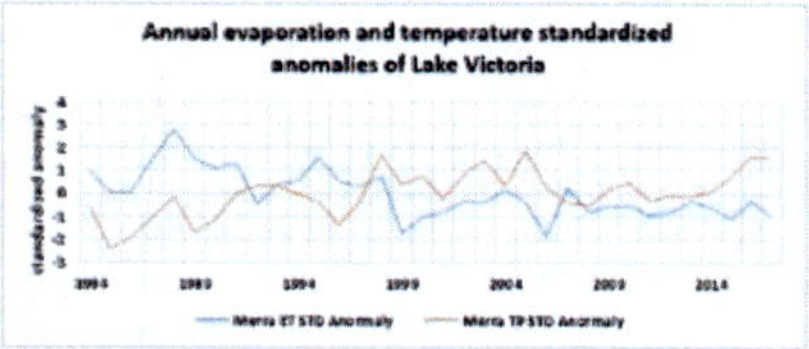

(f) Annual evaporation and temperature standardized anomaly of Lake Victoria for the period 1984-2018

Fig. 8.9 The upper two figures shows the monthly rainfall of 2016–2017 plotted for the seasonal and annual rainfall. Figure 8.9a illustrates that the 1997 ENSO rainfall is higher compared to the 2007 and 2010. Figure 8.9b shows the significant impact of ENSO during the two primary rainy seasons. Figure 8.9c, d show the graphical illustrations of the ENSO induced areal changes. Major areal changes occurred during the 2010 ENSO event relative to the 2007 ENSO event. The lower Fig. 8.9e show the mean water level for a short rainy season (SON; September–November) and a long rainy season (MAM: March-May) period, respectively while Fig. 8.9f show evaporation and temperature anomalies. The linear trend in Fig. 8.9d indicate the annual water level to have declined for the 1984–2018 period. However, this trend, which was obtained using linear fitting is insignificant when tested using Mann-Kendall test. *Source* [14]

8.4.3.2 The Impacts of Global Teleconnection

In 1997, ENSO coupled with Indian ocean dipole (IOD) lasted for the longest period [104] causing unusually high rainfall in East Africa badly damaging socio-economic infrastructure [51, 60, 79]. In this section, the 1997, 2007, and 2010–2011 ENSO rainfall that occurred in East Africa are analysed for their impacts on the lake's dynamics. From Fig. 8.9a, the 1997 rainfall are seen to increase from 36 mm in

February to 261 mm in April. Figure 8.9b indicate an upward trend in the ENSO anomalies from March that remained high throughout 1997. Although rainfall was under 50 mm during Jun–Sep months of 1997, it was still under the influence of ENSO. Rainfall increased in October (145 mm) and November (245 mm) before dropping to 218 mm in December. In 2007, the high rainfall period was from February (98 mm) to May (135 mm) followed by a fluctuation in August that lasted until the end of the year. Rainfall variation was also prominent during the heavy MAM rainy season of 2010 compared to 2007. In general, the MEI anomalies (Fig. 8.9b) indicate the peak of rainfall during the long rainy season to be influenced by ENSO. Compared to 2007 and 2010, rainfall and MEI anomalies of 1997 are seen to be larger during both long (MAM) and short (SON) rainfall seasons. The negative values from June to December reflect the cold phase of ENSO (La Niña) during 2007 and 2010.

The high rainfall during the 2007 and 2010 ENSO caused an increase in the lake's surface area by approximately 0.23% (161 km^2) and 0.42% (294 km^2). Whereas hotspot regions also increased in areas (Fig. 8.9c) by between 0.1 to 0.5 km^2 in 2007, in 2010, maximum change occurred in the Mwanza Gulf (Fig. 8.9d), where the area increased by 13.3 km^2 followed by Birinzi and Emin Pasha Gulfs, which expanded by $>$ 2.5 km^2. The Winam Gulf is least affected by the ENSO rains, and areal changes are less than 1 km^2. The surface area of Lake Victoria extracted from MODIS imagery was likely impacted by its 250 m spatial resolution. More comprehensive future analysis with higher resolution imagery is required to identify more precisely areal changes caused by ENSO events. Due to the limitation of high resolution satellite imagery for the 1990s, the Emin Pasha and Mwanza Gulf hotspots are analyzed using Google Earth Pro (Dec 1996) and Landsat imagery (September 1997). There is just an increase of 0.41 km^2 in region R1 of Emin Pasha, while Emin Pasha (region R2) and Mwanza Gulf decline in surface areas. Surface area changes in hotspots are extremely small. The zoom scale in the digitisation of Google Earth images is also much lower than during the digitisation of Landsat images and the accuracy of the results have been impacted by variable zoom scales, and as such, should be interpreted with caution.

The impacts of high rainfall is observed in the lake's water levels (Fig. 8.9e), where the annual, MAM and SON means of water level fluctuated throughout the study period are plotted. Water levels peaked by 0.9 m in 1998 due to ENSO rainfall of 1997 and high rainfall in 2007 also led to a 0.54 m increase in water level. The mean of MAM and SON found a simultaneous increase in seasonal and annual water levels of the lake. Figure 8.9e shows the water level rise of 1.04 m during long rainy season in 1998. Swenson and Wahr [95] observed the increase of more than 1 m during 1998, while [56] found that water levels rose by 1.7 m due to heavy rains in 1997. Between 2002 and 2006, water levels dropped from 1135.3 m to their lowest level nearly 1134 m (annual). The sharp decline from 2002 to 2006 coincides with Uganda's extension of the Nalubaale dam in Jinja [9, 12, 95] (see Chaps. 4 and 9 for detailed discussion). Annual water level gradually increased from 2010 (1134.6 m) to 2016 (1135.7 m) due to increased rainfall. However, annual water level dropped in 2017 to 1135.3 m. Climate variability and anthropogenic activities (as discussed in Sect. 8.4.4) had significant impacts on the water levels and surface area of Lake

Victoria. Seasonal rainfall and global ENSO teleconnection also influenced the areal behaviour of the lake during the 1984–2018 study period. On the contrast, both temperature and evaporation trends over Lake Victoria computed from MERRA-2 reanalysis data show a uniform near constant behaviour, which is unlikely to have had significance influence on the Lake's physical dynamics (see Fig. 8.9f).

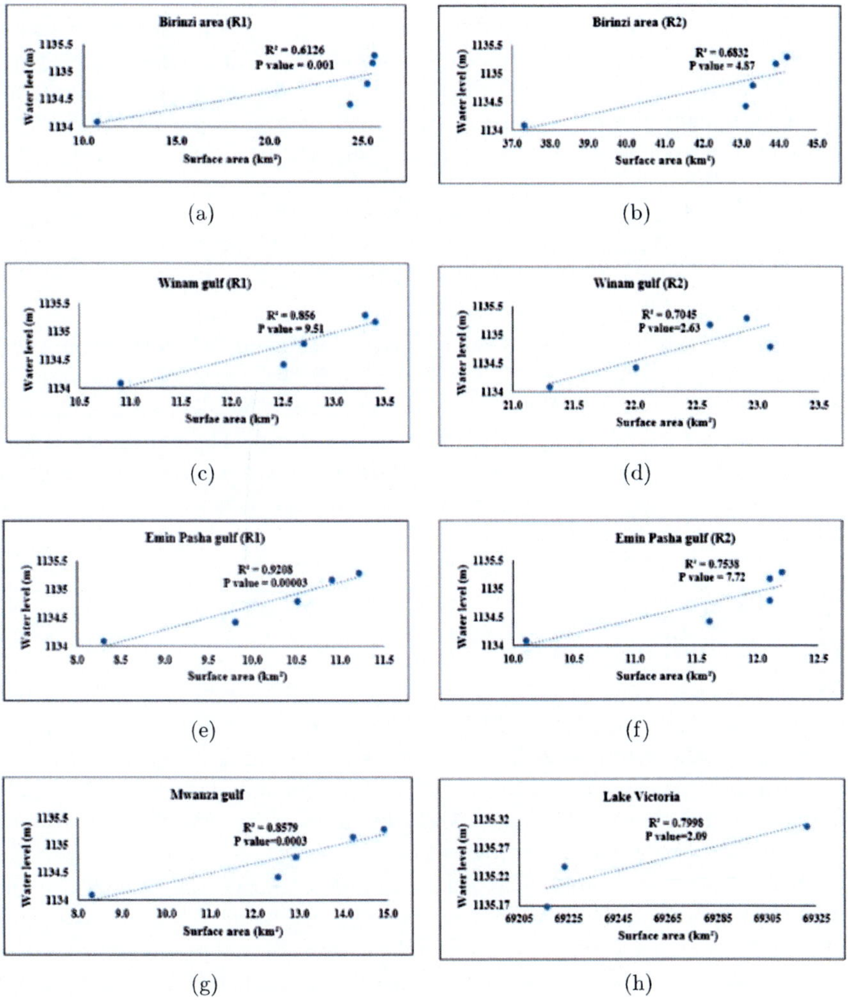

Fig. 8.10 Correlation between annual water level and surface area changes of the Lake and hotspots. *Source* [14]

8.4.4 Anthropogenic Activities and the Dynamics of the Lake

The expansion of the Nalubaale dam greatly impacted on the water levels of Lake Victoria. Remotely sensed Landsat imagery over the lake was not available between 2002 and 2006, therefore, the surface area of each hotspot is calculated from Google Earth Pro imagery using the Ruler tool, see e.g., [89]. Figure 8.5d shows the reduction in the surface area of selected hotspot regions. Tong et al. [98] also observed surface area changes of Lake Victoria during this period. Changes in the surface area in hotspot regions are $< 1.4\,km^2$ for Birinzi, Winam Gulf and Emin Pasha Gulf between 2002 and 2005 while there is a decline of $2.4\,km^2$ in the Mwanza Gulf during this period. Maximum changes are between 2005 and 2006 when regions R1 and R2 of Birnzi decreased by 13.6 and $5.8\,km^2$, respectively. The second highest change occur in Mwanza Gulf where the area receded by $4.2\,km^2$. Surface area in Winam Gulf (region R1) and Emin Pasha Gulf (regions R1 and R2) also reduced by 1.6 and 1.5 km^2, respectively. Figures 8.10 and 8.11a show positive correlations of changes in the hotspot regions with water levels. The correlation coefficient of surface area changes of the hotspots with water level variations is ≥ 0.78 for each region. However, only three hotspot regions; Birinzi (region R1), Emin Pasha (region R1), and Mwanza Gulf are statistically significant ($p < 0.05$). The R^2 of Birinzi (region R1), Emin Pasha Gulf (region R1), and Mwanza Gulf show that 61, 92, and 85% of areal changes, respectively, are explained by the changes in water level. These areal changes of hotspots are highlighted from Google Earth Pro in Fig. 8.11b. The correlation analysis point to the fact that the areal changes of hotspots depend on the increase or decrease in the surface water of the lake.

8.5 Concluding Remarks

This chapter used remotely sensed Landsat, Sentinel-2, Moderate Resolution Imaging Spectro-radiometer (MODIS), Google Earth Pro imagery, Climate Hazards Group InfraRed Precipitation with Station (CHIRPS) rainfall, Multivariate El' Niño-Southern Oscillation (ENSO) Index (MEI) and altimetry data for the past 34 years together with exterior sources (37 published literature from 1969–2018) on the lake's physical parameters to study the dynamics and the associated climate change and anthropogenic impacts. PCA analysis on CHIRPS precipitation was explored to show the impacts of climate variation/change on the lake's physical dynamics. The study used manual digitalization to precisely measure the parameters of the lake. Although the approach was time consuming, it provided more accurate measurements compared to the automated Modified Normalised Difference Water Index (MNDWI) approach which was used to validate the dynamic trends. The results indicated that:

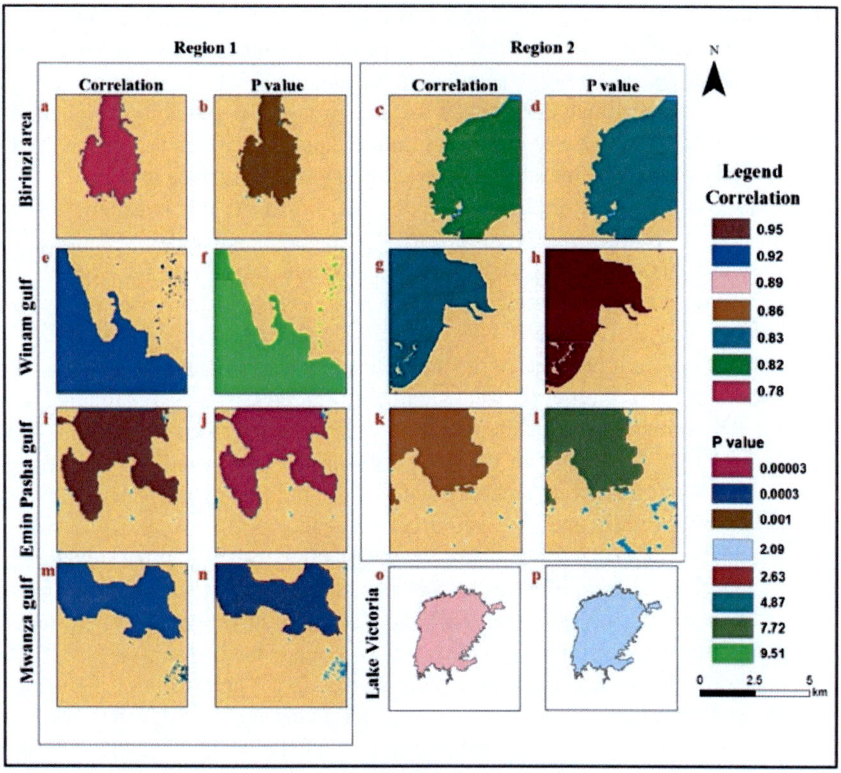

(a)

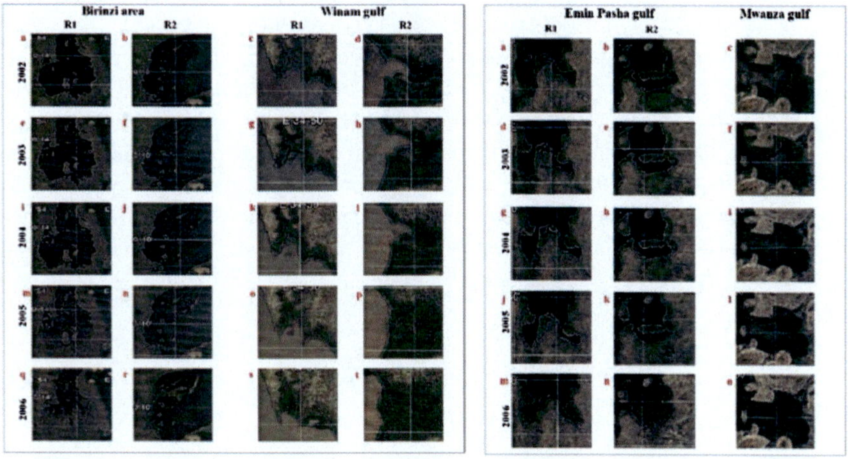

(b)

Fig. 8.11 **a** Regression analysis for the lake and each hotspot's surface area changes with annual water level changes, and **b** visual interpretation of hotspot regions from Google Earth Pro for the 2002–2006 period. *Source* [14]

(i) The mean surface area of Lake Victoria over the past 34 years (1984–2018) is 69,295 km^2 (i.e., 812 km^2 or 1.2% more than the mean of the values quoted in the 37 published literature sampled in this chapter, indicating that the previous studies underestimated the lake's surface areas. The possible explanation of this could be their use of coarse resolution data (e.g., 500 m resolution used by [98]) as well as the methods employed to obtain the surface area.

(ii) The surface area of the lake shrunk by 203 km^2 (−0.3%) between 1984 and 2018. The declining rate was 5.97 km^2/year based on manual digitisation. However, the maximum length and width did not change significantly. The receding rate of the length and width was ~100 m/year and ~70 m/year, respectively, for the 1984–2018 period.

(iii) Specifically, four hotspots of the lake; Birinzi area, Winam Gulf, Mwanza Gulf and Emin Pasha Gulf underwent significant changes. The results found that the vegetation increased around the hotspot regions during the study period, which could be the key factor in the hotspots' surface area receding. The changes in Lake Victoria and its hotspots were likely due to an increasing population, irrigation activities and poor land management.

(iv) Fluctuations in water levels had a downward trend throughout the study period. The 1997 ENSO rainfall was the most catastrophic to East Africa and long rainy seasons also greatly impacted on Lake Victoria's surface area. Due to lack of data, only two hotspots were analyzed to observe the impacts of 1997 ENSO on the hotpsots' surface areas during this period. The results of the analysis showed that there was an increase in water levels and area of the Emin Pasha Gulf. In addition, the lake's surface area increased by 0.23% in 2007 and 0.45% in 2010. The long rainy season of MAM caused an increase of nearly 9 km^2 in surface area in 2017.

(v) The expansion of the Nalubaale dam was another factor that caused changes in water levels and hotspots surface areas between 2002 and 2006.

(vi) This study confirmed that Lake Victoria was influenced by both climate variability (seasonal rainfall and ENSO) and anthropogenic activities. Furthermore, hotspots experienced decrease in areas when water level height decreased, which was confirmed by a regression analysis of hotspot changes and water level heights. For the ENSO analysis, the accuracy of the extraction of surface areas could have been influenced by the coarse resolution of MODIS imagery. Hence, higher resolution imagery is required for a more accurate impact analysis of the ENSO rains on Lake Victoria.

(vii) In this study, we posed the question "*is the Lake dying*?" From the results, it can be concluded that, overall, the lake's surface area has been diminished by 0.3% in 2018 compared to its 1984 value. However, this does not signify that the lake is dying since the changes were largely attributed to hotspots (e.g., Birinzi area, Mwanza Gulf, Emin Pasha Gulf and Winam Gulf), some of which area actually dying.

References

1. Adams WM, Goudie A, Orme AR (eds) (1996) The physical geography of Africa. Oxford University Press, Oxford
2. Akurut M, Willems P, Niwagaba CB (2014) Potential impacts of climate change on precipitation over Lake Victoria, East Africa, in the 21st Century. Water 6(9):2634–2659
3. Anyah RO, Forootan E, Awange JL, Khaki M (2018) Understanding linkages between global climate indices and terrestrial water storage changes over Africa using GRACE products. Sci Total Environ 635:1405–1416. https://doi.org/10.1016/j.scitotenv.2018.04.159
4. Argent R, Sun X, Semazzi F, Xie L, Liu B (2015) The development of a customization framework for the WRF Model over the Lake Victoria basin, Eastern Africa on seasonal timescales. Adv Meteorol. https://doi.org/10.1155/2015/653473
5. Arinaitwe K, Muir DC, Kiremire BT, Fellin P, Li H, Teixeira C, Mubiru DN (2018) Prevalence and sources of polychlorinated biphenyls in the atmospheric environment of Lake Victoria, East Africa. Chemosphere 193:343–350
6. Arinaitwe K, Rose NL, Muir DC, Kiremire BT, Balirwa JS, Teixeira C (2016) Historical deposition of persistent organic pollutants in Lake Victoria and two alpine equatorial lakes from East Africa: Insights into atmospheric deposition from sedimentation profiles. Chemosphere 144:1815–1822
7. Aura CM, Musa S, Yongo E, Okechi JK, Njiru JM, Ogari Z, Ombwa V (2018) Integration of mapping and socio-economic status of cage culture: Towards balancing lake - use and culture fisheries in Lake Victoria, Kenya. Aquac Res 49(1):532–545
8. Awange JL, Ong'ang'a O (2006) Background of Lake Victoria. Lake Victoria: ecology, resources, environment. Springer Science & Business Media, New York, pp 5–16
9. Awange JL, Ogalo L, Bae K-H. Were P. Omondi P, Omute P, Omullo M (2008) Falling Lake Victoria water levels: is climate a contributing factor? Clim Chang 89(3–4):281–297. https://doi.org/10.1007/s10584-008-9409-x
10. Awange JL, Sharifi MA, Ogonda G, Wickert J, Grafarend EW, Omulo MA (2008) The falling Lake Victoria water level: GRACE, TRIMM and CHAMP satellite analysis of the lake basin. Water Resour Manage 22(7):775–796. https://doi.org/10.1007/s11269-007-9191-y
11. Awange J, Forootan E, Kusche J, Kiema J, Omondi P, Heck B, Fleming K, Ohanya S, Goncalves R (2013) Understanding the decline of water storage across the Ramser-Lake Naivasha using satellite-based methods. Adv Water Resour 60:7–23. https://doi.org/10.1016/j.advwatres.2013.07.002
12. Awange J, Anyah R, Agola N, Forootan E, Omondi P (2013b) Potential impacts of climate and environmental change on the stored water of Lake Victoria Basin and economic implications. Water Resour Res 49(12):8160–8173. https://doi.org/10.1002/2013WR014350
13. Awange J, Forootan E, Kuhn M, Kusche J, Heck B (2014) Water storage changes and climate variability within the Nile Basin between 2002 and 2011. Adv Water Resour 73:1–15. https://doi.org/10.1016/j.advwatres.2014.06.010
14. Awange JL, Saleem A, Sukhadiya RO, Ouma YO, Hu K (2019) Physical dynamics of Lake Victoria over the past 34 years (1984–2018): Is the lake dying? Sci Total Environ 658:199–218

15. Awange JL, Palancz B, Völgyesi L (2020) Hybrid imaging and visualization. Employing machine learning with mathematica - python. Springer Nature International, Berlin
16. Baeriswyl P-A, Rebetez M (1997) Regionalization of precipitation in Switzerland by means of principal component analysis. Theoret Appl Climatol 58(1–2):31–41
17. Balirwa JS (2017) Lake Victoria wetlands and the ecology of the Nile tilapia. Routledge
18. Becker M, LLovel W, Cazenave A, Güntner A, Crétaux J-F (2010) Recent hydrological behavior of the East African great lakes region inferred from GRACE, satellite altimetry and rainfall observations. Comptes Rendus Geosci 342(3):223–233. https://doi.org/10.1016/j.crte.2009.12.010
19. Bosilovich MG, Akella S, Coy L, Cullather R, Draper C, Gelaro R, Kovach R, Liu Q, Molod A, Norris P, Wargan K, Chao W, Reichle R, Takacs L, Vikhliaev Y, Bloom S, Collow A, Firth S, Labow G, Partyka G, Pawson S, Reale O, Schubert SD, Suarez M (2015) MERRA-2: initial evaluation of the climate. Technical report series on global modeling and data assimilation, vol 43. NASA/TM-2015-104606/Vol.43
20. Bremner J, Lopez-Carr D, Zvoleff A (2013) Using new methods and data to assess and address population, fertility, and environment links in the Lake Victoria Basin, Population and the environment, Session 247. https://www.iussp.org/en/event/17/programme/paper/6136
21. Busker T, de Roo A, Gelati E, Schwatke C, Adamovic M, Bisselink B, Cottam A (2018) A global lake and reservoir volume analysis using a surface water dataset and satellite altimetry
22. Chamberlain J, Bain C, Boyd D, McCourt K, Butcher T, Palmer S (2014) Forecasting storms over Lake Victoria using a high resolution model. Meteorol Appl 21(2):419–430. https://doi.org/10.1002/met.1403
23. Cheruiyot EK, Mito C, Menenti M, Gorte B, Koenders R, Akdim N (2014) Evaluating MERIS-based aquatic vegetation mapping in Lake Victoria. Remote Sens 6(8):7762–7782
24. Cockerton HE, Street-Perrott FA, Barker PA, Leng MJ, Sloane HJ, Ficken KJ (2015) Orbital forcing of glacial/interglacial variations in chemical weathering and silicon cycling within the upper White Nile basin, East Africa: stable-isotope and biomarker evidence from Lakes Victoria and Edward. Quat Sci Rev 130:57–71
25. Congalton RG (1991) A review of assessing the accuracy of classifications of remotely sensed data. Remote Sens Environ 37:35–46
26. Cunningham MA (2006) Accuracy assessment of digitized and classified land cover data for wildlife habitat. Landsc Urban Plan 78(3):217–228
27. D'Andrimont R, Defourny P (2018) Monitoring African water bodies from twice-daily MODIS observation. GISci Remote Sens 55(1):130–153. https://doi.org/10.1080/15481603.2017.1366677
28. Deus D, Gloaguen R (2013) Remote sensing analysis of lake dynamics in semi-arid regions: Implication for water resource management. Lake Manyara, East African Rift, Northern Tanzania. Water 5(2):698–727. https://doi.org/10.3390/w5020698
29. Dewan A, Corner R, Saleem A, Rahman MM, Haider MR, Rahman MM, Sarker MH (2017) Assessing channel changes of the Ganges-Padma River system in Bangladesh using Landsat and hydrological data. Geomorphology 276:257–279. https://doi.org/10.1016/j.geomorph.2016.10.017
30. Drusch M, Del Bello U, Carlier S, Colin O, Fernandez V, Gascon F, Hoersch B, Isola C, Laberinti P, Martimort P, Meygret A, Spoto F, Sy O. Marchese F, Bargellini P (2012) Sentinel-2: ESA's optical high-resolution mission for GMES operational services. Remote Sens Environ 120:25–36. https://doi.org/10.1016/j.rse.2011.11.026
31. Awange J, Saleem A, Konneh SS, Goncalves RM, Kiema JBK, Hu KX (2019) Liberia's coastal erosion vulnerability and LULC change analysis: post-civil war and Ebola epidemic. Appl Geogr 101:56–67. https://doi.org/10.1016/j.apgeog.2018.10.007
32. Du Y, Zhang Y, Ling F, Wang Q, Li W, Li X (2016) Water bodies' mapping from Sentinel-2 imagery with modified normalized difference water index at 10-m spatial resolution produced by sharpening the SWIR band. Remote Sens 8(4):354. https://doi.org/10.3390/rs8040354
33. Elison MV, Mlaponi E, Musiba MJ, Ngupula GW, Kashindye BB, Kayanda RJ (2018) Changes in the Diet of Synodontis victoriae and Synodontis afrofischeri in Lake Victoria, Tanzanian waters. Afr J Trop Hydrobiol Fish 16(1):10–15

34. Faith JT, Tryon CA, Peppe DJ, Beverly EJ, Blegen N, Blumenthal S, Patterson D (2015) Paleoenvironmental context of the Middle Stone Age record from Karungu, Lake Victoria Basin, Kenya, and its implications for human and faunal dispersals in East Africa. J Hum Evol 83:28–45
35. Feng M, Sexton JO, Channan S, Townshend JR (2016) A global, high-resolution (30-m) inland water body dataset for 2000: first results of a topographic-spectral classification algorithm. Int J Digit Earth 9(2):113–133
36. Forsythe KW, Schatz B, Swales SJ, Ferrato L-J, Atkinson DM (2012) Visualization of lake mead surface area changes from 1972 to 2009. ISPRS Int J Geo-Inf 1(2):108–119. https://doi.org/10.3390/ijgi1020108
37. Funk C, Peterson P, Landsfeld M, Pedreros D, Verdin J, Shukla S, Husak G, Rowland J, Harrison L, Hoell A, Michaelsen J (2015) The climate hazards infrared precipitation with stations - a new environmental record for monitoring extremes. Sci Data 2:150066. https://doi.org/10.1038/sdata.2015.66
38. Gao B (1996) NDWI - A normalized difference water index for remote sensing of vegetation liquid water from space. Remote Sens Environ 58(3):257–266
39. Gilmore S, Saleem A, Dewan A (2015) Effectiveness of DOS (Dark-Object Subtraction) method and water index techniques to map wetlands in a rapidly urbanising megacity with Landsat 8 data. Paper presented at the Research@ Locate'15
40. Global Modeling and Assimilation Office (GMAO) (2015) MERRA-2 tavgM_2d_flx_Nx: monthly mean, time-averaged, single-level, assimilation, surface flux diagnostics V5.12.4. Goddard Earth Sciences Data and Information Services Center (GES DISC, Greenbelt, MD, USA). https://doi.org/10.5067/0JRLVL8YV2Y4
41. Goodchild MF, Hunter GJ (1997) A simple positional accuracy measure for linear features. Int J Geogr Inf Sci 11(3):299–306
42. Grosse G, Schirrmeister L, Kunitsky VV, Hubberten HW (2005) The use of CORONA images in remote sensing of periglacial geomorphology: an illustration from the NE Siberian coast. Permafrost Periglac Process 16(2):163–172. https://doi.org/10.1002/ppp.509
43. Haibo Y, Zongmin W, Hongling Z, Yu G (2011) Water body extraction methods study based on RS and GIS. Procedia Environ Sci 10:2619–2624. https://doi.org/10.1016/j.proenv.2011.09.407
44. Haroon MA, Rasul G (2009) Principal component analysis of summer rainfall and outgoing long-wave radiation over Pakistan. Pak J Meteorol 5(10)
45. Hassan AA, Jin S (2014) Lake level change and total water discharge in East Africa Rift Valley from satellite-based observations. Glob Planet Chang 117:79–90
46. Hu K, Awange JL, Forootan E, Goncalves RM, Fleming K (2017) Hydrogeological characterisation of groundwater over Brazil using remotely sensed and model products. Sci Total Environ 599:372–386. https://doi.org/10.1016/j.scitotenv.2017.04.188
47. Huang S, Li J, Xu M (2012) Water surface variations monitoring and flood hazard analysis in Dongting Lake area using long-term Terra/MODIS data time series. Nat Hazards 62(1):93–100. https://doi.org/10.1007/s11069-011-9921-6
48. Hughes RH, Hughes JS (1992) A directory of African wetlands. Gland, Switzerland, Nairobi, Kenya, and Cambridge. IUCN, UNEP, and WCMC, UK
49. ICRAF (2006) Improved land management in the Lake Victoria Basin: final report on the TransVic project. ICRAF Occasional Paper No. 7. Nairobi. World Agroforestry Centre. Accessed from http://www.worldagroforestry.org/downloads/Publications/PDFS/op14439.pdf on 14 May 2018
50. Idris ZM, Chan CW, Kongere J, Gitaka J, Logedi J, Omar A, Kimura M (2016) High and heterogeneous prevalence of asymptomatic and sub-microscopic malaria infections on islands in Lake Victoria, Kenya. Sci Rep 6:36958
51. Indeje M, Semazzi FH, Ogallo LJ (2000) ENSO signals in East African rainfall seasons. Int J Climatol 20(1):19–46
52. Juma DW, Wang H, Li F (2014) Impacts of population growth and economic development on water quality of a lake: case study of Lake Victoria Kenya water. Environ Sci Pollut Res 21(8):5737–5746

53. Kabenge M, Wang H, Li F (2016) Urban eutrophication and its spurring conditions in the Murchison Bay of Lake Victoria. Environ Sci Pollut Res 23(1):234–241
54. Kayombo S, Jorgensen SE (2006) Lake Victoria: experience and lessons learned brief. International Lake Environment Committee, Lake Basin Management Initiative, Kusatsu, Japan. Accessed from http://www.worldlakes.org/uploads/Victoria_7.15.04.pdf on 23 Apr 2018
55. Kendall RL (1969) An ecological history of the Lake Victoria basin. Ecol Monogr 39(2):121–176. https://doi.org/10.2307/1950740
56. Kizza M, Rodhe A, Xu CY, Ntale HK, Halldin S (2009) Temporal rainfall variability in the Lake Victoria Basin in East Africa during the twentieth century. Theoret Appl Climatol 98(1):119–135. https://doi.org/10.1007/s00704-008-0093-6
57. Khaki M, Awange J (2018) Improved remotely sensed satellite products for studying Lake Victoria's water storage changes. Sci Total Environ 652:915–926
58. Kull D (2006) Connections between recent water level drops in Lake Victoria, dam operations and drought. Accessed from https://www.internationalrivers.org/sites/default/files/attached-files/full_report_pdf.pdf on 19 Aug 2018
59. Laizer LM, Gibson RW, Lukonge E (2018) Seasonal water crises and social dilemmas in semi-arid areas of the lake zone of tanzania. Int J Asian Soc Sci 8(5):213–226
60. Latif M, Dommenget D, Dima M, Grötzner A (1999) The role of Indian Ocean sea surface temperature in forcing East African rainfall anomalies during December-January 1997/98. J Clim 12(12):3497–3504
61. Lopez-Carr D, Pricope NG, Aukema JE, Jankowska MM, Funk C, Husak G, Michaelsen J (2014) A spatial analysis of population dynamics and climate change in Africa: potential vulnerability hot spots emerge where precipitation declines and demographic pressures coincide. Popul Environ 35(3):323–339. https://doi.org/10.1007/s11111-014-0209-0
62. MacIntyre S, Romero JR, Silsbe GM, Emery BM (2014) Stratification and horizontal exchange in Lake Victoria, East Africa. Limnol Ocean 59(6):1805–1838
63. Mailu AM (2000) Preliminary assessment of the social, economic and environmental impacts of water hyacinth in Lake Victoria Basin and status of control. In: ACIAR proceedings. ACIAR (1998), pp 130–139
64. Mertes LA, Dunne T, Martinelli LA (1996) Channel-floodplain geomorphology along the Solimões-Amazon river, Brazil. Geol Soc Am Bull 108(9):1089–1107
65. Mistry VV, Conway D (2003) Remote forcing of East African rainfall and relationships with fluctuations in levels of Lake Victoria. Int J Climatol 23(1):67–89. https://doi.org/10.1002/joc.861
66. Mugimba KK, Chengula AA, Wamala S, Mwega ED, Kasanga CJ, Byarugaba DK, Mutoloki S (2018) Detection of tilapia lake virus (Ti LV) infection by PCR in farmed and wild Nile tilapia (Oreochromis niloticus) from Lake Victoria. J Fish Dis. https://doi.org/10.1111/jfd.12790
67. Musamba EB, Ngaga YM, Boon EK, Giliba RA (2011) Impact of socio-economic activities around Lake Victoria: Land use and land use changes in Musoma Municipality, Tanzania. J Hum Ecol 35(3):143–154
68. Nicholson SE (2017) Climate and climatic variability of rainfall over eastern Africa. Rev Geophys 55(3):590–635. https://doi.org/10.1002/2016RG000544
69. Nyamweya C, Desjardins C, Sigurdsson S, Tomasson T, Taabu-Munyaho A, Sitoki L, Stefansson G (2016) Simulation of Lake Victoria circulation patterns using the Regional Ocean modeling system (ROMS). PLoS ONE 11(3):e0151272. https://doi.org/10.1371/journal.pone.0151272
70. Odada E, Olago D, Ochola W (2006) Environment for development: an ecosystems assessment of Lake Victoria Basin, UNEP/PASS Pan African START Secretariat (PASS), Department of Geology, University of Nairobi, PO Box, 30197. Accessed from https://wedocs.unep.org/bitstream/handle/20.500.11822/7987/environment_development.pdf?sequence=3 on 22 Jun 2018
71. Odada EO, Ochola WO, Olago DO (2009) Understanding future ecosystem changes in Lake Victoria basin using participatory local scenarios. Afr J Ecol 47(s1):147–153

72. Ogallo LJ, JE J (1988) Teleconnection between seasonal rainfall over East Africa and global sea surface temperature anomalies. J Meteorol Soc Jpn Ser. II 66(6):807–822
73. Ogutu-Ohwayo R, Hecky RE, Cohen AS, Kaufman L (1997) Human impacts on the African great lakes. Environ Biol Fishes 50(2):117–131. https://doi.org/10.1023/A:1007320932349
74. Okungu J, Njoka S, Abuodha J, Hecky R, Chebwek N, Okungu J, Ruwa R (2018) An introduction to Lake Victoria catchment, water quality, physical limnology and ecosystem status (Kenyan sector). Accessed from https://www.oceandocs.org/bitstream/handle/1834/7140/ktf0088.pdf?sequence=2&isAllowed=y on 15 Mar 2018
75. Olang L, Fürst J (2011) Effects of land cover change on flood peak discharges and runoff volumes: model estimates for the Nyando River Basin, Kenya. Hydrol Process 25(1):80–89. https://doi.org/10.1002/hyp.7821
76. Oluoch-Otiego J, Oyoo-Okoth E, Kiptoo KKG, Chemoiwa EJ, Ngugi CC, Simiyu G, Opiyo MA (2016) PCBs in fish and their cestode parasites in Lake Victoria. Environ Monit Assess 188(8):483
77. Omute P, Corner R, Awange JL (2012) The use of NDVI and its derivatives for monitoring Lake Victoria's water level and drought conditions. Water Resour Manage 26(6):1591–1613. https://doi.org/10.1007/s11269-011-9974-z
78. Onyutha C, Tabari H, Rutkowska A, Nyeko-Ogiramoi P, Willems P (2016) Comparison of different statistical downscaling methods for climate change rainfall projections over the Lake Victoria basin considering CMIP3 and CMIP5. J Hydro-Environ Res 12:31–45
79. Owiti Z, Ogallo LA, Mutemi J (2008) Linkages between the Indian Ocean dipole and East African seasonal rainfall anomalies. J Kenya Meteorol Soc 2(1)
80. Ouma SO, Ngeranwa JN, Juma KK, Mburu DN (2016) Seasonal variation of the physico-chemical and bacteriological quality of water from five rural catchment areas of lake victoria basin in Kenya. J Environ Anal Chem 3(170):2
81. Pekel JF, Cottam A, Gorelick N, Belward AS (2016) High-resolution mapping of global surface water and its long-term changes. Nature 540(7633):418
82. Phillips J, McIntyre B (2000) ENSO and interannual rainfall variability in Uganda: implications for agricultural management. Int J Climatol: J R Meteorol Soc 20(2):171–182
83. Preisendorfer RW (1988) Introduction. In: Mobley C (ed) Principal component analysis in meteorology and oceanography. Elsevier, Amsterdam, pp 1–9
84. Rabatel G, Gorretta N, Labbé S (2011) Getting NDVI spectral bands from a single standard RGB digital camera: a methodological approach. Paper presented at the Conference of the Spanish Association for Artificial Intelligence. Accessed from https://hal.archives-ouvertes.fr/hal-00648439/document on 21 Aug 2018
85. Reddy GPO, Maji AK, Gajbhiye KS (2004) Drainage morphometry and its influence on landform characteristics in a basaltic terrain, Central India - a remote sensing and GIS approach. Int J Appl Earth Obs Geoinf 6(1):1–16. https://doi.org/10.1016/j.jag.2004.06.003
86. Rzoska J (1976) Lake Victoria, physical features, general remarks on chemistry and biology. The nile, biology of an ancient river. Springer, Dordrecht, pp 167–175
87. Saleem A, Corner R, Awange J (2018) On the possibility of using CORONA and Landsat data for evaluating and mapping long-term LULC: case study of Iraqi Kurdistan. Appl Geogr 90:145–154. https://doi.org/10.1016/j.apgeog.2017.12.007
88. Sene K, Tych W, Beven K (2018) Exploratory studies into seasonal flow forecasting potential for large lakes. Hydrol Earth Syst Sci 22(1):127–141
89. Schloderer G, Bingham M, Awange JL, Fleming KM (2011) Application of GNSS-RTK derived topographical maps for rapid environmental monitoring: a case study of Jack Finnery Lake (Perth, Australia). Environ Monit Assess 180(1–4):147–161. https://doi.org/10.1007/s10661-010-1778-8
90. Schneider SR, McGinnis DF Jr, Stephens G (1985) Monitoring Africa's Lake Chad basin with LANDSAT and NOAA satellite data. Int J Remote Sens 6(1):59–73. https://doi.org/10.1080/01431168508948424

91. Shamsudduha M, Taylor RG, Jones D, Longuevergne L, Owor M, Tindimugaya C (2017) Recent changes in terrestrial water storage in the Upper Nile Basin: an evaluation of commonly used gridded GRACE products. Hydrol Earth Syst Sci 21(9):4533. https://doi.org/10.5194/hess-21-4533-2017
92. Smith KA, Semazzi FH (2014) The role of the dominant modes of precipitation variability over Eastern Africa in modulating the hydrology of Lake Victoria. Adv Meteorol. https://doi.org/10.1155/2014/516762
93. Song Y, Semazzi FH, Xie L, Ogallo LJ (2004) A coupled regional climate model for the Lake Victoria basin of East Africa. Int J Climatol 24(1):57–75. https://doi.org/10.1002/joc.983
94. Stathis D, Myronidis D (2009) Principal component analysis of precipitation in Thessaly region (Central Greece). Glob NEST J 11(4):467–476
95. Swenson S, Wahr J (2009) Monitoring the water balance of Lake Victoria, East Africa, from space. J Hydrol 370(1–4):163–176. https://doi.org/10.1016/j.jhydrol.2009.03.008
96. Taabu-Munyaho A, Nyamweya CS, Sitoki L, Kayanda R, Everson I, Marteinsdóttir G (2014) Spatial and temporal variation in the distribution and density of pelagic fish species in Lake Victoria, East Africa. Aquat Ecosyst Health Manag 17(1):52–61
97. Taylor RG, Todd MC, Kongola L, Maurice L, Nahozya E, Sanga H, MacDonald AM (2013) Evidence of the dependence of groundwater resources on extreme rainfall in East Africa. Nat Clim Chang 3(4):374. https://doi.org/10.1038/NCLIMATE1731
98. Tong X, Pan H, Xie H, Xu X, Li F, Chen L, Luo X, Liu S, Chen P, Jin Y (2016) Estimating water volume variations in Lake Victoria over the past 22 years using multi-mission altimetry and remotely sensed images. Remote Sens Environ 187:400–413. https://doi.org/10.1016/j.rse.2016.10.012
99. Tooth S (2015) Spotlight on google earth as a resource. Geography 100:51–56. Accessed from https://search.proquest.com/docview/1680229076?accountid=10382 on 18 Jun 2018
100. Thiery W, Davin EL, Panitz HJ, Demuzere M, Lhermitte S, Van Lipzig N (2015) The impact of the African Great Lakes on the regional climate. J Clim 28(10):4061–4085
101. van den Thillart G, Wilms I, Nieveen M, Weber RE, Witte F (2018) Hypoxia-induced changes in hemoglobins of Lake Victoria cichlids. J Exp Biol, jeb-177832, https://doi.org/10.1242/jeb.177832
102. Watson CS, Quincey DJ, Carrivick JL, Smith MW (2016) The dynamics of supraglacial ponds in the Everest region, central Himalaya. Glob Planet Chang 142:14–27. https://doi.org/10.1016/j.gloplacha.2016.04.008
103. Widmann M, Schär C (1997) A principal component and long-term trend analysis of daily precipitation in Switzerland. Int J Climatol 17(12):1333–1356. https://doi.org/10.1002/(SICI)1097-0088(199710)17:12<1333::AID-JOC108>3.0.CO2-Q
104. Wolter K, Timlin MS (1998) Measuring the strength of ENSO events: How does 1997/98 rank? Weather 53(9):315–324
105. Wulder MA, Masek JG, Cohen WB, Loveland TR, Woodcock CE (2012) Opening the archive: How free data has enabled the science and monitoring promise of Landsat. Remote Sens Environ 122:2–10. https://doi.org/10.1016/j.rse.2012.01.010
106. Xu H (2006) Modification of normalised difference water index (NDWI) to enhance open water features in remotely sensed imagery. Int J Remote Sens 27(14):3025–3033. https://doi.org/10.1080/01431160600589179
107. Yamazaki D, O'Loughlin F, Trigg MA, Miller ZF, Pavelsky TM, Bates PD (2014) Development of the global width database for large rivers. Water Resour Res 50(4):3467–3480. https://doi.org/10.1002/2013WR014664
108. Ye Q, Zhu L, Zheng H, Naruse R, Zhang X, Kang S (2007) Glacier and lake variations in the Yamzhog Yumco basin, southern Tibetan Plateau, from 1980 to 2000 using remote-sensing and GIS technologies. J Glaciol 53(183):673–676

109. Zhang M-M, Chen F, Tian B-S (2018) An automated method for glacial lake mapping in High Mountain Asia using Landsat 8 imagery. J Mt Sci 15(1):13–24. https://doi.org/10.1007/s11629-017-4518-5
110. Zhou M, Brandt P, Pelster D, Rufino MC, Robinson T, Butterbach-Bahl K (2014) Regional nitrogen budget of the Lake Victoria Basin, East Africa: syntheses, uncertainties and perspectives. Environ Res Lett 9(10):105009. https://doi.org/10.1088/1748-9326/9/10/105009

Chapter 9
Rapid 2002–2006 Fall: Anthropogenic Induced?

...although the droughts of 2004 and 2005 contributed to the lowering of the lake level, if dam operations had adhered to the Agreed Curve, today's lake levels would be around 50 cm higher. There has obviously been an additional factor in the reduction of the lake level ...The secrecy of hydrologic and dam operations data for Lake Victoria and the Victoria Nile is worrying.

D. Kull [21]

9.1 Summary

In 2002–2006, Lake Victoria water level experienced a dramatic fall that caused alarm to water resource managers as to whether the lake was actually dying. Since the lake basin contributes about 20% of the lakes water in form of discharge, with 80% coming from direct rainfall, this chapter presents satellite analysis of the entire lake basin in an attempt to establish the cause of the decline. Gravity Recovery And Climate Experiment (GRACE), Tropical Rainfall Measuring Mission (TRMM) and CHAllenging Minisatellite Payload (CHAMP) satellites are employed in the analysis. Using 45 months of data spanning a period of 4 years (2002–2006), GRACE satellite data are used to analyse the variation of the geoid (equipotential surface approximating the mean sea level) triggered by variation in the stored waters within the lake basin. TRMM Level 3 monthly data for the same period of time are used to compute mean rainfall for a spatial coverage of $25° \times 25°$ (25×25 km) and the rainfall trend over the same period analyzed. To assess the effect of evaporation, 59 CHAMP satellite's occultation for the period 2001 to 2006 are analyzed for tropopause warming. GRACE results indicate an annual fall in the geoid by 1.574 mm/year during the period 2002–2006. This fall clearly demonstrates the basin losing water over the period 2002–2006. TRMM results on the other hand indicate the rainfall over the basin (and directly over the lake) to have been stable during the

J. Awange, *Lake Victoria Monitored from Space*,
https://doi.org/10.1007/978-3-030-60551-3_9

2002–2006 period. The CHAMP satellite results indicate the tropopause temperature to have fallen in 2002 by about 3.9 K and increased by 2.2 K in 2003 and remained above the 189.5 K value of 2002. The tropopause heights showed a steady increase from a height of 16.72 m in 2001 and remained above this value reaching a maximum of 17.59 km in 2005, an increase in height by 0.87 m. Though the basin discharge contributes only 20%, its decline contributed to the fall in the lake waters. Since rainfall over the period remained stable, and temperatures did not increase drastically to cause massive evaporation, the remaining major contributor was the discharge from the expanded Owen Falls dam (i.e., the Kiira dam). This Chapter presents the findings of this investigation.

9.2 The Fall and the Resulting Alarm!

Since the 1960s, Lake Victoria water level has experienced significant fluctuations, see e.g., Fig. 4.1 and also [26, 27]. From 2001 to 2006, however, the lake's water level showed a dramatic fall that alarmed water resource managers as to whether the lake was actually drying up. Kull [21] reported that the lake's levels fell by more than 1.1 m below the 10 year average (see Fig. 4.1).

With the receding of the lake waters, acres of land that were lost to the floods of the 1960s were fast being reclaimed, creating sources of conflicts between man and wildlife. In some beaches, e.g., Usoma in Kenya, wetlands that were once breeding places for fish were dying up, leaving areas of land as playing fields for children and farmland (see, e.g., Fig. 9.1). Ships were now forced to dock deep inside the lake, while the landing bays needed to be extended. Those who directly depended on the lake waters for domestic use were forced to go deeper into the lake to draw water, thus exposing women and children to water-borne diseases and risks of snakes and crocodiles. Water intakes that supplied major towns and cities had to be extended deeper into the lake, thus causing more financial burden to the municipalities that were already strained financially [3].

Lake's surface water, together with groundwater, and soil moisture snow constitute stored water. Changes in the lake's level are directly related to the variation of the water stored in its basin, which contributes 20% inform of river discharge. Since the lake is supplied both by direct rainfall [26, 29] and the river discharges [20], a decrease in stored basin water will contribute to a drop in the lake level. An analysis of the stored water in the Lake Victoria basin in relation to rainfall and evaporation is therefore necessary as a first diagnosis to the fall. It provides water resource managers and planners with information on the state and changing trend of the stored water within the basin. This could be achieved through the use of the latest state-of-the-art GRACE satellites, see, e.g., [15, 37]. Previous methods for studying variation in stored water include, e.g., the Artificial Neural Network [1] and GIS (Geographical Information System) and remote sensing [9, 10, 19, 23]. Besides monitoring the variation in stored water within rivers and lake basins, GRACE satellite products are currently also playing a role of validating models, which have been applied to

(a) (b)

Fig. 9.1 Receding Lake Victoria waters during 2004. The original level of the lake can be seen from the vegetation (e.g., in **a**). It is evident that the lake had receded far deep inside leaving vast land to be used, e.g., for grazing as can be seen from the cow grazing (in **(a)**). This level of receding exposed women and children to dangers of crocodiles and diseases as they ventured deep inside to collect water (in **b**). Needless to say that even the fishermen seen in this photo were not excluded from such dangers

study stored water e.g., [20, 22, 25, 28, 48]. GRACE satellites had been recognized by Hildebrand [18] as having the potential to provide the first space based estimate of terrestrial stored groundwater. Applications of GRACE satellite to monitor and analyse terrestrial water storage changes, e.g., those of Congo, Volta, Nile, Mississippi and Amazon basins, are documented e.g., in [11–13, 30–32, 36, 38, 39]. In this chapter, an in-depth satellite analysis of the lake basin during the period of 2002 to 2006, when the lake waters drastically declined is provided based on the work of Awange et al. [3]. GRACE satellites are used to analyse the variation in the basin's stored water while TRMM [52] and CHAMP satellites are used to analyse rainfall and temperature within the basin respectively.

9.3 Space Diagnostic of the 2002–2006 Water Level Fall

9.3.1 GRACE Satellite Diagnostic

Having been motivated by the potentials of the GRACE satellites (Fig. 9.2), Awange et al. [3] undertook a satellite analysis of the entire lake basin in an attempt to establish the cause of the decline in Lake Victoria's water levels. The GRACE and CHAMP satellites (Fig. 5.2 in Sect. 5.3.3) together with data from the Tropical Rainfall Measuring Mission (TRMM) satellite (Sect. 9.3.2) were employed in the analysis. Using 45 months of data spanning a period of 4 years (2002–2006), the GRACE satellite data were used to analyze the gravity field variations caused by changes in the stored waters within the Lake Victoria basin (Fig. 1.3). As a first step,

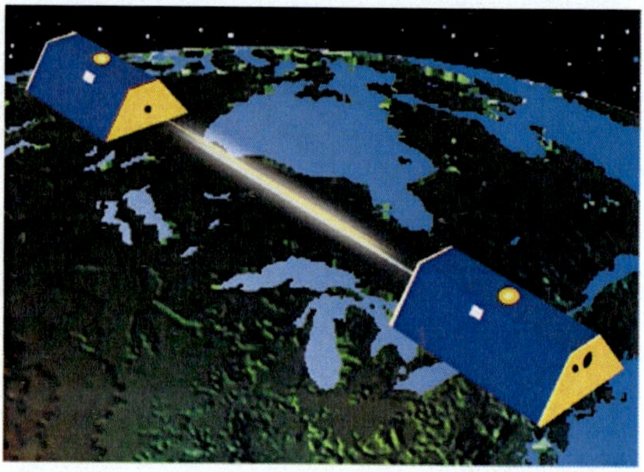

Fig. 9.2 GRACE satellites. *Source* http://earthobservatory.nasa.gov/Study/WeighingWater/printall.php

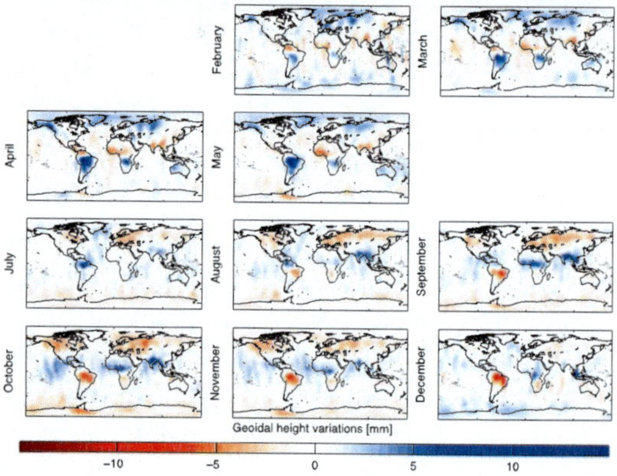

Fig. 9.3 Global monthly geoid variation for 2003. The Amazon basin's signals are compared to those of [39] in Fig. 9.4. The Amazon basins signals from both figures matches. *Source* Awange et al. [3]

to validate the GRACE results, [3] computed the global variation of geoid for the year 2003 (Fig. 9.3) and compared those over the Amazon to the results of [39] in Fig. 9.4. Looking at the Amazon basin in the figures, it is clearly evident that GRACE satellites picked the same signals as those of Tapley et al. [39], and therefore justify its use for analysis of fall in lake Victoria basin.

Following the validation of GRACE results in Figs. 9.3 and 9.4, using Eq. 5.3 on Sect. 5.3.3, monthly geoid variation from GRACE data for the period April 2002 up to April 2006 were then computed. The results are presented in Figs. 9.5, 9.6, 9.7 and 9.8. A comparison of the monthly geoidal variations from 2002 to 2005 gives a clear picture of the variation in the stored water of the Lake Victoria basin. In 2002 (Fig. 9.5), GRACE satellite data were available only for 5 months. For these months, the month of August show decrease in geoid level. The other months of April, November and December indicate a positive variation of between 2–8 mm of the geoid indicating a net gain in mass, i.e., water. In 2003 (Fig. 9.6), January, May, and June recorded a rise in geodal level by 4 mm. The rest of the months indicate values between 1–2 mm.

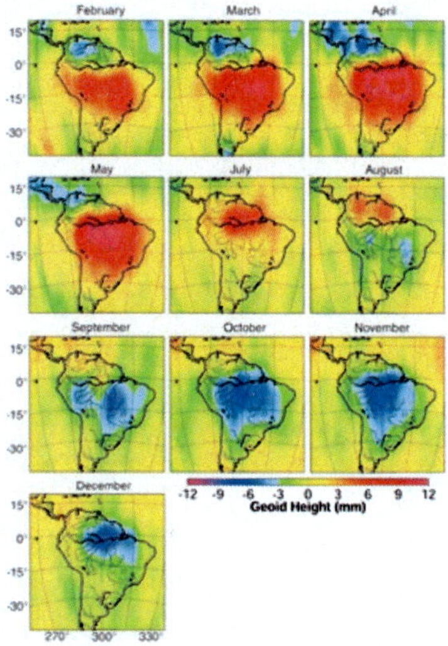

Fig. 9.4 Global monthly geoid variation for 2003 [39]. *Source* Awange et al. [3]

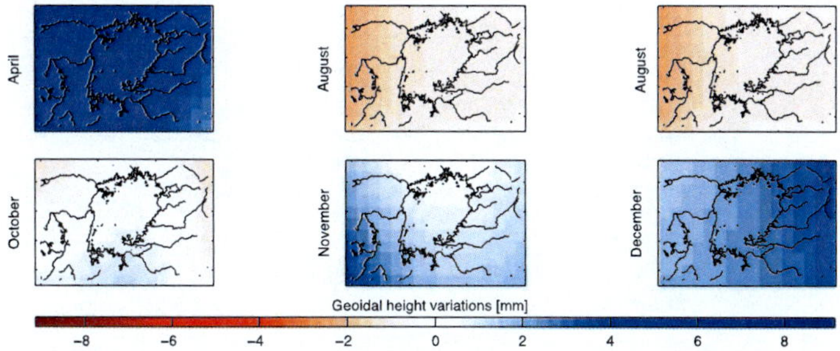

Fig. 9.5 Lake Victoria's stored water changes in 2002. *Source* Awange et al. [3]

Moving on to 2004 (Fig. 9.7), all the months except April indicated a drop in geoid level, thus loss of water. In May, only the North-West part of the basin shows an increase in geoid level of about 1 mm. This is the part of the basin which normally receives high amount of rainfall. July and August also show some 2 mm rise in geoidal level in the Eastern part of the basin. In 2005 (Fig. 9.8), a fall in the geoidal height occurs in all months except July, signifying a further drop in water level within the basin.

In order to visualize the declining trend of Lake Victoria stored water, Awange et al. [3] plotted the changes of the main rainfall season of March-April-May (MAM) in Fig. 9.9 for the period 2002 - 2006 computed from GRACE satellite products. Fig. 9.9 presents the annual variation of the geoid in the lake basin during the high rainy season months of March, April and May (MAM) for the period 2002–2006. The figure clearly indicates decline of stored water in the basin. Since the monthly

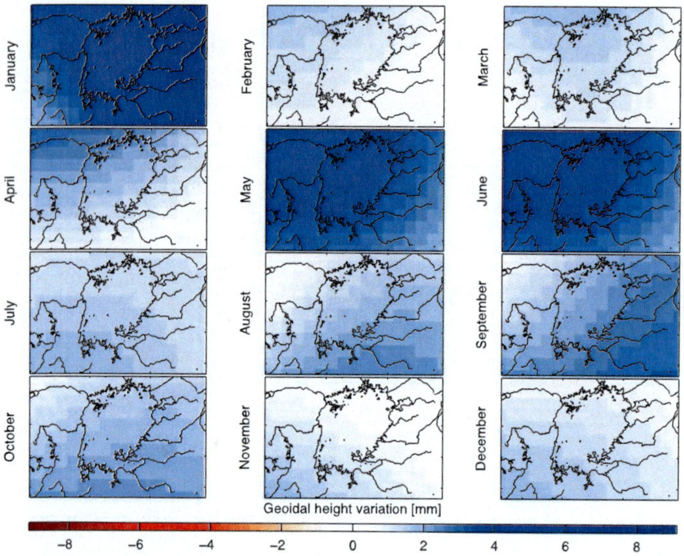

Fig. 9.6 Lake Victoria's stored water changes in 2003. *Source* Awange et al. [3]

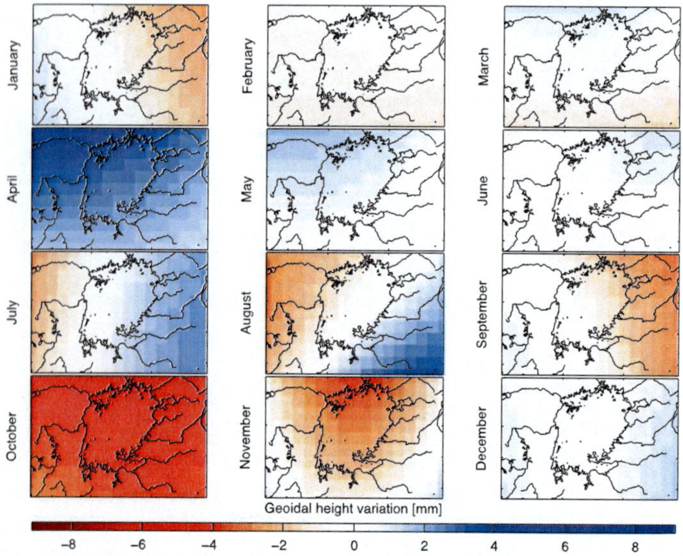

Fig. 9.7 Lake Victoria's stored water changes in 2004. *Source* Awange et al. [3]

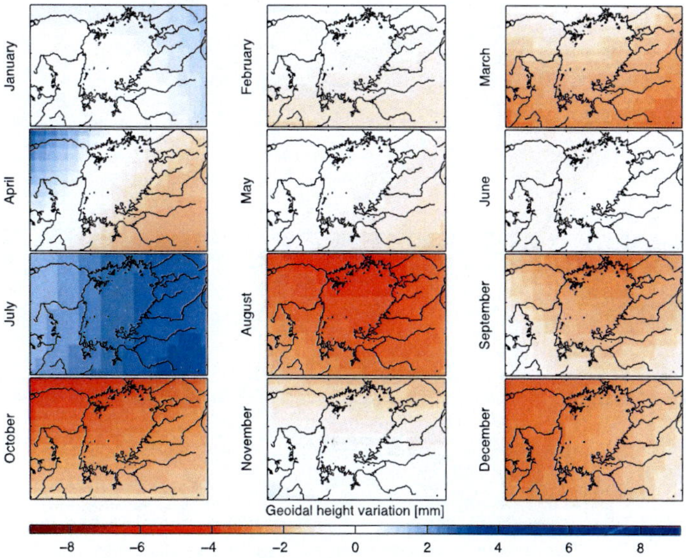

Fig. 9.8 Lake Victoria's stored water changes in 2005. *Source* Awange et al. [3]

geoidal variation are triggered by variation in stored water, then clearly the basin was loosing its stored water. Inter-annual comparison from March 2003 to March 2006 indicate an annual reduction in the basin's stored water as evidenced by a geoidal variation from 2 mm (in March 2003) to -6 mm in March 2006 (i.e., a drop of about 8 mm in the geoid level). From April 2002 to April 2006, a steady decline of the geoid from about 5 mm in 2002 to -3 mm in 2006 is seen, a general reduction of about 8 mm in geoid variation. For the month of May, comparison from 2003 to 2005 also show a drop in the geoidal variation from 5 mm to 1 mm, a reduction of about 6 mm in geoid.

To observe a clear trend of the geoidal variation, a time series graph is plotted in Fig. 9.10. From this figure, the geoidal level is computed to fall at a rate of 0.13 mm/month (1.6 mm/year). Whereas this decline in geoidal level is in mm range, the actual water volume loss caused by such variation for Lake Victoria Basin (LVB) with an area of 258,000 km^2 is significant. Next, we analyse the rainfall pattern over the same period of time. The GRACE results indicate that the basin's total water storage dramatically decreased at a rate of 6.20 mm/month. These changes are expressed in equivalent water thickness (also known as total water storage (TWS)) in Fig. 9.9. For the period 2002–2006, the results indicate a general decline in the lake basin's water level at a rate of 1.83 km^3/month [3].

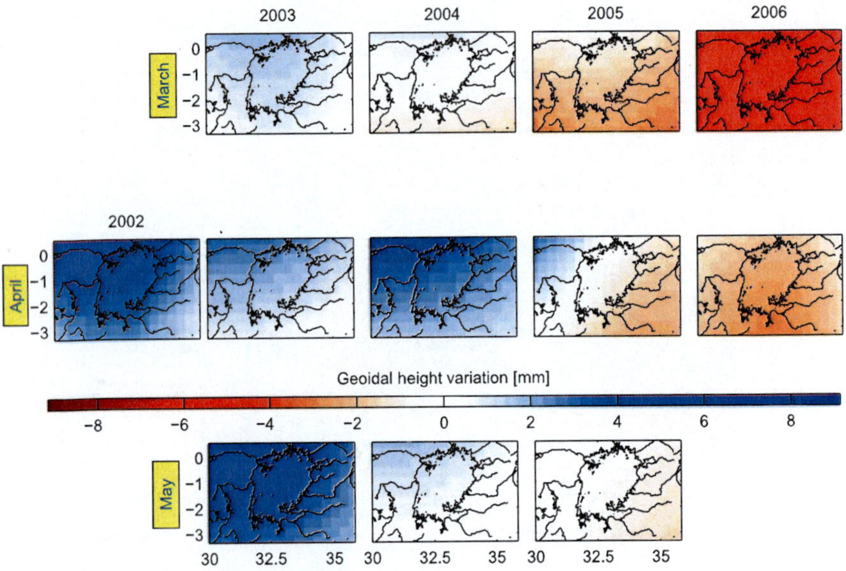

Fig. 9.9 Inter-annual comparison in the geoid during the high rainy season of MAM from 2002–2006. The figure indicates a decline in total water storage over Lake Victoria basin during the period 2002–2006. *Source* Awange et al. [3]

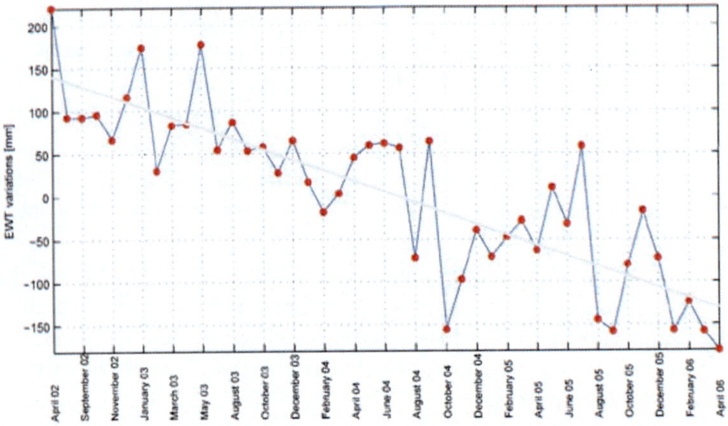

Fig. 9.10 Lake Victoria basin total water storage (equivalent water thickness) changes between 2002–2006, as seen by the GRACE satellite. The figure indicates that the GRACE satellites observed a general decline in the lake's basin waters over this period (cf. Fig. 5.5 in Sect. 5.5.2 obtained from satellite altimetry). *Source* Awange et al. [3]

9.3.2 TRMM Satellite Diagnostic

Due to the extreme difficulty of obtaining surface rainfall data for the entire basin and also directly over the lake, this study used the TRMM data [2, 14, 34, 35, 41, 49] are adopted. TRMM satellite, jointly managed by National Aeronautics and Space Administration (NASA) of the United States and Japan Aerospace Exploration Agency (JAXA) was designed to monitor and study tropical rainfall. It has the advantage of providing four dimensional (X, Y, Z, t) rainfall and latent heat data over inaccessible areas such as the ocean, un-sampled terrains, etc. The primary rainfall parameters on TRMM are the TRMM Microwave Imager (TMI), the Precipitation Radar (PR) and the Visible and Infrared Radiometer System (VIRS) [49].

Monthly TRMM rainfall data are sampled for the lake basin at a spatial resolution of 25 km $\times$ 25 km from 2002 to 2006 [52]. In Fig. 9.11, annual comparison for the high rainfall seasons of MAM for the period 2002–2006 is provided. The figure indicates the direct rainfall on the lake to have been intense during the month of April throughout 2002 to 2006 (i.e., between 200-300 mm) but no clear pattern emerges to suggest any reduction in the basin rainfall. To provide a clear picture, we plotted a time series basin mean monthly rainfall for the entire period 2006-2006 (Fig. 9.12). From the figure, no clear trend is seen to suggest a drastic reduction in rainfall over the basin during this period.

The rainfall data from TRMM as previously mentioned is an average over a 25 km $\times$ 25 km spatial coverage [52]. EAC [16] surface rainfall data indicated a similar trend to the TRMM data. Though no drastic reduction in rainfall is noted in Fig. 9.12, EAC [16] however suggest that the amount of the rainfall during this period is relatively small compared to previous years, and thus a long term reduction in rainfall could also have contributed to the fall in the lake level. This calls for an analyses of draught versus the lake level, i.e., [4] which is presented in Chap. 10.

9.3.3 CHAMP Satellite Diagnostic

Radio occultation with Global Positioning System (GPS) takes place when radio signals from a transmitting GPS satellite, setting or rising behind the Earth's limb, are received by a GPS receiver aboard a Low Earth Orbiting (LEO) satellite. The meteorological products of occultation comprise the air pressure P, air temperature T and the water vapour pressure P_w. By measuring the occulting GPS signal due to the effect of the atmosphere, CHAMP satellite (LEO) is capable of providing accurate tropospheric measurements to sub-kelvin accuracy, see e.g., [5–8, 17, 24, 33, 40, 42–46]. More details on CHAMP satellites are presented in Sect. 5.6.

Temperature profiles computed are presented in CHAMP level 3 data discussed in details by Wickert [44]. Using the entire spectrum of CHAMP level 3 data, those of Lake Victoria basin (Longitude: 30° E–36° E; Latitude: 1° N–3° 20'S) were selected. Figure 9.13 presents the position of occultation within the lake basin for

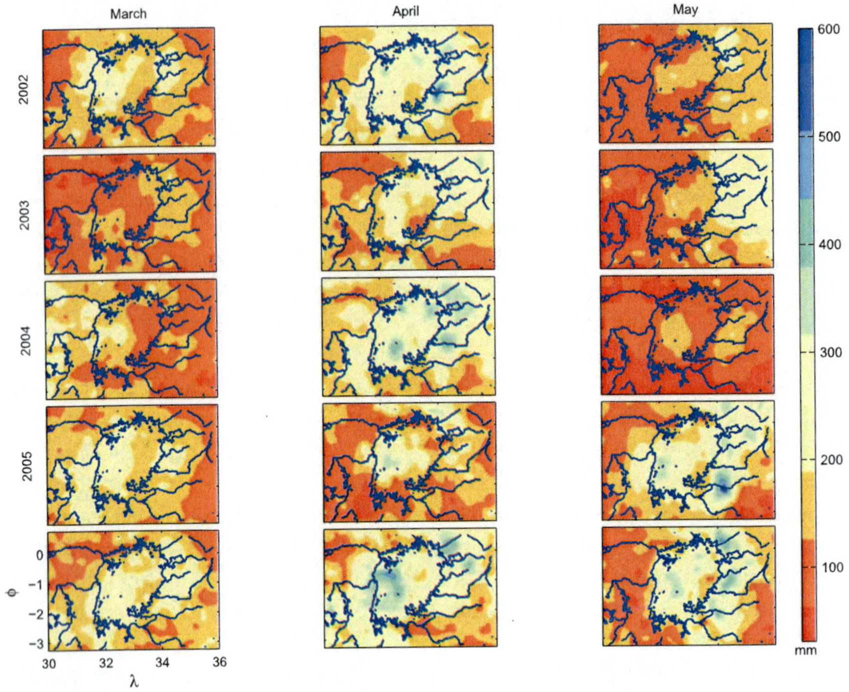

Fig. 9.11 Inter-annual comparison of rainfall over the Lake Victoria basin during the high rainy season of MAM from 2002–2006 using TRMM results. *Source* Awange et al. [3]

the period 2001–2006. In total, 53 occultations took place directly within the lake basin and its neighbourhood.

Whereas Fig. 9.13 show the position of the occultations to be well distributed within the basin, an understanding of the occultations, which occurred every month for the period 2001 to 2006 is presented in Fig. 9.14. Though the occultation data within the lake basin is sparse compared to other mid-latitude regions, nonetheless, the data is sufficient to provide an indication on any significant change in tropopause temperature over the study period.

For each occultation, temperature profile graphs are plotted and the tropopause temperature and height selected (see, e.g., Fig. 9.15). In this study, the tropopause is defined according to the World Meteorology Organization (WMO) definition [47]. For occultation number 195, which occurred at 1.9°S, 32.7°E on 25th of July 2004 from 21h 20m 27s for example, the temperature profile are as plotted in Fig 9.15.

In order to obtain a meaningful deduction, time series for tropopause temperatures and height for the study periods were computed (Figs. 9.16 and 9.17). Annual means of the basin's tropopause temperatures and heights depicted in Figs. 9.18 and 9.19 respectively are thereafter computed. From the time series Figs. (9.16 and 9.17), a

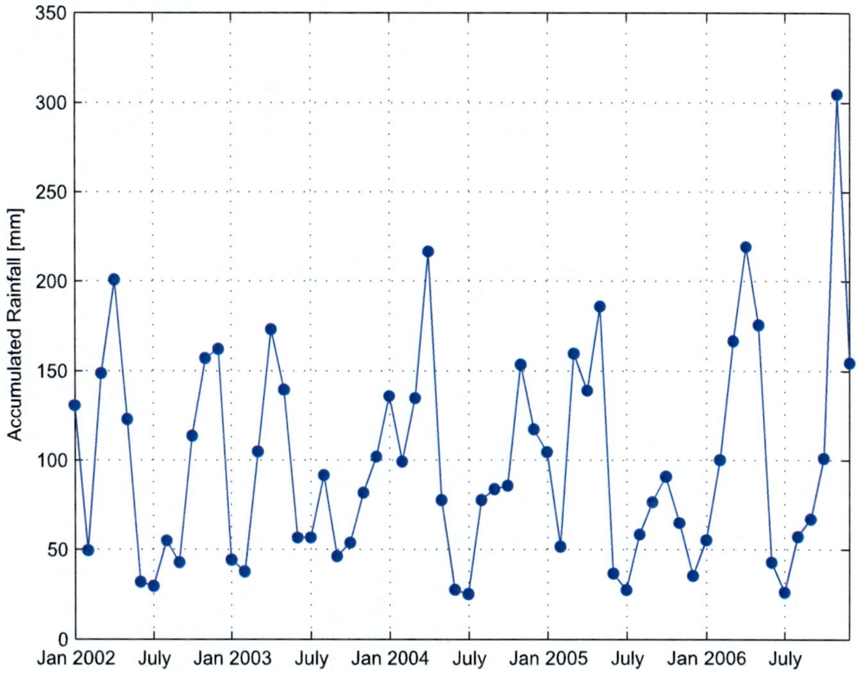

Fig. 9.12 Time series of rainfall 2002–2006 for the lake Victoria basin as observed by the TRMM satellite. *Source* Awange et al. [3]

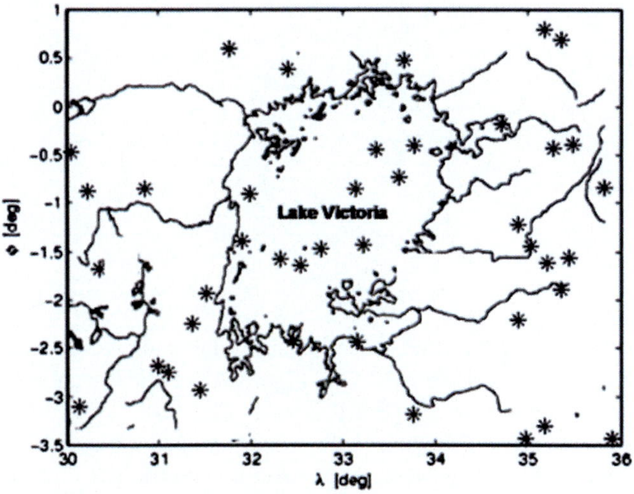

Fig. 9.13 Occultations within the lake basin 2001–2006. *Source* Awange et al. [3]

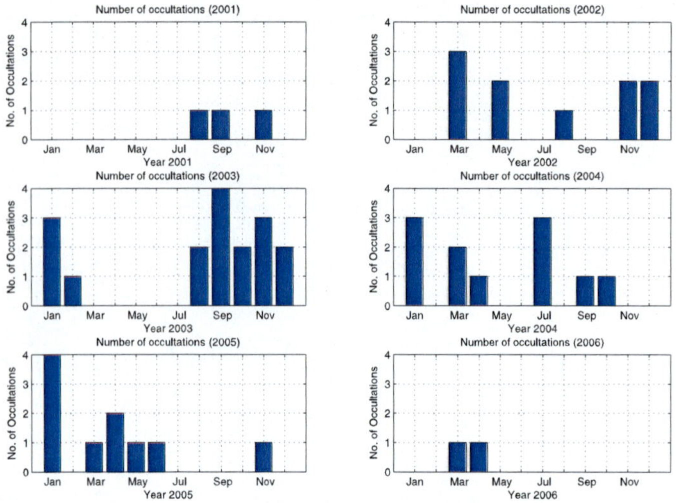

Fig. 9.14 Monthly number of occultations for the period 2001–2006. *Source* Awange et al. [3]

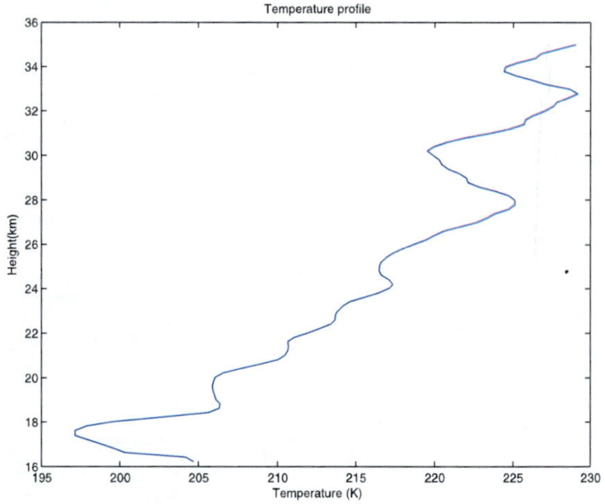

Fig. 9.15 Temperature profile for occultation 195 of 25th July 2004. *Source* Awange et al. [3]

uniform trend is maintained where the tropopause temperatures are between 188–194 K, and the heights between 16.7–17.7 m. Figure 9.18 indicates an annual decrease in the tropopause temperature by 3.9 K from 2001 to 2002. Thereafter, there was an increase of 2.2 K in temperature in the following year. Though there was a drop from 2003 to 2005 by about 1.8 K, an increase is noticed thereafter up to 2006. A significant point to note from the figure however, is the fact that these temperatures have remained relatively above the lowest value of 2002 by more than 0.4 K. This point is corroborated by the heights data of Fig. 9.19, where an increase is noticed from 2001 to 2002 of about 0.58 m. This is followed by a subsequent fall in height by 0.33 m in 2003. Thereafter, the tropopause heights increase by 0.62 m to reach

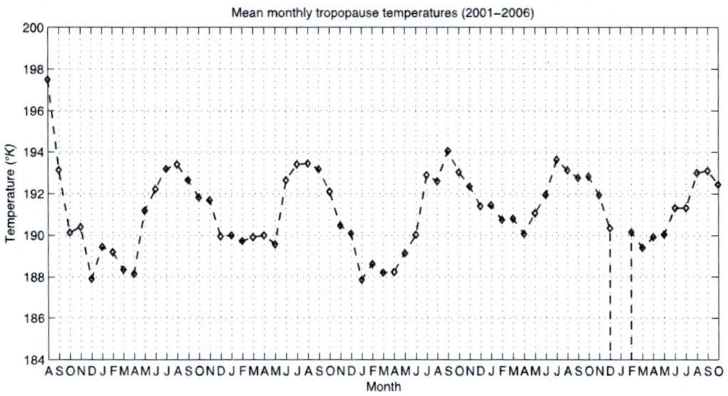

Fig. 9.16 Time series tropopause temperatures. *Source* Awange et al. [3]

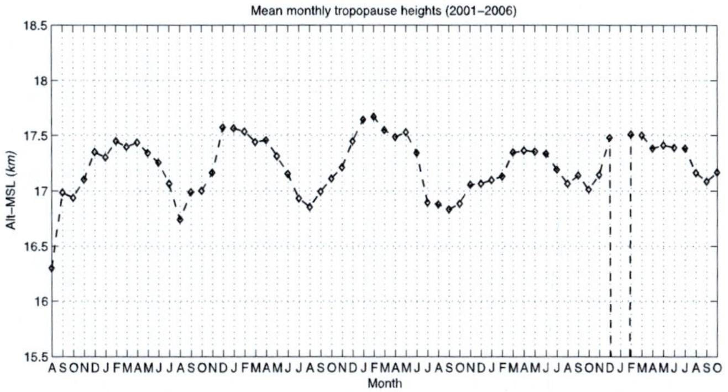

Fig. 9.17 Time series tropopause heights. *Source* Awange et al. [3]

the highest value in 2005. Significant to note is the fact that the heights remain above the lowest value of 2001 by more than 0.87 m.

9.3.4 The 2002–2006 Rapid Fall: The Cause

TRMM Level 3 monthly precipitation data for the same period of time were used to compute mean rainfall at a spatial resolution of $0.25^\circ \times 0.25^\circ$ (25×25 km), as shown in Fig. 9.11, from which the rainfall trends were analyzed (Fig. 9.12). TRMM rainfall data over Africa has been validated, e.g., in Awange [50]. To assess the effects of evaporation, and in the absence of evaporation data, GNSS remote sensing data (59 CHAMP satellite occultations) for the period 2001 to 2006 were analyzed to define if tropopause warming took place (see the approach in [8]). The results indicated that the tropopause temperature fell in 2002 by about 3.9 K and increased by 2.2 K in 2003 and remained above the 189.5 K value of 2002. The tropopause heights

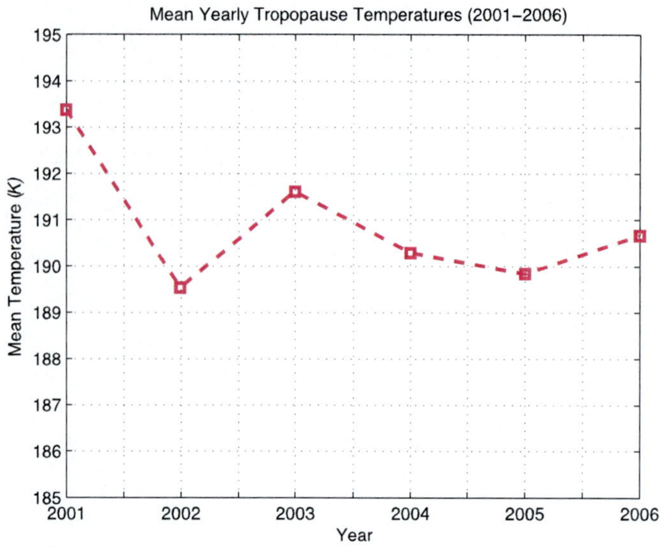

Fig. 9.18 Mean annual tropopause temperatures for the Lake basin from 2001 to 2006. *Source* Awange et al. [3]

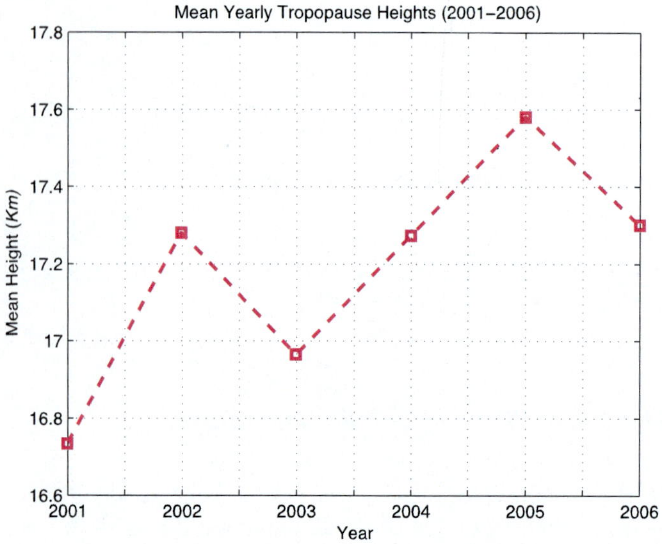

Fig. 9.19 Mean annual tropopause heights for the Lake basin from 2001 to 2006. *Source* Awange et al. [3]

showed a steady increase from a height of 16.72 m in 2001 and remained above that value reaching a maximum of 17.59 km in 2005, an increase in height by 0.87 m. Temperatures did not, therefore, increase drastically to cause massive evaporation. TRMM results indicated the rainfall over the basin (and directly over the lake) to have been stable during this period (see Figs. 9.11 and 9.12). Since rainfall over the period remained stable, and temperatures did not increase drastically to cause increased evaporation, the remaining culprit for the 2002–2006 fall was suspected to be discharge from the expanded Owen Falls dam (i.e., Kiira dam). Awange et al. [3] concluded, thanks to the space satellites missions (GRACE, CHAMP and TRMM), that the fall in Lake Victoria's water level between 2001 and 2006, was due to human impact on the basin's environment (i.e., expanded dam) as opposed to natural factors.

In a related work, Swenson and Wahr [51] used satellite gravimetric and altimetric data to study trends in water storage and lake levels of multiple lakes in the Great Rift Valley region of East Africa for the years 2003–2008. GRACE total water storage estimated by Swenson and Wahr [51] corroborated the findings of Awange et al. [3] that the lake's water level had declined by as much as 60 mm/year, while their altimetric data indicated that levels in some large lakes in the East African region dropped by as much as 1–2 m. Swenson and Wahr [51] concluded that the largest decline occurred in Lake Victoria and, like Awange et al. [3], attributed this to the role of human activities, both pointing fingers to Uganda and concurring with Daniel Kull (see Chap. 4) and also a British Broadcasting Cooperation BBC's report.[1]

Both the findings of Awange et al. [3] and Swenson and Wahr [51] provide evidence that the GRACE satellites could be used to provide independent means of assessing the relative impacts of climate and human activities on the balance of stored waters of Lake Victoria that does not depend on in-situ observations, such as dam discharge values, which may not be available to the public domain.

9.4 Concluding Remarks

The chapter has demonstrated that the lake levels fluctuate annually and seasonally with significant drops during drought seasons and rapid rise during periods of above average rainfall in the region. Lake Victoria water levels are very sensitive to climatic factors and as such, any long-term analysis should incorporate climate parameters, which forms the subject of this chapter. During the short period between 2001 and 2006, studies with satellite datasets, e.g., [3] revealed no major influence of precipitation on the falling water level. With 80% of Lake Victoria water coming from direct rainfall, changes in the lake level are directly related to the variation in the water stored in its basin, which contributes around 20% in the form of river discharge into the lake. A decrease in stored basin water was therefore suspected to

[1]http://news.bbc.co.uk/2/hi/africa/4696240.stm.

contribute to the drop in the lake level. An analysis of the stored water in the Lake Victoria Basin (LVB) in relation to rainfall and evaporation was therefore necessary as a first diagnosis step. This would provide water resource managers and planners with information on the state and changing trend of the stored water within the LVB. Such basin scale observations could only be achieved through the use of satellites such as Gravity Recovery and Climate Experiment (GRACE) discussed in Chap. 5. Conventional methods for studying variations in stored water such as "boots on the ground" in-situ based, Artificial Neural Network, GIS (Geographical Information System) or traditional image-based remote sensing could not diagnose the problem, see e.g., [3]. The following facts were established following the satellite analysis of the Lake basin:

(a) From the GRACE analysis of the geoidal (gravity field) variation of the lake basin, a fall of the geoid level at a rate of 1.6 mm/year was observed. This signifies an annual loss of the basin's stored water during the period of 2002 to 2006. Since the basin catchment discharge contributes about 20% of the Lakes hydrology, the reduction of the basins stored water contributed to the decline of the 20% input to the lake and thus reduction in the Lake's waters.
(b) Reduction in rainfall over the basin during the period from 2002 to 2006 from TRMM data was not so significant to trigger rapid fall in the lake level. However, EAC [16] observes (from surface data) a general decline in the rains on the lake and its basin in recent years. A draught analysis of rainfall versus the lake levels spanning climatological time frame (30 years) is presented in Chap. 10.
(c) CHAMP satellite data indicated there was an increase of 2.2 K in temperature in the from 2002 to 2003 and that the temperatures have remained relatively above the lowest value of 2002 by more than 0.4 K. Though the CHAMP satellite data used was sparse, EAC [16] also note an increase in surface data by 1°C. This amount of increase in temperature could have contributed to evaporation but not at a scale to cause rapid decline of the lake.
(d) The increased withdrawal of the water from the lake basin could only be attributed to the expansion of the Owen Falls (now consisting of the original Naluabaale Dam and the new Kiira Dam extension) as pointed out in EAC [16, 21].

GRACE analysis of lake Victoria's basin has thus helped to reveal the trend in the decline of the basin's stored water. This however could have contributed to a reduction of the 20% discharge into the lake. Since the TRMM rainfall analysis does not show a drastic drop in rainfall over the same period and the CHAMP satellite does not show massive increase in temperature to cause significant evaporation, the main ***culprit is the expanded Owen Falls dam***, one of the conclusion pointed out in EAC [16, 21]. The GRACE, TRIMM and CHAMP satellites thus offers objective and unbiased means for future monitoring water storage changes within Lake Victoria's basin.

References

1. Altunkaynak A (2006) Forecasting surface water level fluctuations of Lake Van by artificial neural networks. Water Resourc Manag 21:399–408
2. Amitai E, Marks DA, Wolff DB, Silberstein DS, Fisher BL, Pippitt JL (2006) Evaluation of radar rainfall products: lessons learned from the NASA TRMM validation program in Florida. J Atmosph Ocean Technol 23:1492–1505
3. Awange JL, Sharifi M, Ogonda G, Wickert J, Grafarend EW, Omulo M (2008a) The falling Lake Victoria water levels: GRACE, TRIMM and CHAMP satellite analysis of the lake basin. Water Resour Manag 22:775–796. https://doi.org/10.1007/s11269-007-9191-y
4. Awange JL, Ogallo L, Kwang-Ho B, Omondi P, Omulo M (2008b) Falling Lake Victoria water levels: is climate a contributing factor? Clim Change 89:281–297. https://doi.org/10.1007/s10584-008-9409-x
5. Awange JL, Grafarend E (2005) Solving algebraic computational problems in geodesy and geoinformatics. The answer to modern challenges. Springer, Berlin
6. Awange JL, Fukuda Y, Takemoto S, Wickert J, Aoyama Y (2004) Analytic solution of GPS atmospheric sounding refraction angles. Earth, Planets Space 56:573–587
7. Awange JL (2012) Environmental monitoring using GNSS. Global navigation satellite system. Springer, Berlin
8. Awange JL (2018) GNSS environmental sensing. Revolutionizing environmental monitoring. Springer, Berlin
9. Awange JL, Kiema JBK (2013) Environmental geoinformatics. Monitoring and management. Springer, Berlin
10. Awange JL, Kiema JBK (2018) Environmental geoinformatics. Extreme hydro-climatic and food security challenges: exploiting the big data, 2nd edn. Springer International Publishers, Berlin
11. Awange JL, Forootan E, Kuhn M, Kusche J, Heck B (2014) Water storage changes and climate variability within the Nile Basin between 2002 and 2011. Adv Water Resour 73:1–25. https://doi.org/10.1016/j.advwatres.2014.06.010
12. Ndehendehe C, Awange J, Agutu N, Kuhn M, Heck B (2016) Understanding changes in terrestrial water storage over West Africa between 2002 and 2014. Adv Water Resour 88:211–230. https://doi.org/10.1016/j.advwatres.2015.12.009
13. Crowley JW, Mitrovica JX, Bailey RC, Tamisiea ME, Davis JL (2006) Land water storage within the Congo Basin inferred from GRACE satellite gravity data. Geophys Res Lett 33:L19402. https://doi.org/10.1029/2006GL027070
14. Dinku T, Anagnostou EN (2006) TRMM calibration of SSM/I algorithm for overland rainfall estimation. J Appl Meteorol Climatol 45:875–886
15. Dunn C, Bertiger W, Bar-Sever Y, Desai S, Haines B, Kuang D, Franklin G, Harris I, Kruizinga G, Meehan T, Nandi S, Nguyen D, Rogstad T, Thomas JB, Tien J, Romans L,Watkins M, Wu S-C, Bettadpur S, Kim J (2003) Instrument of GRACE - GPS augments gravity measurements. GPS World, 1 February 2003
16. EAC (East African Community) (2006) Lake Victoria basin commission. Special report on the decline of water levels of Lake Victoria, EAC Secretariat, Arusha, Tanzania
17. Gurbunov ME, Gurvich AS, Bengtsson L (1996) Advanced algorithms of inversion of GPS/MET satellite data and their application to the reconstruction of temperature and humidity, Rep. No. 211, Max-Plunk-Institut für Meteorologie, Hamburg, Germany
18. Hildebrand PH (2005) Toward an improved understanding of the global fresh water budget. A symposium on living with a limited water supply 85th AMS annual meeting, 9-13 January, San Diego
19. Kalivas DP, Kollias VJ, Karantounias G (2003) A GIS for the assessment of the spatio-temporal changes of the Kotychi lagoon, western Peloponnese, Greece. Water Resour Manag 17:19–36
20. Khaki M, Awange JL (2019) Improved satellite remotely sensed products for studying climate-induced changes over Lake Victoria. Science of the total environment, vol 652, pp 915–926. https://doi.org/10.1016/j.scitotenv.2018.10.279

21. Kull D (2006) Connections between recent water level drops in Lake Victoria, Dam operations and drought. http://www.irn.org/programs/nile/pdf/060208vic.pdf. Cited 18th May 2006
22. Loukas A, Mylopoulos N, Vasiliades L (2006) A modeling system for the evaluation of water resources management strategies in thessaly. Water Resources Management, Greece. https://doi.org/10.1007/s11269-006-9120-5
23. Mendoza ME, Bocco G, Bravo M, Granados EL, Osterkamp WR (2006) Predicting water-surface fluctuation of continental lakes: A RS and GIS based approach in central Mexico. Water Resour Manag 20:291–311
24. Melbourne WG, Davis ES, Duncan CB, Hajj GA, Hardy K, Kursinski R, Mechan TK, Young LE, Yunck TP (1994) The application of space borne GPS to atmospheric limb sounding and global change monitoring. JPL Publication, pp 94–18
25. Mylopoulos N, Mylopoulos Y, Tolikas D, Veranis N (2006) Groundwater modeling and management in a complex lake-aquifer system. Water Resour Manag 21:469–494
26. Nicholson SE (1998) Historical fluctuations of lake Victoria and other lakes in the Northern Rift valley of East Africa. In: Lehman JT (ed) Environmental change and response in East African lakes. Kluwer, Dodrecht, pp 7–35
27. Nicholson SE (1999) Historical and modern fluctuations of lakes Tanganyika and Rukwa and their relationship to rainfall variability. Clim Change 41:53–71
28. Owes TY, Taimeh AY (1996) Evaluation of small basin water harvesting system in arid region of Jordan. Water Resour Manag 10:21–34
29. Phoon SY, Shamseldin AY, Vairavamoorthy K (2004) Assessing impacts on Lake Victoria basin, Africa. In: 30th WEDC international conference, Vientiane, Lao PDR
30. Ramillien G, Cazenave A, Brunau O (2004) Global time variations of hydrological signals from GRACE satellite gravimetry. Geophys J Int 158:813–826
31. Ramillien G, Frappart F, Cazenave A, Güntner A (2005) The variation of land water storage from an inversion of 2 years GRACE geoids. Earth Planet Sci Lett 235:283–301
32. Rodell M, Chen J, Kato H, Famiglietti JS, Nigro J, Wilson CR (2006) Estimating groundwater storage changes in the Mississippi River basin (USA) using GRACE. Hydrogeol J 15(1):159–166. https://doi.org/10.1007/s10040-006-0103-7
33. Steiner AK (1998) High resolution sounding of key climate variables using the radio occultation technique. Dissertation, Institute for Meteorology and Geophysics, University of Graz, No. 3
34. Sanderson VL, Kidd C, McGregor GR (2006) A comparison of TRMM microwave techniques for detecting the diurnal rainfall cycle. J Hydrometeorol 7:687–704
35. Shige S, Sasaki H, Okamoto K, Iguchi T (2006) Validation of rainfall estimates from the TRMM precipitation radar and microwave imager using a radiative transfer model: 1. Comparison of the version-5 and -6 products. Geophys Res Lett 33:L13803. https://doi.org/10.1029/2006GL026350
36. Smith AB, Walker JP, Western AW, Ellet KM (2005) Using ground-based gravity measurements to monitor changes interrestial water storage. Engineers Australia 29th hydrology and water resource symposium, 21–23, February 2005, Canberra
37. Tapley B, Reigber C (2004) GRACE (Gravity recovery and climate experiment). McGraw-Hill Yearbook of Science and Technology
38. Tapley BD, Bettadpur S, Watkins MM, Reigber CH (2004a) The gravity recovery and climate experiment: mission overview and early results. Geophys Res Lett 31:L09607. https://doi.org/10.1029/2004GL019920
39. Tapley BD, Bettadpur S, Ries JC, Thompson PF, Watkins MM (2004b) GRACE measurements of mass variability in the Earth system. Science 305: 503–505
40. Tsuda T, Heki K, Miyazaki S, Aonashi K, Hirahara K, Tobita M, Kimata F, Tabei T, Matsushima T, Kimura F, Satomura M, Kato T, Naito I (1998) GPS meteorology project of Japan-Exploring frontiers of geodesy-. Earth Planets Space 50(10):i–v
41. Viltard N, Burlaud C, Kummerow CD (2006) rain retrieval from TMI brightness temperature measurements using a TRMM PR-based database. J Appl Meteorol Climatol 45:455–466
42. Vorob'ev VV, Krasil'nikova TG (1994) Estimation of the accuracy of atmospheric refractive index recovery from Doppler shift measurements at frequencies used in the NAVSTAR system. Phys Atmos Oceans 29:602–609

43. Wickert J, Beyerle G, Schmidt T, Marquardt C, König R, Grundwaldt L, Reigber C (2003) In: Reigber C, Lühr H, Schwintzer P (eds) First CHAMP mission results for gravity, magnetic and atmospheric studies. Springer, Heidelberg
44. Wickert J, Schmidt T, Beyerle G, König R, Reigber C, Jakowski N (2004) The radio occultation experiment aboard CHAMP: operational data analysis and validation of vertical atmospheric profiles. J Met Soc Jpn, Spec issue 'Appl GPS Remote Sens Meteorol Related Fields', 82(1B):381–395
45. Wickert J, Schmidt T, Beyerle G, Michalak G, König R, Heise S, Reigber C (2006a) GPS radio occultation with CHAMP and GRACE: recent results. In: Foelsche U, Kirchengast G, Steiner A (eds) Atmosphere and climate: studies by occultation methods. Springer, Berlin, pp 3–16
46. Wickert J, Schmidt T, Beyerle G, Heise S, Reigber C (2006) Global atmospheric sounding with GPS radio occultation aboard CHAMP. In: Rummel R, Reigber C, Rothacher M, Boedecker G, Schreiber U (eds) Observation of the earth system from space, pp 55–67, Springer, Berlin
47. WMO (1986) Atmospheric ozone; Technical report 16, World Meteorological Organisation, Geneva, Switzerland
48. Xu CY, Singh VP (2004) A review on monthly water balance models for water resources investigations. Water Resour Manag 12:20–50
49. Kummerow C, Barnes W (1998) The tropical rainfall measuring mission (TRMM) sensor package. J Atmos Ocean Technol 15:809-817
50. Awange JL, Hu K, Khaki M (2019) The newly merged satellite remotely sensed, gauge and reanalysis-based multi-source weighted-ensemble precipitation: evaluation over Australia and Africa (1981–2016). Scie Total Environ 670:448–465. https://doi.org/10.1016/j.scitotenv.2019.03.148
51. Swenson S, Wahr J (2009) Monitoring the water balance of Lake Victoria, East Africa, from space. J Hydrol 370(1–4):163–176. https://doi.org/10.1016/j.jhydrol.2009.03.008
52. TSDIS [TSDIS-F] (2006) Tropical rainfall measuring mission science data and information system. Interface control specification between the tropical rainfall measuring mission science data and information system (TSDIS) and the TSDIS science user (TSU) TSDIS-P907 volume 4: File specifications for TRMM products - Level 2 and Level 3 release 6.07, code 902

Chapter 10
Rapid 2002–2006 Fall: Climate Induced?

It is unclear what hydrological viewpoint, when taken in context of the full hydrologic record and the inherent variability of climate and hydrology, would assume a high likelihood of continued high lake levels and flows. –D. Kull [18].

10.1 Summary

In Chaps. 4 and 9, we saw that Lake Victoria water level had receded at an alarming rate, with fingers pointed at the expanded "Kiira" hydroelectric power station in Uganda. However, since the lake receives 80% of its refill through direct rainfall and only 20% from the basin discharge, climatic contributions cannot be ignored, since the 80% water is directly dependant on it. It is therefore necessary to investigate climatic contribution to the declining Lake Victoria water level observed over a long period, i.e., 30 years based on the work of Awange et al., [3]. This chapter extends on Chap. 9, which examined the possible causes of the fall in Lake Victoria within the 2002–2006 period, contributing towards our knowledge of the hydrology of the lake by investigating the climatological contribution to the falling Lake Victoria water levels. This is achieved by relating the lake levels to climatic indicators, e.g.., precipitation and drought.

Use is made of 30 years period anomalies for rainfall, river discharge and lake level changes of stations within Lake Victoria basin to analyse linear and cyclic trends of climate indicators in relation to Lake levels. Linear trend analysis using the Student's t-test indicate a decreasing pattern in rainfall anomalies, with the slope being statistically similar to those of water levels at both Kisumu, Maziba and Jinja stations for the same period of time (1976–1999), thus showing a strong correlation. On the other hand, cyclic trend analysis using Discrete Fourier Transform (DFT) shows cyclic period of water level to coincide with those of droughts and rainfall.

J. Awange, *Lake Victoria Monitored from Space*,
https://doi.org/10.1007/978-3-030-60551-3_10

The strong relationship between climatic indicators of drought and rainfall on the one hand, and lake levels on the other hand signifies the need to incorporate climate information in predicting, monitoring and managing lake level changes.

10.2 Introductory Remarks

Lake Victoria's water levels have been monitored to infer the impacts of climate change, see e.g.., [24] and anthropogenic, e.g.., [1]. Previous interests in Lake Victoria's water balance traces its roots back to the work of [10]. Other numerous subsequent works that evaluated the lake's water balance have been in the context of further understanding the hydrology of the Nile River. Nicholson and Yin [23] derived a water balance model for Lake Victoria that simulates lake level changes using only the over-lake rainfall as input. Linear regression was applied to model discharge based on rainfall and lake levels only. The model was used to calculate the end of year lake levels from 1931 to 1994, given an initial lake level value in the year 1930. This model indicated that the lake level fluctuated according to the over-lake rainfall with very high values for 1961–1964. This confirms that fluctuations of Lake Victoria and other East African Lakes are driven predominantly by rainfall [24, 34] and as such is impacted by change in climate. The lake typically recharges during the "short rains of September, October and November (OND)" and "long rain of March, April and May (MAM)" seasons, but the amount of recharge dependents on seasonal rainfall amounts as well as water demand at the Uganda's power utility.

Prior to 1910, rainfall measurements were made at few stations of Lake Victoria's catchments, though historical information on the Lake's levels extends back to the late eighteenth century [34]. Records of Nile flow permitted a rough estimation of the lake levels going back several centuries [25]. Though tidal gauge data exist in East Africa, they are inadequate to provide a critical analysis of the Lake water level. In an attempt to circumvent this shortfall, Awange et al. [1] employed TRMM (Tropical Rainfall Measuring Mission), GRACE (Gravity Recovery and Climate Experiment) and CHAMP (CHAllenging Minisatellite Payload) satellite data and revealed a reduction in the 20% basin discharge during the period 2002–2006, with little change in the basin rainfall over the same period. Awange et al. [1] pointed an accusing finger at the expanded Owen Falls hydropower complex, now consisting of the original Naluabaale Dam and the new Kiira Dam extension as a possible culprit in the decline in water over the 2002–2006 period (see detailed discussion of this study in Chap. 9). Although Awange et al. [1] noted no significant change in rainfall over the 5-year period, EAC [12] had pointed out that a long-term reduction in rainfall (i.e., climatic effects) could also have contributed to the fall in Lake Victoria water level over the 2002–2006 period. This was also affirmed by [18], see details in Chap. 4. This is because 80% of the total catchments of Lake Victoria rely on direct rainfall while the remaining 20% comes from the river and underground discharges.

The influence of climatic change on water level was noticed, e.g.., by Magadza [20] who analysed the sensitivity of major rivers of the African continent. Magadza [20] examined changes in Zimbabwe's main water storage facilities during the period of 1991–1992 drought cycles, and established that the storage had dwindled to less than 10% of its installed capacity. Jallow et al. [13] and Li et al. [19] also studied the impact of climate change on water level. In their study of the flow of the Gambian river, [13] indicated that, with an estimated error of about 8%, the Gambia river flow was very sensitive to climate change. Based on the results of river flow responses and vulnerability analysis, climate variables alone were found to cause a 50% change in runoff in the Gambia river catchments [13]. Li et al. [19] noted that primarily climate indicators of precipitation and temperature influenced the fluctuation of Lake Qinghai water levels. In general, Manneh [21] points out that a 1% change in rainfall results in a 3% change in runoff, which in-turn reduces the Lake's recharge. While lake level fluctuations have been shown to track drought episodes, e.g.., [8], the concern here is the 202–2006 falling of the Lake Victoria levels, hence the need to investigate the contribution of climate change besides other environmental and man-made factors. Mistry and Conway [22] investigated the climatological factors responsible for the rise in the lake level, and found out that there was a significant correlation between the Lake rainfall series and the Lake levels. They also pointed out that there was a time lag of 1–2 years between rainfall episodes and the water level peaks of the lake. Since the rainfall series are based on land-based observations, and the Lake itself is roughly one quarter of the whole basin, the lake level variability is partially explained by the over-lake rainfall.

10.3 Data Exploration

The Lake Victoria Basin (LVB, e.g.., Fig. 1.3), described elaborately in [4, 35], experiences a hot and humid equatorial climatic condition that is modified by the effects of altitude, relief and influence of the Lake [17]. Surface water temperatures range between 23.5°C and 29.0°C and the major rivers discharging into it are: Sio, Nzoia, Yala, Nyando, Sondu Miriu, Mogusi, and Migori from the Kenyan side of the Lake; Kagera from the Ugandan side runs all the way from Rwanda and empties its waters in Uganda. To analyse the contribution of climatic change on the fall of Lake Victoria's water level, the following data were used [3].

Rainfall data (1961–1999): Kisumu station (latitude 0°6'S, longitude 34°45'E and 1500m above sea level) was selected because of its position in relation to other stations in the region (see Fig. 1.3). There are a number of other meteorological stations within the LVB, but all are found in the same (Lake Victoria) climatological zone with homogeneous anomalies. Past studies have shown significant homogeneity in the patterns of rainfall anomalies in East Africa including Lake Victoria basin, resulting into common use of single rain gauge location to represent large areas [2, 7, 27, 28]. Several authors have also noted that large scale moisture transported by monsoonal winds enhance basin precipitation significantly, e.g.., [36]. Thus lake Victoria does

Table 10.1 Seasonal rainfall anomalies for the Kisumu meteorological station. Source: IGAD Climate Prediction and Applications Centre Nairobi

YEAR	DJF	MAM	JJA	SON
1961	–	−1.8	2.6	6.2
1962	4.3	3 .2	−0.2	0.7
1963	1.7	1.4	−1.0	−2.1
1964	1.4	0.5	1.1	−1.1
1965	−0.5	1.7	−3.0	2.1
1966	1.2	0.2	0.8	−2.3
1967	−3.1	−1.1	−3.7	4.0
1968	0.1	1.1	1.6	1.2
1969	3.3	−1.8	−1.8	−0.9
1970	1.5	−0.3	−2.1	−2.1
1971	−2.0	2.2	−0.8	−0.9
1972	−0.3	−0.6	2.5	1.9
1973	0.7	−1.0	−2.6	−2.1
1974	−3.1	−0.3	1.8	−0.8
1975	−2.1	0.9	0.0	−1.4
1976	−0.2	−1.5	0.2	1.0
1977	−0.1	2.5	1.4	−1.3
1978	2.3	1.4	1.7	0.8
1979	1.9	−0.3	−0.7	−0.8
1980	−0.5	−0.3	−1.0	−1.4
1981	−2.0	0.4	0.8	−2.9
1982	−1.2	−3.7	2.7	5.0
1983	−1.9	−1.7	−0.4	−0.1
1984	−1.5	−3.1	3.2	0.9
1985	−0.1	2.0	0.5	−2.6
1986	−0.6	1.4	−0.6	−0.5
1987	0.1	−1.7	−0.2	1.5
1988	0.0	1.1	−0.5	0.1
1989	0.2	0.2	−1.9	0.2
1990	0.9	−1.0	−0.3	−2.5
1991	−0.2	0.8	−2.2	0.4
1992	−2.3	−1.8	4.0	0.4
1993	1.6	−0.4	0.3	−2.9
1994	−2.2	3.5	−0.4	0.1
1995	−0.7	0.9	−0.9	1.2
1996	0.0	1 .5	−0.2	1.6
1997	−1.9	−0.3	−1.1	−1.4
1998	6.7	−1.4	−1.0	−2.1
1999	−1.5	1.7	2.3	−0.2

Table 10.2 Seasonal and annual averages for L. Victoria water level for the Kibos staton From 1965 to 2000. Lake level units are in cm. Source: Ministry of Water, Kenya

YEAR	December–February	March–May	June–August	September–November	Annual average
1965	1244.8	1293.4	1275.5	1249.4	1268.2
1966	1287.7	1296.4	1270.2	1267.5	1276.4
1967	1262.7	1213.4	1062.1	1158.5	1172.1
1968	1250.3	1250.3	1273.9	1230.1	1251.8
1969	1269.8	1282.4	1286.1	1243.2	1270.9
1970	1261.6	1243.1	1291.1	1242.5	1259.0
1971	1257.9	1241.6	1247.0	1238.3	1242.0
1972	1248.4	1241.0	1240.8	1158.8	1221.0
1973	1239.2	1255.6	1245.8	1228.2	1244.1
1974	1232.7	1221.5	1246.3	1223.9	1228.5
1975	1217.2	1220.4	1214.3	1223.3	1218.3
1976	1218.6	1221.2	1224.7	1207.3	1215.0
1977	1212.6	1228.1	1237.9	1218.4	1225.1
1978	1216.1	1266.8	1266.6	1249.4	1256.4
1979	1264.2	1317.4	1319.9	1253.9	1290.0
1980	1251.9	1252.8	1141.6	1106.0	1183.7
1981	1218.3	1208.5	1220.8	1205.0	1208.9
1982	1205.8	1205.4	1217.1	1198.3	1206.1
1983	1219.4	1230.4	1228.1	1227.2	1228.9
1984	1222.2	1224.5	1209.1	1180.0	1207.0
1985	1196.1	1192.4	1211.8	1195.8	1195.5
1986	1159.2	1189.6	1196.8	1167.1	1177.5
1987	1187.2	1205.1	1208.8	1186.2	1196.6
1988	1182.8	1196.1	1180.9	1206.0	1195.0
1989	1197.4	1233.6	1225.4	1206.1	1217.4
1990	1208.1	1259.7	1179.9	1223.0	1217.9
1991	1193.0	1241.3	1246.3	1234.0	1231.2
1992	1214.5	1209.1	1219.1	1197.0	1208.9
1993	1220.9	1216.5	1217.8	1170.7	1203.1
1994	1194.9	1185.7	1193.4	1193.0	1192.2
1995	1205.1	1197.7	1166.6	1083.8	1164.5
1996	1212.0	1205.7	1234.8	1231.7	1219.7
1997	1203.0	1220.7	1088.2	1190.7	1175.2
1998	1199.1	1316.3	1328.6	1264.1	1277.6
1999	1253.5	1267.8	1284.3	1267.2	1273.9
2000	1246.7	1241.6	1222.6	1204.2	1229.4

Table 10.3 Table of calculated seasonal and annual Lake Victoria water level anomalies for Kibos station

YEARS	MAM	JJA	SON	DJF	Annual
1961	−2.6	−2.6	−2.2	−2.7	−2.4
1962	0.1	0.5	0.8	−0.2	0.4
1963	1.2	1.6	1.4	0.9	1.4
1964	2.2	2.2	2.3	1.9	2.2
1965	1.7	0.9	1.0	0.7	1.3
1966	1.8	0.8	1.4	2.2	1.0
1967	−0.7	−3.0	−1.2	1.3	0.3
1968	0.4	0.9	0.5	0.9	1.0
1969	1.3	1.1	0.8	1.6	1.2
1970	0.2	1.2	0.8	1.3	1.2
1971	0.2	0.4	0.7	1.2	0.5
1972	0.1	0.3	−1.2	0.8	0.2
1973	0.6	0.4	0.5	0.5	0.4
1974	−0.4	0.4	0.4	0.3	0.3
1975	−0.4	−0.2	0.3	−0.2	−0.1
1976	−0.4	0.0	0.0	−0.2	−0.2
1977	−0.2	0.2	0.2	−0.4	0.4
1978	0.9	0.7	1.0	−0.3	0.7
1979	2.4	1.7	1.1	1.4	1.1
1980	0.5	−1.5	−2.5	1.0	0.4
1981	−0.8	−0.1	−0.1	−0.2	−0.1
1982	−0.9	−0.2	−0.3	−0.6	−0.2
1983	−0.2	0.0	0.4	−0.2	−0.2
1984	−0.3	−0.3	−0.7	−0.1	−0.4
1985	−1.3	−0.3	−0.3	−1.0	−1.0
1986	−1.3	−0.5	−1.0	−2.3	−1.2
1987	−0.9	−0.3	−0.6	−1.3	−0.9
1988	−1.2	−0.8	−0.1	−1.4	−0.5
1989	−0.1	0.0	−0.1	−0.9	−0.4
1990	0.7	−0.8	0.3	−0.6	0.2
1991	0.2	0.4	0.6	−1.1	0.2
1992	−0.8	−0.1	−0.3	−0.3	−0.5
1993	−0.6	−0.1	−0.9	−0.1	−0.8
1994	−1.5	−0.6	−0.4	−1.0	−1.5
1995	−1.1	−1.1	−3.1	−0.7	−1.0
1996	−0.9	0.2	0.6	−0.4	−0.6
1997	−0.4	−2.5	−0.5	−0.7	−1.0
1998	2.3	1.9	1.3	−0.9	1.1
1999	0.9	1.1	1.4	1.0	0.5
2000	0.2	−0.1	−0.1	0.8	−0.5

Table 10.4 Seasonal discharge averages for R. Nzoia and R. Yala for the period 1962–1999. The discharge units are given in m^3/s. Source: Ministry of Water, Kenya

YEARS	Nzoia Dec–Feb	Nzoia Mar–May	Nzoia Jun–Aug	Nzoia Sep–Nov	Yala Dec–Feb	Yala Mar–May	Yala Jun–Aug	Yala Sep–Nov
1962	62.4	62.4	126.3	121.4	73.0	53.9	65.1	52.9
1963	53.8	138.7	123.2	73.0	26.0	53.3	52.0	31.1
1964	90.5	69.9	136.2	127.4	29.7	34.9	50.3	51.8
1965	62.4	62.4	49.2	33.5	16.4	17.9	13.0	16.1
1966	14.3	44.5	44.7	77.8	11.7	29.7	23.0	25.3
1967	12.6	57.6	150.3	81.6	6.9	18.8	35.0	36.1
1968	37.7	95.1	118.4	48.8	23.8	49.7	52.3	33.9
1969	24.0	25.8	48.7	40.8	23.0	21.0	27.0	26.0
1970	29.5	53.0	113.6	85.8	13.8	29.0	40.6	39.6
1971	31.9	43.6	101.5	76.7	12.4	20.6	45.6	43.8
1972	41.8	45.3	93.4	60.1	16.1	18.6	41.8	41.1
1973	32.7	19.4	86.8	67.8	27.5	18.8	37.5	40.7
1974	27.6	24.8	64.1	66.6	11.6	17.8	33.3	32.2
1975	11.3	34.5	132.9	147.6	9.3	17.7	38.6	54.3
1976	22.5	28.0	77.0	47.7	13.9	19.3	31.6	28.1
1977	19.2	102.0	156.4	169.4	12.3	28.3	49.9	51.0
1978	78.5	122.0	153.3	126.5	29.4	51.0	50.6	46.1
1979	45.7	48.5	80.8	29.3	28.2	29.2	52.2	27.9
1980	12.0	35.8	52.6	29.8	10.5	19.3	29.2	20.5
1981	10.5	92.8	97.7	92.5	6.4	33.7	36.2	35.9
1982	23.3	55.2	110.7	85.7	19.8	17.3	33.9	36.8
1983	68.9	38.2	93.5	171.1	30.8	19.2	29.9	52.8
1984	29.0	26.7	50.4	42.9	16.9	13.6	24.9	18.3
1985	10.4	79.3	92.4	50.3	9.7	28.8	37.4	27.2
1986	12.4	34.9	40.7	26.8	10.2	18.6	18.2	15.5
1987	8.7	24.8	47.9	29.8	7.8	16.2	22.8	18.8
1988	12.1	62.5	121.5	135.9	12.7	37.0	37.0	50.1
1989	22.3	41.3	69.6	54.0	15.6	25.7	32.2	32.1
1990	33.5	56.6	76.9	73.1	21.9	43.6	32.5	22.6
1991	23.9	42.5	95.8	72.4	11.9	18.5	33.2	22.9
1992	16.1	31.6	99.3	117.0	10.0	15.9	40.3	53.2
1993	32.2	43.2	72.9	31.7	18.9	18.5	28.1	20.0
1994	10.2	47.0	160.9	96.3	7.1	26.1	49.0	41.2
1995	33.3	37.1	61.3	91.9	19.6	24.0	33.7	31.9
1996	28.5	44.6	95.8	70.3	15.7	23.3	37.2	38.3
1997	21.6	48.4	65.0	59.1	14.3	24.1	23.1	26.0
1998	102.4	50.9	133.5	146.6	51.8	35.1	50.4	44.7
1999	31.9	58.8	91.9	77.7	15.4	28.1	29.0	37.0

not generate its own climate through precipitation-evapotraspiration-reprecipitation recycling only [5, 30]. Based on this information, the rainfall anomalies of Kisumu station (Table 10.1) represented the entire region.

Lake level data: Lake level data from 3 stations within LVB were used. These were: Kibos in Kenya (1965–2000), see Table 10.2, Jinja in Uganda (1976–2005), and Maziba in Uganda (1956–2003). The computed lake level anomalies for the Kibos station are given in Table 10.3.

River discharge: Stream flows are usually used as surrogates rather than rainfall (precipitation) in a climate change impact assessment, as it is easier to detect climate change in runoff than precipitation, since changes in precipitation are usually amplified in runoff [11]. Furthermore, stream flow data may be perceived as representing the integrated effects of the spatial variability of precipitation within the catchment. Therefore, the stream flow data may provide as much information as precipitation time series derived from several rainfall stations in the catchment [11]. Discharge measurements from two rivers, i.e., rivers Yala and Nzoia from Kenya for the period (1962–1999) were considered (see Table 10.4). River Nzoia is the longest with the largest catchment basin in the region while river Yala is the second longest river [16]. To obtain the river discharge, the speed of river is multiplied by the cross sectional area and the units are given in m^3/s.

10.4 Climatological Analysis

The daily records of data d_i (rainfall, water level and discharge) were averaged to get the monthly means using

$$X_{DM} = \frac{1}{n}\sum_{n=1}^{n} d_i, \tag{10.1}$$

where n denotes the number of days in a month. Seasonal means were then obtained by averaging the monthly means for the three months of a particular season, i.e.,

$$X_{SM} = \frac{1}{3}\sum_{m=1}^{3} X_{DM}, \tag{10.2}$$

while for annual means, the average of the 12 monthly means in Eq. 10.1 were taken for a given year, i.e.,

$$X_{AM} = \frac{1}{12}\sum_{m=1}^{12} X_{DM}. \tag{10.3}$$

In order to achieve comparison of the data, the rainfall, river discharge, and lake level data were normalized (standardized) using

$$Z = \frac{X_t - X_{30}}{\sigma}, \quad (10.4)$$

where Z is the standardized value, X_t is the observed value at a particular time (e.g.., X_{SM} or X_{AM}), X_{30} the 30-year mean for the parameter and σ is the standard deviation. X_{30} is given by

$$X_{30} = \frac{1}{30}\sum_{m=1}^{30} X_{AM}. \quad (10.5)$$

The 30-years-mean was in accordance with World Meteorological Organization (WMO) requirement that this be the standard used to define the rainfall "normal" of a region [2]. Using Eqs. 10.1–10.5 drought, linear and cyclic trend analysis were performed as discussed below.

10.4.1 Determination of Drought Seasons and Years

Using rainfall data computed from Eq. 10.3, the normal precipitation $N = X_{30}$ - typically considered a 30-year mean is obtained from Eq. 10.5. The normal precipitation is then used to compute the percentage of normal P_N (quartile) using

$$P_N = \frac{X_A}{N} \times 100\%, \quad (10.6)$$

where X_A is the actual precipitation. Using the percentage of normal from Eq. 10.6, Drought Severity Index (DSI) is determined by considering all observations which are less than 25% (first quartile) of the ranked historical records to be dry, while those which are more than 75% (third quartile) are considered wet (Table 10.5). Four seasons common in the region are December-January-February (DJF), March-April-May (MAM), June-July-August (JJA) and September-October-November (SON). MAM is the 'long-rains' season and SON the 'short-rains' season in the region. Drought seasons and years were identified by observing the years and seasons in which the anomalies were below the drought threshold line (i.e., DSI). These drought years are then used in the analysis process to compare those times of drought in the region and the Lake Victoria levels.

Table 10.5 Values of drought severity classified in percent of normal

Percentage (%)	Description of condition
<75	Wet
25–75	Near normal
> 25	Drought

10.4.2 Trend Analysis

Under trend analysis, linear trends of the anomalies with respect to the common epochs of the data were plotted and the slope of the line computed using *polyfit* function of Matlab software. The slopes of linear trends from rainfall, water level and discharge data for the common period 1976–1999 were then tested for statistical significance using the Student's t-test. Assuming that each data follows the normal distribution with sampled mean and standard deviation, let a_1 and a_2 be the estimated slopes from the first order least squares method and σ_1 and σ_2 the standard deviations of a_1 and a_2, respectively. The unbiased combined (pooled sampled) standard deviation of two datasets is given, e.g.., by [14] as

$$S = \sqrt{\frac{\sigma_{a1}^2 + \sigma_{a2}^2}{n}}, \tag{10.7}$$

where n is the number of samples with the degree of freedom of $n - 1$. If the samples have different sizes, the combined standard deviation of the datasets is given as

$$S = \sqrt{\frac{(n_1 - 1)\sigma_{a1}^2 + (n_2 - 1)\sigma_{a2}^2}{n_1 + n_2 - 1}\left(\frac{1}{n_1} + \frac{1}{n_2}\right)}, \tag{10.8}$$

where n_1 and n_2 are the numbers of each samples with the degree of freedom of $n_1 + n_2 - 2$. The null and alternative hypotheses are then formulated as:

$$\begin{aligned} H_0 &: a_1 = a_2 \\ H_1 &: a_1 \neq a_2, \end{aligned} \tag{10.9}$$

and H_0 is the null hypothesis that the slope of a_1 equals that of a_2, while H_1 is the alternate hypothesis. Using Student's t-distribution, and S from either Eq. 10.7 or 10.8 depending on the sample sizes, we express the test value as

$$t = \frac{|a_1 - a_2|}{S}, \tag{10.10}$$

and test it against the test criterion (tc) at 99% confidence level. The null hypothesis is accepted if $t < tc$ or rejected otherwise. In case the Null hypothesis is accepted, i.e., the slopes a_1 and a_2 are statistically equal, then the occurrences are said to be related.

Annual anomalies are too general and cannot give a clear impression of drought on seasonal basis. It is possible to experience drought in one season in a year, which can result to severe impacts. This cannot be reflected if annual anomalies are used in case there was much rainfall in another season of the same year since the rainy season would compensate for the drought season. Because of this, it is important to analyse drought on seasonal basis along with annually anomaly analysis. Cyclic analysis, i.e., periods of year-to-year recurrences of droughts, were investigated using Discrete Fourier Transform (DFT; see, e.g.., [15, 31, 32] for alternate wavelet analysis approach). Discrete Fourier Transform of the N seasonal data, (dj), are expressed as

$$\begin{aligned} X(k) &= \sum_{j=1}^{N-1} d(j)e^{-(\frac{2\pi jk}{N})i} \\ &= \sum_{j=1}^{N-1} d(j)[cos(\tfrac{2\pi jk}{N}) + isin(\tfrac{2\pi jk}{N}), \end{aligned} \tag{10.11}$$

where $k = 0 \ldots (N-1)$ [9, 29]. Its power spectrum, P, is given as the mean of the absolute values of $X(k)$ and the maximum frequency that can be recovered without aliasing, i.e. the Nyquist frequency (fc), is given as

$$fc = \frac{1}{2\Delta}, \tag{10.12}$$

where Δ is the sampling interval. In our cases, the Nyquist frequency is 0.5 and this means that the minimum cyclic trend can be found with the seasonal data of 2 year.

10.5 Climate Impacts on Lake Victoria

10.5.1 Drought Years from the Rainfall Anomalies

In Table 10.6 and Fig. 10.1, drought years from annual rainfall anomalies of the region are identified. It was noted that floods are also recurrent within the lake basin. Some of the extreme drought and floods have been linked to El Nino southern oscillation (ENSO), Indian Ocean Dipole, and many other large scale climate systems [27]. Prominent cycles that have been detected in rainfall over eastern Africa including lake Victoria basin are 2.0–2.5 years, 2.7–3.3 years with less prominent cycles at 3.5–4.4 years, 5.0 years, 6.0–6.5 years, 7–8 years and 10–11 [26, 28]. Because of the El Ñino rains of 1998, the 1990s were not particularly dry, but significant droughts did occur in 1990 and in 1997 (Fig. 10.1).

The anomalies computed from Eq. 10.4 indicate years that received below and above normal rainfall. The DSI in Fig. 10.2 shows the seasonal rainfall anomalies for all the four seasons for the Lake Victoria Basin (LVB) region. For the same year, the figure indicates that one season can receive too much rainfall than the others. A good indication of this is in 1997, which has been highlighted as a drought year. This resulted from the below 'normal' rains that were received in the first three seasons of the year. In the last month of 1997, which is part of the 1998 DJF season, the El Ñino rains were experienced in the region. The year 1997 is ranked as a drought year since the El Ñino rains started in the last month of that year. General patterns of drought identified for the Lake Victoria seasons, e.g.., in [3] were:

- During the hot dry season (DJF), severe drought can be expected approximately every 7–8 years.

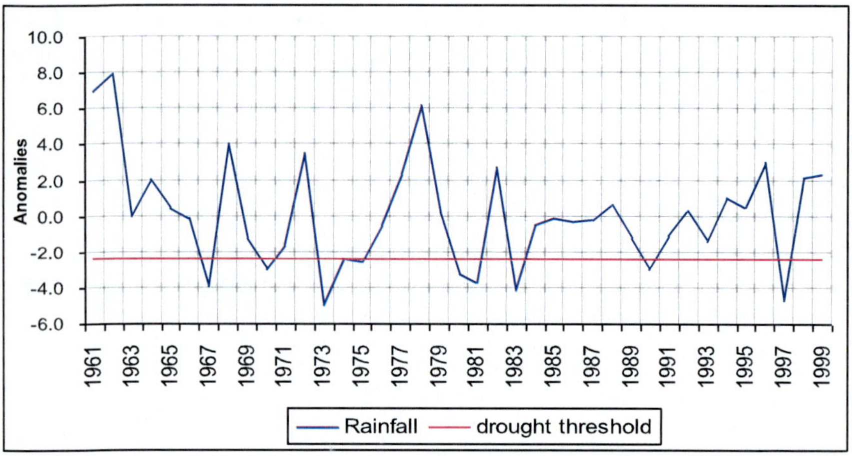

Fig. 10.1 Time series of annual rainfall anomalies for Lake Victoria. Source: Awange et al. [3]

Table 10.6 Seasonal drought years

Seasons	Drought years
DJF	1967, 1971, 1974, 1975, 1981, 1982, 1983, 1984, 1992, 1994, 1997, 1999
MAM	1961, 1967, 1969, 1976, 1982, 1983, 1984, 1987, 1992, 1998
JJA	1963, 1965, 1967, 1969, 1970, 1973, 1980, 1989, 1991, 1995, 1997, 1998
SON	1963, 1966, 1970, 1973, 1975, 1977, 1980, 1981, 1985, 1990, 1993, 1997, 1998

- During the long rainy wet season (MAM), severe drought can be expected every 5–8 years.
- During the dry season (JJA) there is no clear cycle of drought events.
- During the short rainy wet season (SON), severe drought can be expected every 3–4 years.

10.5.2 Linear Trend Analysis

Annual rainfall anomalies' linear trend is plotted in Fig. 10.3, while those of lake levels are plotted in Figs. 10.4 and 10.5 for Kibos and Jinja stations. Figures 10.3, 10.4 and 10.5 for rainfall and lake levels show annual declining linear trends. Figure 10.6 presents a linear trend for Nzoia river. The linear trends are tested according to Eqs. 7 and 9 at 99% confidence level. Table 10.7 presents the slope data, standard deviations of the slopes and the sample sizes for water level, rainfall and discharge data types for Kibos, Jinja, Kisumu, Nzoia and Maziba stations. Tables 10.8 and 10.9

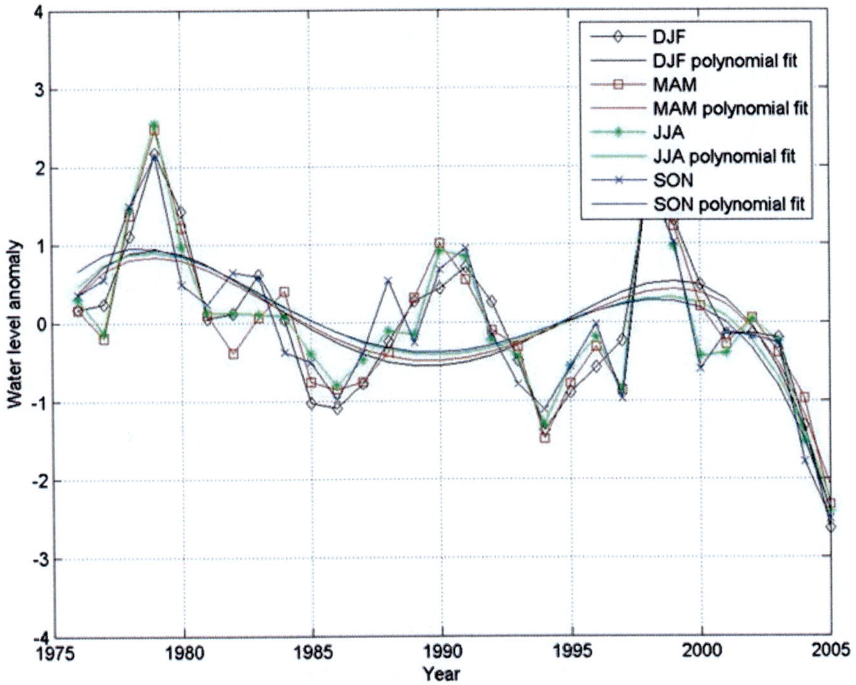

Fig. 10.2 Seasonal anomalies of Jinja water level from 1976 to 2005 and their fourth order polynomial fits. Source: Awange et al. [3]

show the results of the statistical analyses. From the analysis, it is noted that the null hypothesis is accepted in the entire test, indicating that the slopes were significantly the same. The results indicate the acceptance of the null hypothesis and imply that the decline in rainfall contributes to the fall in water level. A plot of the rainfall anomalies together with the drought threshold (DSI) in Fig. 10.1 indicate a general declining trend in the lake water levels since 1965 after it had attained a rise as a result of much El Ñino rains that were received in the region during the period between 1961 and 1964. Some years, which recorded large annual lake levels fall, e.g.., 1967, 1980 and 1983 (Figs. 10.4 and 10.5) coincide with the annual drought year periods (Fig. 10.1), where the rainfall anomalies were below drought thresholds.

10.5.3 Cyclic Trend Analysis

Though the analysis of annual rainfall anomalies and annual lake level averages gives some general relationship overview, seasonal anomalies are also examined for periodic cyclic trends. Seasonal lake levels time series are plotted and smoothed using the Polyfit function of Matlab software and is presented in Fig. 10.2 for Jinja

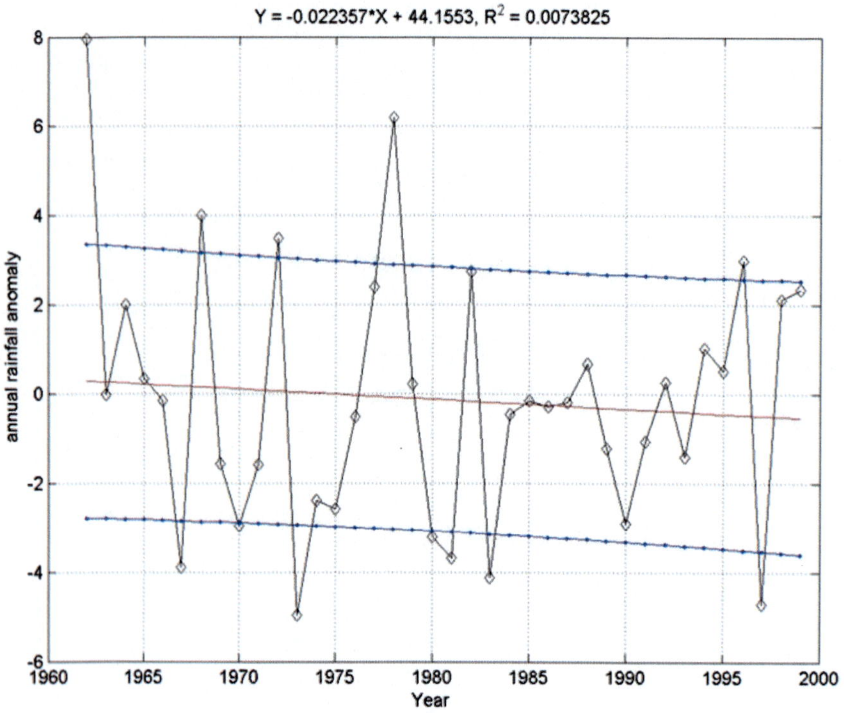

Fig. 10.3 Annual rainfall anomaly in Kisumu station from 1962 to 1999. The first order line fit results and R^2 values are presented in the top of each plot. Source: Awange et al. [3]

station, which shows a cyclic trend for the seasons DJF, MAM, JJA and SOD. Figure 10.2 shows seasonal variation of rainfall anomalies for all the four seasons for Lake Victoria region (Table 10.10). The seasons with very low lake levels are; JJA (1967), SON (1972), SON (1980), DJF (1986), SON (1995) and JJA (1997), some of which were identified as drought years in Fig. 10.1. Lake level rose in all seasons from 1961, reaching its peak in 1964. This is attributed to the above-normal rainfall in the region from 1961–1964, which caused the Lake to rise with an unexpected 2.5 m [33].

From Figs. 10.2, 10.4, and 10.5, a general trend in declining lake level is evident. As in the first 20 years 1961–1979, in most of the seasons, the lake levels were above the threshold line and yet from 1980–2000, most of the seasons had below normal' lake levels. For the same years, the figure indicates that one season can receive too much rainfall compared to the others. A good indication of this is in 1997, which has been highlighted as a drought year. This resulted from the below normal' rains that were received in the first three seasons of the year. In the last month of 1997, which is part of the 1998 DJF season, the El'Nino rains were experienced in the region. The year 1997 is ranked as a drought year yet El'Nino rains started in the last month of that year. This cyclic pattern is analysed using Discrete Fourier Transformation,

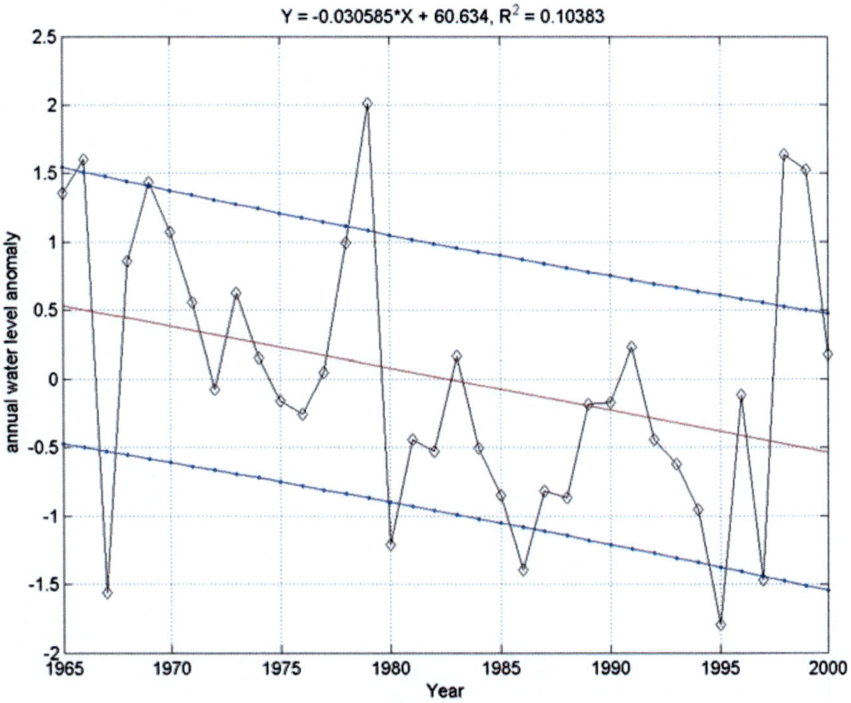

Fig. 10.4 Annual water level anomaly in Kibos station from 1965 to 2000. Source: Awange et al. [3]

which shows the dominant peaks in the signals and provides the cyclic period (Figs. 10.7 and 10.8). The periods of these peaks for rainfall and water level at Jinja are compared to the drought periods of Awange et al. [2], Table 10.7. The results of Table 10.11, which compares the drought cycles from Awange et al. [2] and those obtained from Fourier Transformation of rainfall and water level data, are interesting. The results indicate closeness in the periods of drought cycles and those of less rainfall and drop in water level. Climate change, indicated by variation in rainfall and drought therefore contribute to the fall in lake Victoria's water level in the long term.

Lake levels anomalies calculated using Eq. 10.5, plotted in Figs. 10.2, 10.4, and 10.5 shows the fluctuations, and identify seasons and years when the Lake recorded very low levels. The figures indicate a general falling linear trend of the lake levels since 1965 after it had attained a rise because of much rain in the region during 1961–1964. Though the Lake levels show a general falling trend, there are very significant falls in certain years like in 1967, 1980–1981, 1986, 1997, 2000 and 2005 for Jinja station. Drought in the region would also have an impact on the river flows in the LVB. Therefore, for evidence of drought impact on the lake levels, the stream flows of rivers in the basin are considered.

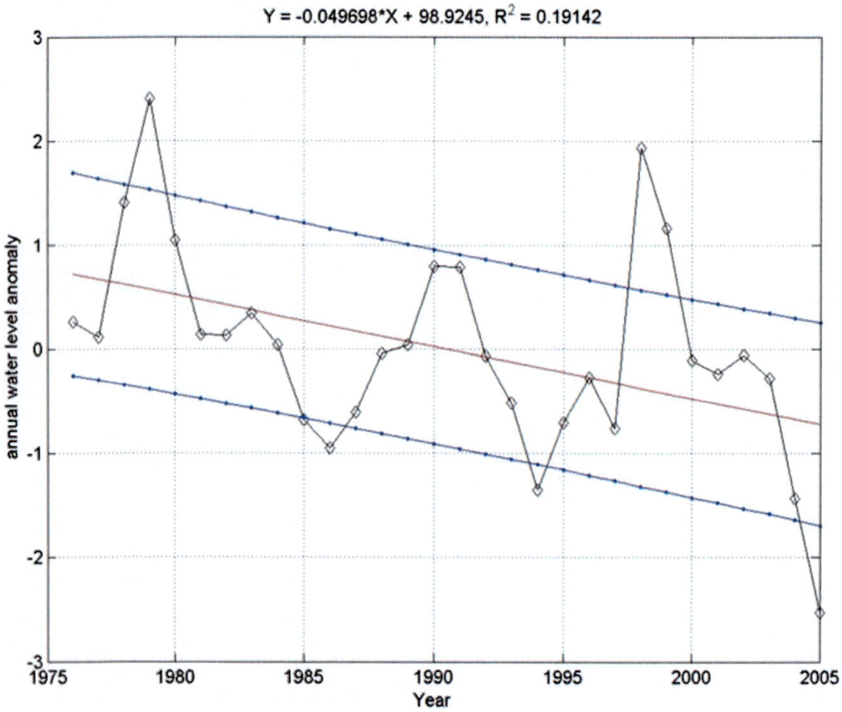

Fig. 10.5 Annual water level anomaly in a station in Uganda from 1976 to 2005. Source: Awange et al. [3]

Statistical analyses (Tables 10.8 and 10.9) indicate the impact of drought on the Lake water level. This is also evidenced by the declining trend (Fig. 10.6) of the discharge of river Nzoia, which contributes to the 20% refill of the Lake. The explanation for the years when the Lake level was increasing and the flow of the river decreasing as in 1968/1969 (Figs. 10.4 and 10.6) could be due to the rainfall falling directly on the lake. The years when the river flow was increasing and the lake level falling as in 1994 could be due to rains on the catchment and not over the Lake or the rate of lake water withdrawal could have been faster than the expected input from both the over-lake rainfall and the river flows (see also Fig. 10.5 for Jinja station). Since the flow of a river is sensitive to the rainfall of the region, similarities in the trend patterns of the river and lake levels indicate that rainfall in the basin have greater impact on the lake levels. Any reduction in the river contributes to the general reduction of the 20% lake recharge from the basin. For the droughts of the late 1960s and early 1980s, both the lake levels and the river discharge show a declining trend. They both show a rise as a result of the 1961/1962 rains and the 1997/1998 El Ñino rains.

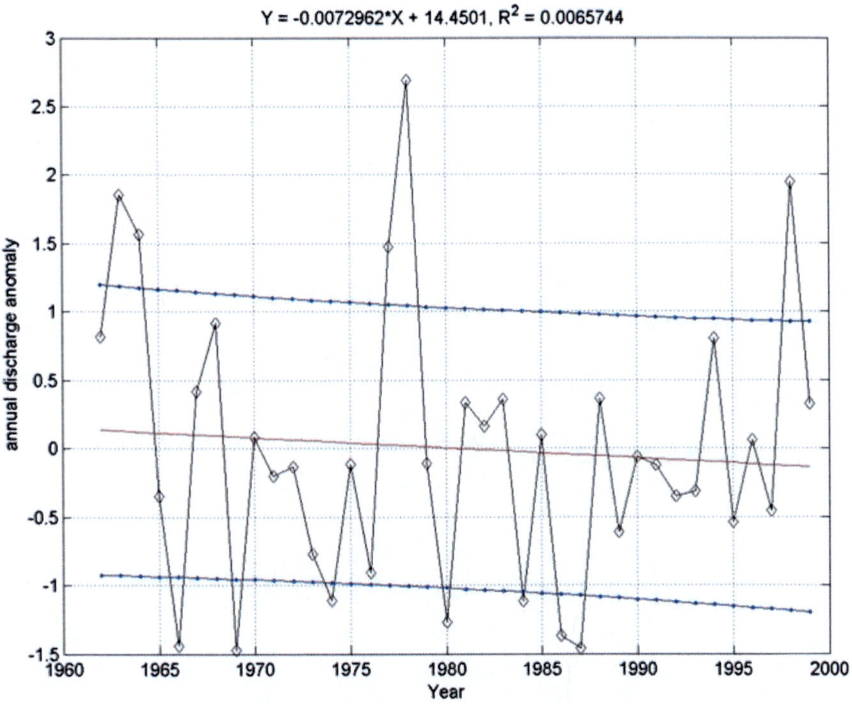

Fig. 10.6 Annual discharge anomaly in river Nzoia station from 1962 to 1999. Source: Awange et al. [3]

Table 10.7 Estimated slopes and their SD of rainfall, water level and discharge data from four stations in Kenya and Uganda from 1976 to 1999

Item	Kibos (water level)	Jinja (water level)	Kisumu (rainfall)	Nzoia (discharge)	Maziba (water level)
Slope a_i	−0.0081	−0.0296	−0.0046	−0.0012	−0.0351
Std_ a_i	0.0301	0.0270	0.0766	0.0302	0.0326
Number of data	24	24	24	24	16

Drought in the Lake Victoria basin, therefore, contributes to sharp drops of the lake levels. The magnitude of the impact of drought on the lake levels is determined by the period of the drought. Prolonged droughts in the region have long-term impacts on the water levels of the Lake, while short droughts have short-lived impacts on the lake levels since the lake regain its water level when rains resume. Though it has been noticed that drought in the lake region contributes to the sharp drops of the lake levels, it can also be noticed that the lake levels indicate a reducing trend in its levels since 1965 and this is not only as a result of drought in the region.

Table 10.8 Student's t-test results of the estimated slopes of water level and discharge data from Kibos, Jinja and river Nzoia

Item	Kibos versus Jinja (water level)	Kibos versus Nzoia (discharge)	Kibos versus Maziba (discharge)
Test value (t)	2.6049	0.7928	2.6890
critical value (tc)	3.7676	2.8073	2.7116
Accept or reject H_0	Accept	Accept	Accept

Table 10.9 Student's t-test results of the estimated slopes of rainfall, water level and discharge data from Kisumu, Kibos, Jinja and river Nzoia

Item	Kisumu (Rainfall) versus Kibos (water level)	Kisumu (Rainfall) versus Jinja (water level)	Kisumu (Rainfall) versus Nzoia (discharge)	Kisumu (Rainfall) versus Maziba (water level)
Test value (t)	0.2083	1.5079	0.2023	1.4996
critical value (tc)	2.8073	3.7676	2.8073	2.7116
Accept or reject H_0	Accept	Accept	Accept	Accept

Table 10.10 Seasonal mean and SD for Lake Victoria water levels

Seasons	Mean (X)	SD (σ)
DJF	1,224.3	28.8
MAM	1,235.9	34.5
JJA	1,225.7	54.7
SON	1,209.2	40.8

Table 10.11 Cyclic trends of rainfall (Kisumu, 1962–1999) and water level (Jinja, 1976–2005) found by both Awange et al., [3] ($C_A wange$) and the method by the Fourier analysis ($C_{Fourier_rainfall}$, $C_{Fourier_waterlevel}$)

Item	C_{Awange}(year) Kisumu	$C_{Fourier-rainfall}$ (year) Kisumu	$C_{Fourier-waterlevel}$ (year) Jinja
DJF	7∼8	∼8.3	∼10.0
MAM	5∼8	∼2.9, ∼2.6, ∼4.8, ∼8.3	∼10.0
JJA		2.0	∼4.0, ∼10.0, ∼16.7, ∼50.0
SON	3∼4	∼2.4	∼4.0, ∼10.0, ∼16.7, ∼50.0

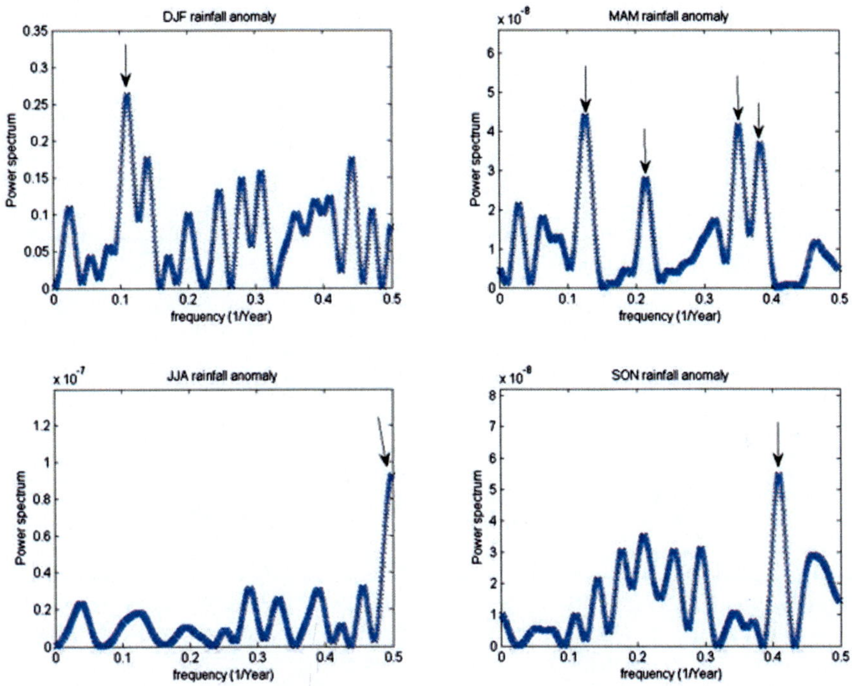

Fig. 10.7 Seasonal anomaly of Kisumu rainfall from 1962 to 1999 and its power spectrum. The strong modes are indicated by an arrow. Source: Awange et al. [3]

10.6 Concluding Remarks

This chapter has demonstrated that the lake levels fluctuate annually and seasonally with significant drops during drought seasons and rapid rise during periods of above average Fig. 10.8. Lake Victoria water levels are very sensitive to climatic factors and as such, any long-term analysis should incorporate climate parameters. During the short period between 2002 and 2006, studies with satellite datasets [1] revealed no major influence of precipitation on the falling water level. In case of long-term fluctuation of the Lake Victoria water level, however, climatic factors play a significant role. Further analysis incorporating more data, e.g.., rainfall and discharges from river Kagera, which could not be obtained during the preparation of this contribution needs to be incorporated to enhance the conclusion.

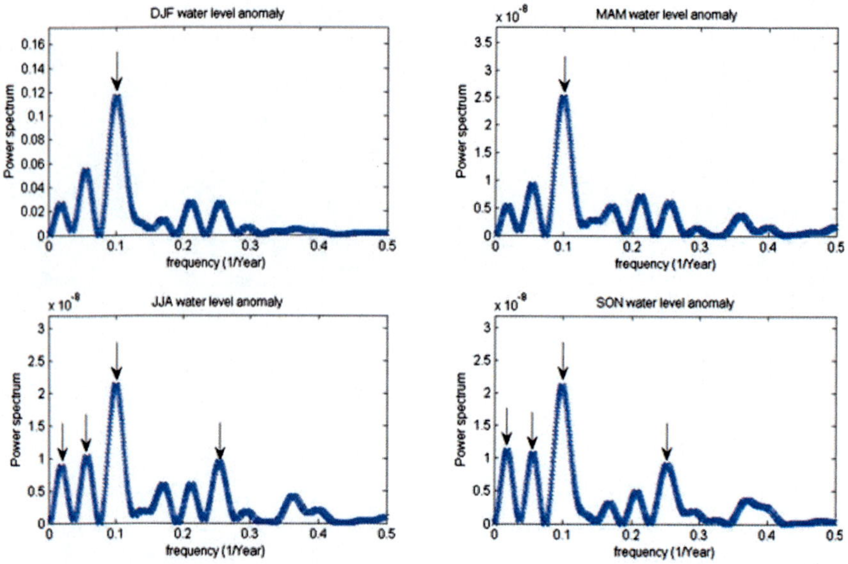

Fig. 10.8 Seasonal anomaly of Jinja water level from 1976 to 2005 and its power spectrum. The strong modes are indicated by an arrow. The strong modes are indicated by an arrow rainfall in the region. Source: Awange et al. [3]

References

1. Awange JL, Sharifi M, Ogonda G, Wickert J, Grafarend EW, Omulo M (2008a) The falling lake Victoria water levels: GRACE, TRIMM and CHAMP satellite analysis of the lake basin. Water Res Manag 22:775–796. https://doi.org/10.1007/s11269-007-9191-y
2. Awange JL, Aluoch J, Ogallo L, Omulo M, Omondi P (2007) An assessment of frequency and severity of drought in the Lake Victoria region (Kenya) and its impact on food security. Clim Res 33:135–142
3. Awange JL, Ogallo L, Kwang-Ho B, Omondi P, Omulo M (2008b) Falling lake Victoria water levels: is climate a contributing factor? Clim Change 89:281–297. https://doi.org/10.1007/s10584-008-9409-x
4. Awange JL, Ong'ang'a O (2006) Lake Victoria: ecology, resources and environment. Heidelberg, Springer
5. Anyah RO, Semazzi FHM, Xie L (2006) Simulated physical mechanisms associated climate variability over lake Victoria East Africa. Mon Wea Rev 134:3588–3609
6. Aseto O, Ong'nga O (2003) Lake Victoria (Kenya) and its environs: resource, opportunities and challenges. Africa Herald Publishing House, Kendu Bay, Kenya
7. Basalirwa CPK, Ogallo LA, Mutua FM (1993) The design of a regional minimum raingauge network. Water Res Dev 9:411–424
8. Beaudoin AB (2002) On the identification and characterization of drought and aridity in postglacial paleoenvironmental records from the northern great plains. Géographie Physique et Quaternaire 56(2):229-246 E-SCAPE Contribution 3. Note: Volume dated 2002, but published in 2004
9. Bracewell RN (1965) The Fourier transform and its applications. McGraw-Hill
10. Brooks CEP (1923) Variations in the levels of central African lakes Victoria and Albert. Geophys Mem Lond 20:334–337

11. Chiew FHS, McMahon TA (1996) Trends in historical streamflow records. In: Jones JAA, Liu C, Woo MK and Kung HT (Eds), Regional Hydrological response to climate change. Kluwer Academic Publishers. pp 63–68
12. EAC (East African Community) (2006) Lake Victoria basin commission. Special report on the 385 decline of water levels of Lake Victoria. EAC Secretariat, Arusha, Tanzania
13. Jallow BP, Barrow MKA, Leatherman SP (1996) Vulnerability of the coastal zone of the Gambia to sea level rise and development of response options. Clim Res 6:165–177
14. Johnson RA, Wichern DW (2002) Applied multivariate statistical analysis, 5th edn. Prentice Hall, New Jersey
15. Kaiser G (1994) A friendly guide to wavelets. Birkhäuser, Boston
16. Kirugara D, Nevejan N (1996) Identification of pollution sources in the Kenyan part of the Lake Victoria catchment area. Kenya-Marine-and-Fisheries-Research-Institute, pp 78
17. Kite GW (1981) Recent changes in level of Lake Victoria. Hydrol Sci Bull Oxford 26:233–243
18. Kull D (2006) Connections between Recent water Level Drops in Lake Victoria, Dam operations and drought. http://www.irn.org/programs/nile/pdf/060208vic.pdf. Accessed 18th May 2006
19. Li XY, Xu HY, Sun YL, Zhang DS, Yang ZP (2007) Lake-level change and water balance analysis at Lake Qinghai, West China during recent decades. Water Res Manag 21:1505–1516. https://doi.org/10.1007/s11269-006-9096-1
20. Magadza CHD (1996) Climate change: some likely multiple impacts in southern Africa. In: Downing TE (ed) Climate change and world food security. Springer, Heidelberg, Germany, pp 449–483
21. Manneh A (1997) Vulnerability of the water resources sector of The Gambia to climate change. In: Republic of The Gambia: Final Report of The Gambia/U.S. Country Study Program Project on Assessment of the Vulnerability of the Major Economic Sectors of The Gambia to the Projected Climate Change. Banjul, The Gambia, (unpublished)
22. Mistry VV, Conway D (2003) Remote forcing of East African rainfall and relationships with fluctuations in levels of Lake Victoria. Int J Climatol 23:67–89
23. Nicholson SE, Yin X, Ba MB (2000) On the feasibility of using Lake water balance model to infer rainfall: an example from Lake Victoria. Hydrol Sci J 45:76–96
24. Nicholson SE (1999) Historical and modern fluctuations of lakes Tanganyika and Rukwa and their relationship to rainfall variability. Clim Change 41:53–71
25. Nicholson SE (1998) Historical fluctuations of lake Victoria and other lakes in the Northern Rift valley of East Africa. In: Lehman JT (ed) Environmental change and response in East African lakes. Kluwer, Dodrecht, pp 7–35
26. Ogallo LA (1984) Temporal fluctuations of seasonal rainfall patterns in East Africa. Mausam 35:175–180
27. Ogallo LA (1988) Relationships between seasonal rainfall in East Africa and the southern oscillation. J Climatol 8:31–43
28. Ogallo LA (1993) Dynamics of the East African climate. Proc Indian Acad Sci (Earth Planet Sci) 102:203–217
29. Press WH, Flannery BP, Teukolsky SA, Vetterling WT (1989) Numerical Recipes, The art of scientific computing (FORTRAN version). Cambridge University Press
30. Song Y, Semazzi FHM, Xie L, Ogallo LA (2004) A coupled regional climate model for lake Victoria basin East Africa. Mon Wea Rev 134:3588–3609
31. Torrence C, Compo GP (1998) A practical guide to wavelet analysis. Bull Am Meteorol Soc 79:61–78
32. TorrenceC Webster PJ (1999) Interdecadal changes in the ENSO - Monsoon system. J Clim 12:2679–2690
33. United States Department of Agriculture (2005) Low Water Levels Observed on Lake Victoria. Production Estimates and Crop Assessment Division, Foreign Agricultural Service, USDA
34. Yin X, Nicholson SE (2002) Interpreting annual rainfall from the levels of Lake Victoria. J Hydrometeorol 3:406–416

35. Awange JL, Saleem A, Sukhadiya RM, Ouma YO, Kexiang H (2019a) Physical dynamics of Lake Victoria over the past 34 years (1984–2018): is the lake dying? Sci Total Environ 658:199–218
36. Khaki M, Awange J (2019) Improved remotely sensed satellite products for studying Lake Victoria's water storage changes. Sci Total Environ 652:915–926

Chapter 11
Climate Change and Its Economic Implications

The major economic sectors of Lake Victoria Basin (LVB) that are subjected to first-order impact of climatic change are: water resources, ecosystems and fishery, agriculture, energy, transportation, infrastructure and communications, and public health and labor productivity. The second-order economic impact of climatic change are such as lingering food shortages, energy poverty, malnutrition and impaired learning ability, and gradual loss of ecosystems that previously supported economic and social life of inhabitants."

—N. Agola [8]

11.1 Summary

The changing climatic patterns and increasing human population within the Lake Victoria Basin (LVB), together with overexploitation of water for economic activities call for assessment of water management for the entire basin. Based on the work of [8], this chapter focuses on the analysis of a combination of available in-situ climate data, Gravity Recovery And Climate Experiment (GRACE), Tropical Rainfall Measuring Mission (TRMM) observations, and high resolution Regional Climate simulations during recent decade(s) to assess the water storage changes within LVB that may be linked to recent climatic variability/changes and anomalies. Trend analysis, principal component analysis (PCA), and temporal/spatial correlations were employed to explore the associations and co-variability among LVB stored water, rainfall variability, and large scale forcing associated with El-Niño/Southern Oscillation (ENSO) and Indian Ocean Dipole (IOD). Potential economic impacts of human and climate-induced changes in LVB stored water were also explored.

Overall, observed in-situ rainfall from lake-shore stations showed a modest increasing trend during the recent decades. The dominant patterns of rainfall data from the TRMM satellite estimates suggest that the spatial and temporal distribution

J. Awange, *Lake Victoria Monitored from Space*,
https://doi.org/10.1007/978-3-030-60551-3_11

of precipitation have not changed much during the period of 1998–2012 over the basin consistent with in-situ observations. However, GRACE-derived water storage changes over LVB indicate an average decline of 38.2 mm/yr for 2003–2006, likely due to the extension of the Owen Fall/Nalubaale dam, and an increase of 4.5 mm/yr over 2007–2013, due to two massive rainfalls in 2006–2007 and 2010–2011. The temporal correlations between rainfall and ENSO/IOD indices during the study period, based on TRMM and model simulations, suggest significant influence of large scale forcing on LVB rainfall, and thus stored water. The contributions of ENSO and IOD on the amplitude of TRMM-rainfall and GRACE-derived water storage changes, for the period of 2003–2013, are estimated to be ~2.5 cm and ~1.5 cm, respectively.

11.2 Introductory Remarks

Freshwater, the most fundamental natural resource for human beings, is required in abundance for drinking, agriculture and all forms of socio-economic development. Its stored potential (surface, groundwater, soil moisture, ice, etc.) is increasingly facing challenges from climate change as well as anthropogenic activities. That current and future climate change is expected to significantly impact the fresh water systems including rivers, streams and lakes, in terms of flow and direction, timing, volume, temperature and its inhabitants has been documented in numerous publications [19, 55]. Changes in the freshwater system, both in terms of quality and quantity, resulting from both natural climate variability (e.g., rainfall patterns) and change, and other anthropogenic influences such as excessive water withdrawals and construction of dams for hydropower generation in the upstream will have significant consequences on the ecosystem and the people depending on them [55]. The conditions are expected to get worse for hugely populated basins such as Lake Victoria Basin (LVB) [26].

Lake Victoria, the second largest freshwater body on Earth, is a source of freshwater and livelihood for more than 42 million people living around it [9, 18] and indirectly supports another 340 million people along the Nile Basin [11, 65] being the source of the White Nile. Lake Victoria Basin (LVB, Fig. 2.2) constitutes an area of 193,000 km^2 and extends over Burundi (7.2%), Kenya (21.5%), Rwanda (11.4%), Tanzania (44%) and Uganda (15.9%) [9, 18]. The basin acts as a constant source of water to the lake through its massive catchment area and its ability to influence the regions' seasonal rainfall. In the last decade, however, the stored waters within LVB have come under immense pressure from climate change and anthropogenic factors that resulted in significant fluctuations. However, the lake level remained above average since the early 1960s [48, 49] till the early 2000s. Discharge estimates from the lake for the period 1950–2005 show that the net balance between recharge and discharge remained relatively stable over the estimation period [58]. A decreasing trend in the lake's level in the past decade as shown, e.g., by [15, 35, 59, 67], however, is attributed equally to over-abstraction and natural climate change such as evaporation [16, 58, 66, 67].

Lake Victoria Basin (LVB) is characterized by modified equatorial type of climate with substantial rainfall occurring throughout the year, particularly over the lake's surface, to semiarid type characterized by intermittent droughts over some near-shore regions [4]. The seasonal rainfall over the basin is further characterized by a bimodal cycle, just like most areas of East Africa, and is controlled mainly by the north-south migration of Inter Tropical Convergence Zone (ITCZ), a quasi-permanent trough that occurs over Lake Victoria [7] due to locally induced convection, orographic influence and land-lake thermal contrast, which modulates rainfall pattern over the lake and hinterlands. The large-scale precipitation over the lake is mainly initiated from the easterly/southeasterly (Indian Ocean) monsoon flow that transports maritime moisture into the interior of East Africa. The humid Congo air mass has also been linked to significant rainfall amounts received over the western and northwestern parts of the lake [7]. Large-scale winds over the Lake Basin are mainly easterly trades most of the year. Superimposed on this basic flow regime are the south-easterly (SE) or north-easterly (NE) monsoons that are mostly driven towards, and often converge over, the ITCZ location. The strength of the monsoons also depends on the sub-tropical anticyclones over the Arabian Sea (Arabian high pressure cell) and southwestern Indian Ocean (Macarene high pressure cell).

In terms of inter-annual variability, LVB climate is characterized by periodic episodes of anomalously wet/dry conditions with some of the memorable events including the 1961/62 and 1997/98 floods that left behind a huge trail of damage to property and infrastructure. The 1961/62 floods were associated with a strong zonal SST gradient over the equatorial Indian Ocean and mid-troposphere westerly flow from Tropical Atlantic [1, 3, 5]. It is noteworthy that 1997/98 floods coincided with one of the warmest ENSO episodes (strongest El Niño) of the last century as well as very strong IOD mode. Hence, the inter-annual variability of the LVB is also closely linked to the SST anomalies over the global ocean basins.

On the one hand, climate Change influences rainfall and temperature patterns thereby affecting LVB's stored water. This is attributed to the fact that more than 80% of LVB's water source is derived directly from the seasonal precipitation [18] and almost an equivalent amount of the precipitation is lost to evaporation [68, 74]. The temperature in the LVB region is projected to increase by 3–4 °C by the end of this century without much change in the rainfall regime, leading to a significant downward trend in the Lake's net basin supply as a result of enhanced evaporation [68] as well as increased water temperatures. Impacts of climate change on LVB have been reported, e.g., in [39, 58, 66, 67].

On the other hand, on anthropogenic influence on LVB, [74] characterized most of the LVB's catchment areas as semi-arid zones, with exception of areas close to the lake, and hence the catchments ability to discharge water into Lake Victoria is expected to decrease as a result of increased abstraction demand for agricultural and industrial activities. This, in addition to declining lake water quantity and quality due to increasing population will thus have serious impacts on the regional water requirement, domestic food supplies, and global food trade [17, 27, 31].

Combined, the impacts of both climate change and other anthropogenic factors on LVB's total water storage (TWS) is having a toll on the economic as well as the environment of the region. For instance, there are already signs of declining fish trades [27] and access to fresh water in the LVB leading to environmental scarcity [20, 46]. Change of fish community and loss of phytoplankton [26, 27] are some impacts of climate change and anthropogenic influences on the lakes water quality, questioning the quality and health of the food. Lake Victoria's outflow is determined by the "agreed curve" (see Chap. 4) drawn between Egypt and Uganda, which also determines the level of hydropower generation. The current and more alarming anthropogenic stress is the increasing demand for power as a result of increasing population in the basin area [45, 58]. The impact of hydropower plants along the Nile river are found to be largest during the drought seasons (or years) and is therefore, expected to put more pressure on the lake with increasing hydropower plants [26, 45]. Recent studies on climate variability and change over the LVB and fluctuations of Lake Victoria levels show some worrying scenes of drought patterns and receding lake levels, which are both attributed to natural climate change and increasing human influence [14–16, 67, 68, 74]. Thus, it is very important to monitor the basin's hydrological cycle using the up-to-date technology and methods to inform the policy-makers and politicians, who play the most important role in managing the regional water resource. All these poses a significant environment and economic challenge to the East African region as a whole, leading to various levels of domestic and interstate conflicts, see e.g., [20].

This chapter examines the changes of total water storage (surface, groundwater and soil moisture) caused by climate variability and extremes during the period (2003–2013) over LVB and the potential economic impacts. To achieve this, freely available global high resolution satellite data sets of Tropical Rainfall Measuring Mission (TRMM) rainfall estimates and Gravity Recovery and Climate Experiment (GRACE) time-variable gravity fields [60, 69, 70] coupled with outputs from various regional climate models (RCMs) in addition to analysis of observed *in-situ* rainfall data over specific stations within the lake's perimeter are employed to study trends of climate over the basin.

A brief overview of the various data sets and methods used to investigate the impacts of climate variability and extremes on stored water potential of LVB are discussed in what follows. These include observed in-situ data, Gravity Recovery And Climate Experiment (GRACE) and Tropical Rainfall Measuring Mission (TRMM). The next subsection gives brief highlights on each dataset used.

11.2.1 Rainfall Data (1960–2012)

Monthly observed in-situ precipitation data for stations along Lake Victoria Basin (see, Fig. 11.1) were employed in this analysis. There are a number of other meteorological stations within the Lake Victoria basin, but only those representatives of their climatological zones with homogeneous anomalies were used. The annual rainfall total was computed through accumulation of the monthly observed data. These data sets were first subjected to quality control and homogeneity tests, see e.g., [54, 56],

Fig. 11.1 Lake Victoria Basin and in-situ rainfall stations (red) used in the study (*Source* [33] modified in [8])

before being analyzed. The slopes of linear trends from the annual rainfall total for the common period 1921–2012 were computed using least-squares regression analysis while statistical significance was assessed using Student's t-test [16]. Linear regression model was applied to the accumulated annual rainfall total for various stations used for the study.

11.2.2 Tropical Rainfall Measuring Mission (TRMM)

The rainfall measurements employed in this work are a product derived largely from observations made by the Tropical Rainfall Measuring Mission or TRMM [36]. TRMM products have been employed in a number of studies of African precipitation where they have been found to be adequate when compared with ground truth observations [12, 13, 47, 52]. The product employed in this work is referred to as the *TRMM and Other Precipitation Data Set* (denoted as 3B43), and covers the period 1998 to 2013. 3B43 provides monthly rainfall (average hourly rate) between latitudes 50 °N/50 °S over a $0.25° \times 0.25°$ grid. It is derived not only from TRMM instruments, but also a number of other satellites and ground-based rain-gauge data. Over time, the products produced from the TRMM observations are updated as the processing techniques and methods for integrating the different data sets are improved upon. In this work, the latest version, number 7, which has been found to be a significant improvement over the previous version 6 owing to such changes as the use of additional satellites and a superior means of incorporating rain gauge information from the Global Precipitation Climatological Centre [25, 29] was used.

11.2.3 Gravity Recovery and Climate Experiment (GRACE)

The Gravity Recovery And Climate Experiment (GRACE) is a United States (National Aeronautics and Space Administration, NASA) and German (Deutsche Zentrum für Luft- und Raumfahrt, DLR) space mission which has been providing products that describe the temporal variation of the Earth's gravity field arising from mass movements within the Earth's system. Level 2 time-variable gravity field products of GRACE have been frequently used to study the Earth's water storage variations [15]. This study uses the latest release five (RL05) monthly GRACE solutions, provided by the German Research Centre for Geosciences (GFZ, [21]), covering 2003–2013.

For computing monthly total water storage (TWS) fields over the LVB basin, the following items are considered:

1. GRACE level-02 products contain correlated errors among higher order spherical harmonics, known as the north-south striping pattern in spatial domain [37]. In order to remove stripes, the de-correlation filter of DDK3 [38] to the GFZ-RL05 solutions was applied. The filtered solutions can also be downloaded from http://icgem.gfz-potsdam.de/ICGEM/TimeSeries.html. Evaluation of the DDK filter for computing correct water storage variations is addressed e.g., in [73].
2. Residual gravity field solutions with respect to the temporal average of 2003–2013 were computed.
3. The residual coefficients were then convolved with a basin function, while considering the basin boundary of Fig. 11.1. For computing the basin function, a uniform mass distribution with the value of one inside the LVB basin and no mass outside the basin (S1 = 1, is a uniform mass in the basin) was assumed. Then, the uniform mass was transformed into spherical harmonics. The obtained coefficients were filtered with the same DDK3 filter as was applied for GRACE products.
4. In order to account for leakages (see e.g., [23, 24]), the total surface mass of the basin was calculated from the basin function coefficients (S2, synthesized uniform mass in the LVB basin). The ratio of S1/S2 reflects the effect of the truncation of the spherical harmonics as well as signal attenuation due to filtering GRACE products over LVB. More discussion of the leakage problem can be found, e.g., in [34].
5. The derived ratio is multiplied by coefficients in item 2 and the results were transformed into $0.5^\circ \times 0.5^\circ$ TWS maps within LVB, following [71].

11.2.4 CRU Data

The University of East Anglia Climate Research Unit (CRU) gridded observational data comprises of 1200 monthly observed climate from 1901 to 2000. CRU data are derived from gauge observations over land areas only and are interpolated on a regular grid of $0.50^\circ \times 0.50^\circ$ [42]. The data sets contain five climatic variables including precipitation, surface temperature, diurnal temperature range (DTR), cloud cover and vapor pressure. In this chapter, only monthly mean surface temperature and precipitation are utilized to complement the available station-based observations.

11.2.5 Regional Climate Simulations

The results of simulated rainfall climatology are presented from four state-of-the-art high resolution Regional Climate Models [a random sample from the Coordinated Regional Downscaling Experiment (CORDEX)] a group of models being used in CORDEX (http://wcrp-cordex.ipsl.jussieu.fr/). CORDEX Africa Project (http://start.org/cordex-africa/about/) used different RCMs to simulate rainfall over the whole Africa domain. The four RCMs data from the CORDEX archive used in constructing simulated climatology over the LVB were WRF, MPI, CRCM5, and PRECIS. The data is from 1989 to 2008 (20 years). The spatial resolution for RCMs-CORDEX is 50 km and for analysis, data was extracted for the LVB domain stretching from 31 °E to 36 °E, and 4 °S to 2 °N. Details on these RCMs are explained in [50].

Given the importance of rainfall in the water balance of the LVB (see Sect. 1.1.3), in this chapter, we only concentrate on comparing the model vs observations (TRMM 3B43-V7 and CRU). Evaluation was undertaken on how the model simulates the impact of large-scale forcings on the seasonal and interannual variability of LVB rainfall (i.e., influence of IOD and ENSO during the years 2005 and 2006, respectively). In order to understand the IOD and ENSO influence, spatial correlations between Nino3.4 and IOD indices for both model and observed (TRMM) data were also computed. Knowing the temporal pattern of ENSO and IOD from the indices, their contributions were co-estimated considering linear trends as well as the annual and semi-annual components in the TRMM-derived rainfall and GRACE-derived TWS changes from 2003 to 2013.

11.3 Analysis of Climate and Economic Implications

11.3.1 Rainfall Variability Analysis

The trend analysis results for precipitation over the Lake Victoria Basin (LVB) are shown in Fig. 11.2. Stations located within the LVB generally showed

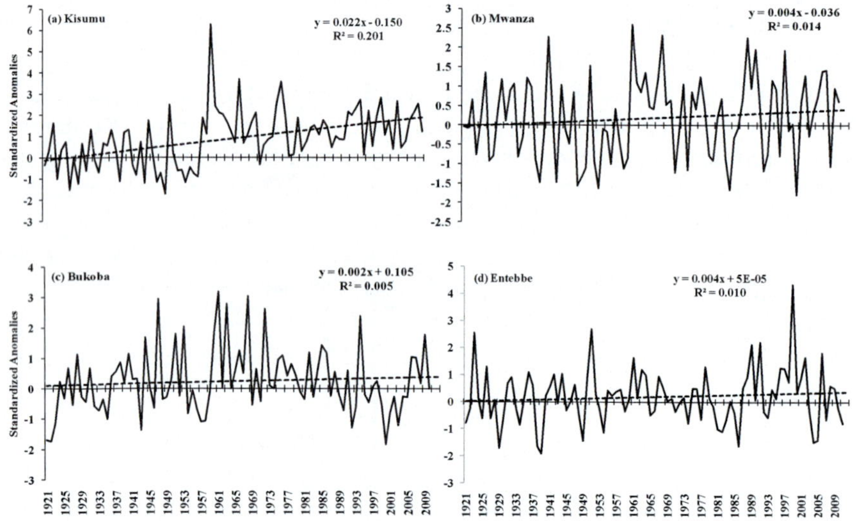

Fig. 11.2 Rainfall trends for some stations in the Lake Victoria Basin. *Source* [8]

modest increase in rainfall trends (e.g., see, Fig. 11.2 a, b, c and d). The increase in trends shown by these stations are, however, not significant at 95% confidence level when Student t-test is applied. Further, Principal Component Analysis (PCA, [10, 57]) analysis of TRMM data are computed to isolate the dominant spatial and temporal patterns of rainfall variability over the LVB during the recent years. TRMM rainfall estimates are here preferred given the more complete spatial coverage, albeit over a relatively short period. To extract the period with relatively more rainfall, the rainfall values of each monthly grids are summed up and presented with respect to their corresponding months in Fig. 11.3. Impacts of the EL-Niño Southern Oscillation phenomenon can be seen e.g., in 2006–2007 and 2011–2012.

Applying PCA on rainfall data of LVB, four dominant EOFs and PCs are established and shown in Fig. 11.4. EOF1 and PC1 (representing 63% of total variance of the rainfall) show a superposition of the annual and seasonal variabilities. The amplitude of the signal in some years such as that of 2007 is amplified as a result of El-Niño. EOF2 and PC2 representing 13% of total rainfall are also related to the annual variation with the same dipole structure of the annual TWS changes in Fig. 11.7. A lag of one-month between PC2 of TRMM and PC2 of TWS changes was obtained. PC3 shows a summation of inter-annual changes and a linear trend over the basin. Considering the structure of EOF3, which is negative over the north west and positive over the southeast, rainfall rate of −2.0 and 2.8 mm/yr respectively are estimate over them, for the period of 2003–2013. The derived trends, however, were not statistically significant. The fourth mode of PCA on rainfall changes (EOF4 and PC4) are not interpreted since the temporal pattern is quite noisy and they represent only 3% of variance in rainfall.

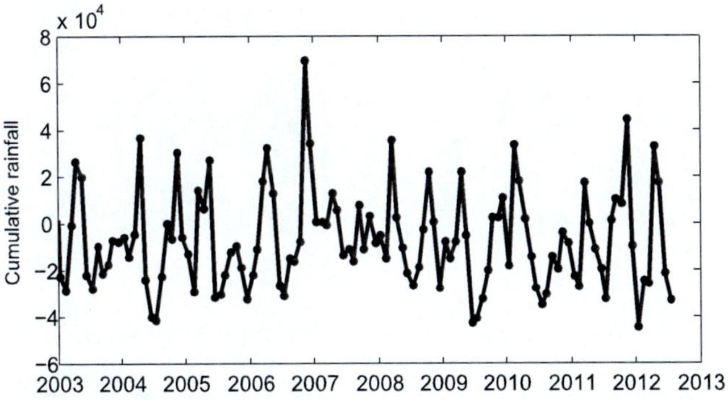

Fig. 11.3 An overview of the cumulative rainfall, derived from each month of TRMM data over LVB, for the period of 2003– 2013. *Source* [8]

11.3.2 Simulated Climatology of LVB (1989–2008)

The observed bimodal rainfall pattern over the LVB (31.5 °E–34 °E; 2.5 °S–1 °N) is well reproduced by three of the four CORDEX Regional Climate Models (RCMs) as shown in Fig. 11.5. However, the MPI RCM captures the bimodal rainfall regime but underestimates the peaks during MAM and OND seasons. This level of RCMs differences (uncertainties) in reproducing the LVB spatial and temporal mean patterns of precipitation presents a challenge in using numerical (theoretical) modeling techniques to understand climate-hydrology connections as well as water level/storage variability over LVB. The RCMs inability to reproduce variability of some peculiar rainfall features of the LVB climate has been linked to incomplete representation/parameterization of localized convective and boundary layer processes that exert significant influence on the spatio-temporal distribution of LVB rainfall [4, 6, 63, 64].

In Fig. 11.6, the Canadian Regional Climate Model version 5 (CRCM5), compared to TRMM estimates, overestimates over-lake seasonal rainfall amounts for both MAM and OND seasons. On the other hand, the PRECIS model as well as the other two models (not shown) consistently simulate drier conditions over the LVB; in some places underestimating the rainfall totals by nearly 100% of the observed (TRMM) seasonal total, especially during the March–May (MAM). However, the CRCM5 captures the OND seasonal mean rainfall pattern quite well compared to TRMM, and also consistent with the dominant EOF loadings of TRMM in Fig. 11.4. The PRECIS model also reproduces the observed spatial distribution of rainfall during OND although the simulated center of rainfall maximum is over the northeastern quadrant of the Lake as opposed to southwestern and western quadrants as in TRMM estimates and CRCM5 simulation.

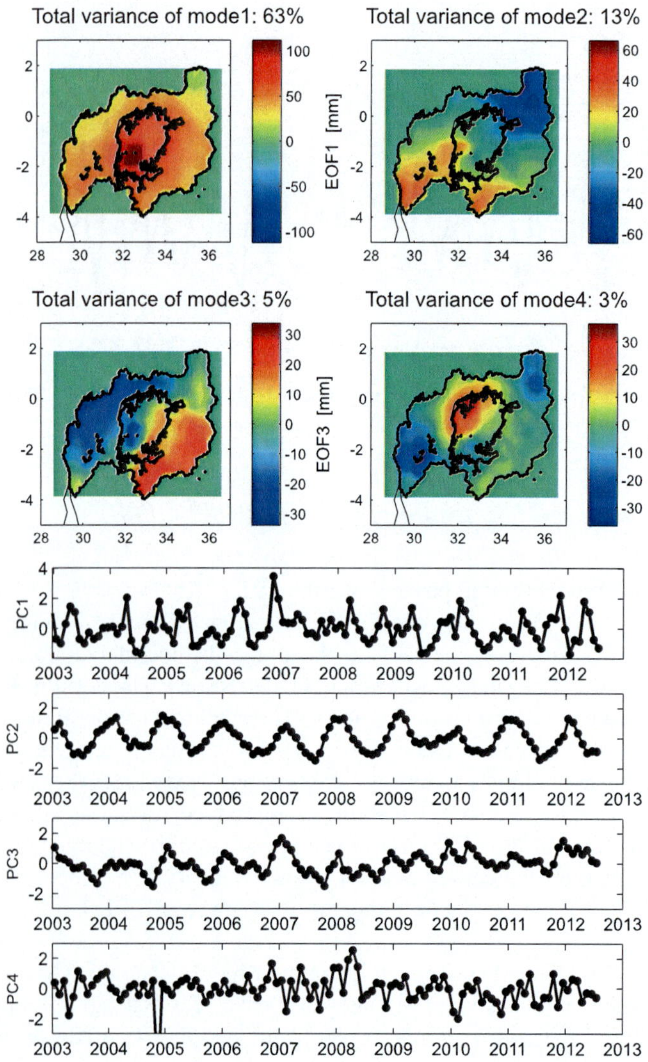

Fig. 11.4 PCA decomposition of rainfall changes derived from TRMM, over LVB. EOFs are rainfall anomaly maps and PCs are their corresponding unit-less temporal patterns. *Source* [8]

11.3.3 GRACE Total Water Storage over LVB

PCA analysis was then employed on TWS to examine whether the observed and simulated patterns of climate variability discussed in the previous section are consistent with the water storage variability derived from GRACE data. As a result, its first two dominant EOFs and PCs are shown in Fig. 11.7, where EOF1 and PC1

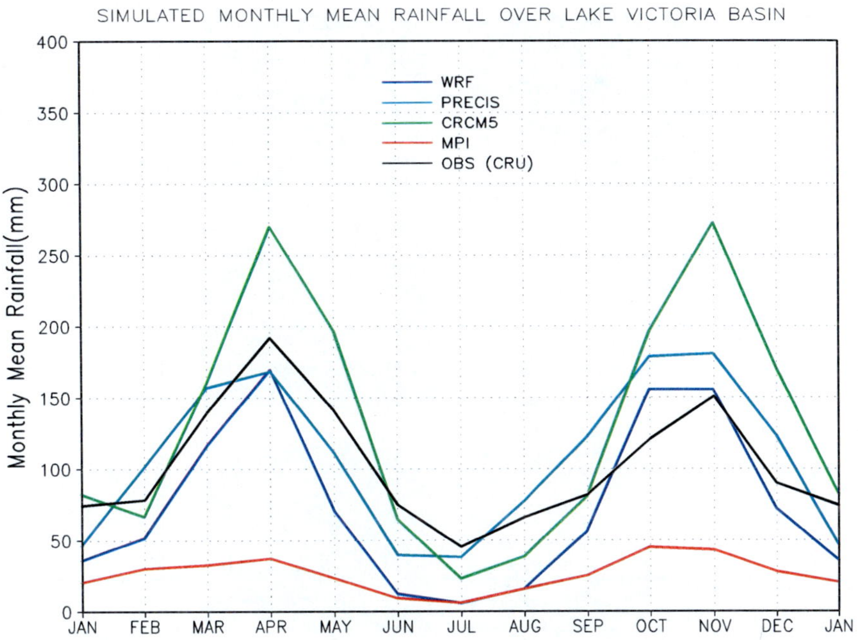

Fig. 11.5 Mean annual cycle of precipitation (mm) over Lake Victoria Basin (31.5E-34E, -2.5S-0.5N). *Source* [8]

represents 82% of total variance in TWS changes and EOF2 and PC2 represents 14%. EOF1 shows a strong anomaly all over the basin, while its corresponding PC1 shows the dominant trend of the basin. Using a linear regression, an average mass decline of 38.2 and increase of 4.5 mm/yr over the LVB, respectively was established for the periods of 2003–2007 and 2007–2013. EOF2 shows a spatial north-south dipole structure, which as PC2 indicates, corresponds to the annual changes of TWS over the basin. The TWS decline of 2003–2007 is attributed to the extension of the Owen Falls (Nalubaale) dam as stated e.g., in [15, 67]. The positive rate of 2007–2013 is likely due to the positive impact of El Niño in the years 2007 and 2013. This result is supported by rainfall analysis of Sect. 11.3.1.

11.3.4 Inter-annual Variability: Influence of ENSO and IOD

Some previous studies over equatorial eastern Africa (including LVB) have shown that local forcings modulate regional climate by either amplifying or suppressing the anomalies triggered by perturbations in the large-scale circulations that are propagated through global teleconnections such as El-Niño/Southern Oscillation and east-west sea surface temperature (SST) gradient over equatorial Indian Ocean

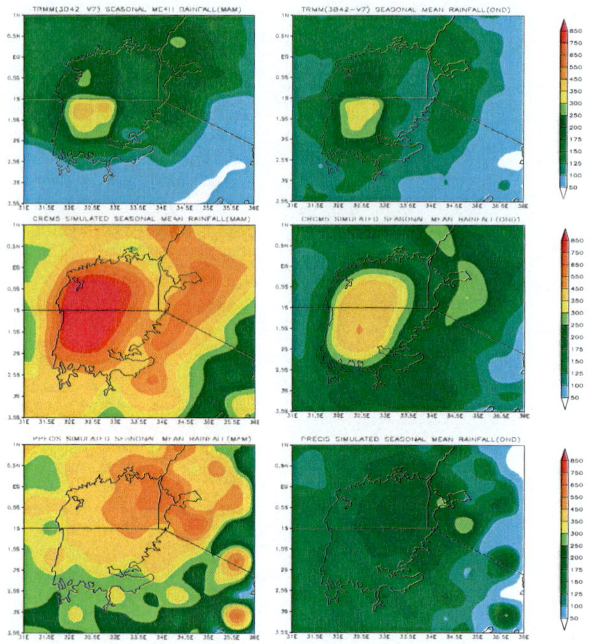

Fig. 11.6 Spatial pattern of seasonal mean rainfall (mm) over LVB. Left panels (March–May season); Right panels (October–December season). *Source* [8]

[i.e. IOD mode: [2, 30, 53, 61, 62], among others]. ENSO and IOD have thus been indicated as significant triggers of some of the past extreme LVB rainfall anomalies (floods and droughts).

In this chapter, we show in Fig. 11.8 the observed and simulated rainfall anomalies during 2005 and 2006, associated with fairly strong La Niña and El Niño/IOD conditions respectively. Generally, the apparent ENSO influence on the spatial variability of LVB rainfall is manifest, with more widespread below normal rainfall amounts during the OND season (2005) and the opposite during 2006 season (based on 1989–2008 average). Over-lake rainfall is more depressed during La Niña (2005), but there is a modest increase during El Niño years (2006 and 2010), although TRMM estimates show significant increases over the western and northern quadrants of the Lake. This feature is clearly reproduced by all the four CORDEX models, compared to TRMM estimates. Given the recent improvements in ENSO prediction, with lead times over 6 months, the apparent link between LVB rainfall and ENSO can have very practical application for LVB water resources availability and governance.

In Fig. 11.9, we show the spatial correlations between ENSO (Nino3.4 index) and LVB TRMM on the one hand, and simulated monthly rainfall totals on the other hand during the OND season. In October (Fig. 11.9, top), statistically significant correlation between Nino3.4 and TRMM (3B43-V7) during 1998–2008 is observed over the western parts of the Lake as well as the northeastern shores (Winam Gulf and

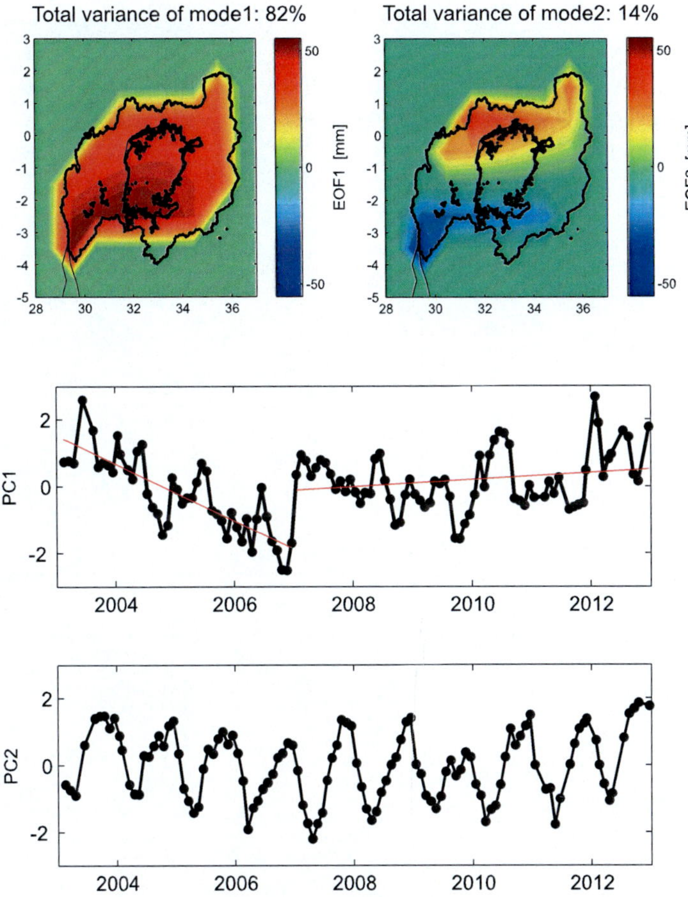

Fig. 11.7 PCA decomposition of TWS changes over LVB. EOFs are counted as anomaly maps that show the spatial distribution of TWS changes within the basin. Corresponding PCs are temporal variations which are scaled with their standard deviations to be unit-less. *Source* [8]

surrounding areas). In contrast, significant r-values between nino3.4 and simulated rainfall tend to be more widespread, especially over the northern sector of the Lake. Similar correlation patterns are derived from TRMM during November (Fig. 11.9. middle), but nino3.4 index correlation with the simulated rainfall show very weak correlations ($r \sim 0$), especially over the lake surface. The spatial correlation pattern in December (Fig. 11.9, bottom) for both TRMM and model are somehow similar to the pattern in October (Fig. 11.9, top).

A conspicuous similarity in the monthly spatial correlation patterns between IOD and rainfall (Fig. 11.10), and those shown in Fig. 11.9 is unmistakable. This apparently implies that co-occurrence of IOD and ENSO events exert significant influence

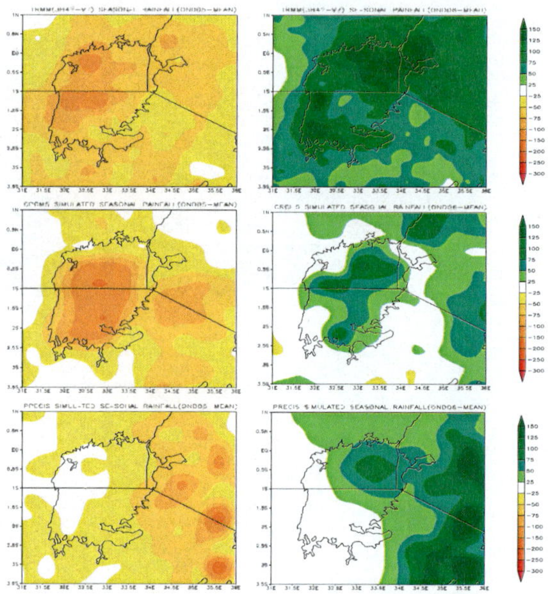

Fig. 11.8 Left: Spatial pattern of 2005 (La Nina), and Right: 2006 (El Nino) seasonal rainfall anomalies (mm) from long term mean over LVB. *Source* [8]

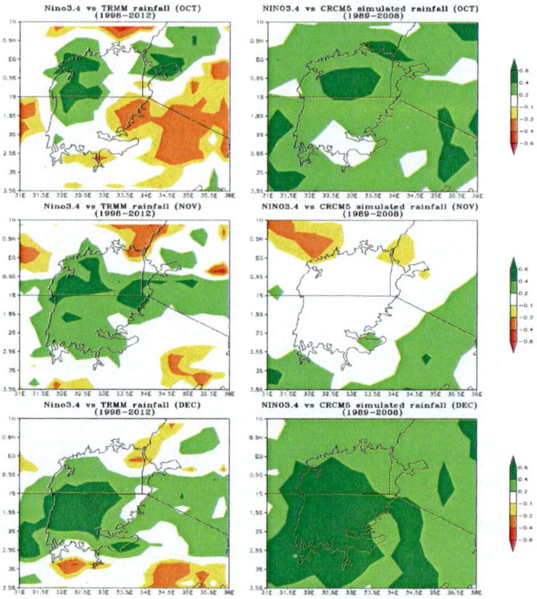

Fig. 11.9 Spatial correlation between ENSO (nino3.4 index) and monthly rainfall over LVB (r = 0.44 significant at 0.05 confidence level). Top (October), Middle (November) and bottom (December). *Source* [8]

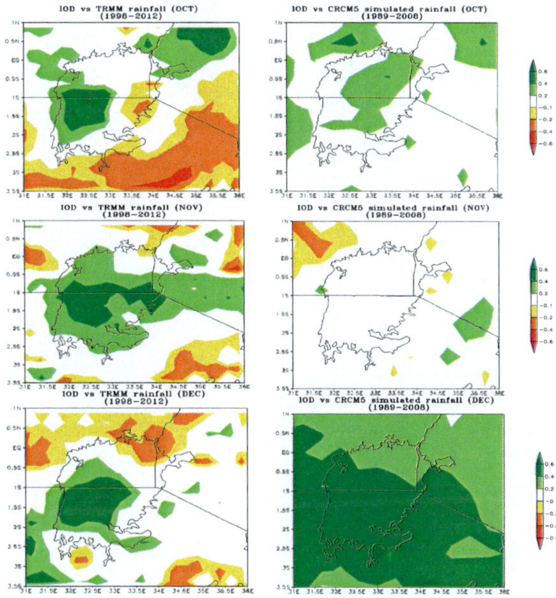

Fig. 11.10 Spatial correlation between IOD (index) and monthly rainfall over LVB (r = 0.44 significant at 0.05 confidence level). Top (October), Middle (November) and bottom (December). *Source* [8]

on LVB rainfall, and hence significantly influence climate-sensitive socio-economic activities (see, Sect. 11.4 over the lake and its hinterland).

In order to estimate the impact of ENSO and IOD on the variability of rainfall and thus stored water, it is assumed that the normalized temporal patterns of the nino3.4 and IOD indices as known. Then, their contributions are co-estimated, beside a linear trend as well as the annual and semi-annual components, in the variability of TRMM-rainfall and GRACE-TWS, over 2003–2013. Thus, it is assumed that the dominant temporal behavior of the rainfall and TWS changes is represented by $[a, b.t, c.sin(2\pi t), d.cos(2\pi t), \; e.sin(4\pi t), f.cos(4\pi t), g.\bar{E}(t - \phi_{ENSO}), h.\bar{I}(t - \phi_{IOD})]$, where t is time in year (2003–2013), $\bar{E}$ and $\bar{I}$ respectively contain the normalized ENSO and IOD indices and ϕ_{ENSO} and ϕ_{IOD} are the phase lags in year between the indices and the rainfall/TWS time series. The contributions of the components a, b, c, d, e, f, g are co-estimated using a least squares procedure. The correlation between nino3.4 and IOD indices and rainfall time series are found to be maximum when the lag is zero. Therefore, the normalized ENSO $\bar{E}$ and IOD $\bar{E}$ indices without considering any time lags, i.e. $\phi_{ENSO} = \phi_{IOD} = 0$ are considered for the rainfall. The estimated coefficients for g and h are summarized in Fig. 11.11. The magnitude of ENSO and IOD over 2003–2013 reached 25 mm whereas the magnitude of the annual ($\sqrt{c^2 + d^2}$) and semi-annual components ($\sqrt{e^2 + f^2}$) were 70 and 50 mm, respectively. The same procedure was repeated for TWS time series while considering a lag of one month for both ENSO and IOD ($\phi_{ENSO} = \phi_{IOD} = 1/12$). This selection is due to the fact that a delay of around one to two months exists between rainfall changes and TWS changes as was discussed under rainfall variability analysis. The corresponding coefficients are summarized in Fig. 11.12. The magnitude of

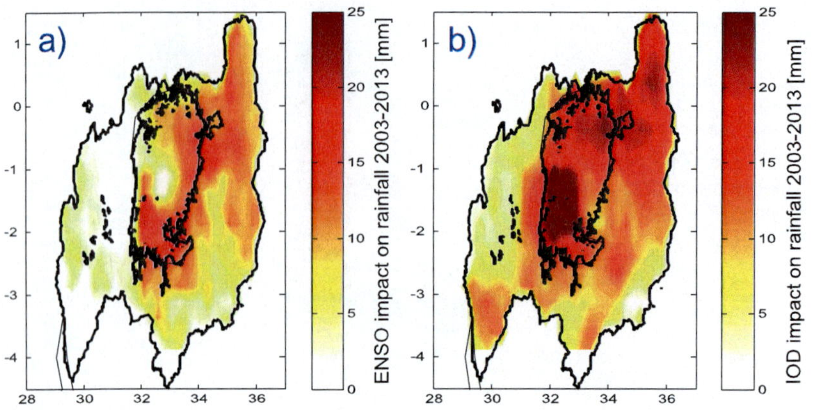

Fig. 11.11 Contribution of ENSO (left) and IOD (right) on the TRMM-derived rainfall variability of LVB

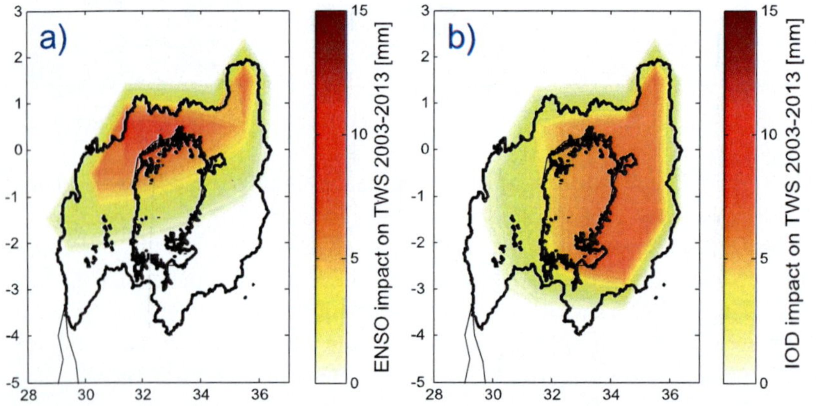

Fig. 11.12 Contribution of ENSO (left) and IOD (right) on the GRACE-derived TWS variability of LVB. *Source* [8]

their contribution reached 15 mm, over the period of 2003–2013. This is relatively less than what was observed in TRMM-rainfall in Fig. 11.11. Considering the simple water balance equation (see Sect. 5.1), where TWS is equal to precipitation minus evaporation minus runoff, when a phenomenon like ENSO happens, the amplitude of precipitation increases. One should, however, also consider that consequently, the amplitude of evaporation and runoff will increase and to some extent cancel out a part of the extra input water.

11.4 Economic Implications of Climate on LVB Water

This section provides an overview assessment of the economic impact of climate change linked to changes in stored water potential of Lake Victoria Basin as discussed in Sect. 11.3.3. It is important to point out that impact of climatic change on economic activities is systemic, thus quite complex and cannot be reduced to only monetary metrics for a single time period. Invariably, the economic impact of climatic change can be categorized as first-order impact, and second order impact. The first order impact can be noticed right after a major extreme climatic event occurs, such as drought or floods (e.g., the El Niño rains of 2007, Fig. 11.3). The second-order impacts are linked to climatic variations in the LVB that happens over protracted length of time or erratic happenings such as unpredictable rainy and dry seasons, which do not correspond to, or altogether disrupt planned-economic activities. In addition, lingering economic effects often happen in an incremental patter over protracted periods of time.

Equally important, is the need to understand the complex link between economic and social variables, which when subjected to climatic change, then engenders negative outcomes, both in the short and long term. At the center of economic impact assessment overview is also the heavy dependence of majority of the LVB population on certain economic activities, and therefore negative impact on such activities due to climatic change must be perceived within this reality. For instance, 80% of the LVB population is engaged in small-scale agricultural production and livestock farming, while fishing directly or indirectly support the livelihood of about 3 million people [22, 41, 51]. The population of LVB depends on wood biomass for 90% of their energy requirement [40].

It is difficult to arrive at precise monetary figures when making assessment of economic impact of climatic change in the LVB. This is because costs extend well beyond non-economic sectors in the eco-system, but have indirect negative bearing on economic activities in the LVB. Compounding the difficulty of measuring precise economic impact is the sheer lack of accurate statistical data of the gross domestic product (GDP) of the LVB. Lake Victoria Basin Commission (LVBC) officials give conflicting GDP figures of $30 billion, and 40 billion for 2011 and 2012, respectively in various presentations [32, 43]. Knowing the accurate GDP can be helpful in estimating the economic impact of changes in stored water potential of Lake Victoria due to climate change. We can then know percentage decrease or increase in GDP that may have resulted from such variability. Hence the overview assessment of economic impact given here is restricted to giving the correlating economic impact to distinctive climatic events - drought, floods and erratic seasonal rainfall patterns within spatial dimension.

The major economic sectors that are subjected to first-order impact of climatic change are: water resources, ecosystems and fishery, agriculture, energy, transportation, infrastructure and communications, and public health and labor productivity. The second-order economic impact of climatic change are such as lingering food shortages, energy poverty, malnutrition and impaired learning ability, and gradual loss of ecosystems that previously supported economic and social life of inhabitants. The 1997/98 El Niño floods (see, Fig. 11.2) caused damage to buildings, roads, communications systems, crops, and in addition to costs of treating diseases [44]. This type of damage has immediate and lingering future costs. Taking the costs of replacement of infrastructure, one can assess immediate costs for all damaged structures, in addition to lost value due to impaired infrastructure, cost of treating diseases, and lost productivity due to diseases and inability to move and communicate freely.

Likewise, the drought spawned by La Niña between October 1998 to 2000 led to massive crop and livestock loss, decreased hydro-electric power station outputs, water shortage and contamination-related diseases [44]. Awange et al. [17] found a link between highly variable climate pattern in the LVB to the frequency and severity of droughts and food insecurity in the region or parts of it. A commissioned research by United States Agency for International Development (USAID) conducted by International Resource Group's [28], gives some conservative estimates of cost of climate change for LVB at about $6.5 billion for the year 2005, in period in which LVB level dropped (see, Fig. 11.7 and also [15]). This study gives the GDP of the LVB at around $ 31.4 billion, thus the cost of climate change impact stands at almost 21% of the region's GDP for the single year. Even more surprising result of this study is the huge cost of public healthcare, which claims 4.4% of LVB GDP. Huge costs in healthcare are related to the elevated incident of malaria, diarrheal diseases and malnutrition, all of which have direct link either to drought or floods [72]. The economic impact overview assessment here depicts great exposure of the LVB's economic activities to adverse impact of climate change. However there is need for accurate data from which reliable monetary cost of the impact of climate change can be measured and therefore allowing for cost-effective adaptation mechanisms to be planned and implemented.

11.5 Concluding Remarks

In this chapter, decadal water storage changes over the basin derived from monthly GRACE, TRMM and RCM products were analyzed. The PCA results from both GRACE and TRMM together with in-situ data analyzed showed a general increase in rainfall and water volume over Lake Victoria Basin (LVB).

Overall it has been confirmed that there has been a modest increase in rainfall and stored water over the basin during the last decade. This is captured by in-situ-observed data obtained from lake-shore stations, TRMM and GRACE satellite remote sensing. TRMM data suggest that rainfall conditions have not changed much during the study period (1998–2013) over the basin while GRACE-TWS indicates average mass decline of 38.2 mm/yr for the period 2003–2007 and increase of 4.5 mm/yr for 2007 to 2013 over the LVB. This decline has been attributed to expansion of the Owen Falls/Nalubaale Dam, at Jinja Uganda in earlier investigations by Awange et al. [15] and Swenson and Wahr [67].

Futhermore, the four high-resolution regional climate model simulations analysed clearly reproduced the broad spatial and temporal patterns of precipitation over the LVB, as well as El Nino and La Nina linked anomalous wet and dry conditions during the recent decades. However, only two (CRCM5 and PRECIS) of the four RCMs capture the observed spatial distribution of rainfall over the LVB, and this is likely to compromise their ability to depict the correct (GRACE) water stored over the LVB.

The economic impact assessment of LVB depicts great exposure of the LVB's economic activities to adverse impact of climate change, specifically its impact on stored water.

References

1. Anyamba EK (1984) Some aspects of the origin of rainfall deficiency in East Africa. In: On GATE (ed) Proceeding of the WMO regional scientific conference. WAMEX and Tropical Meteorology, Dakar, Senegal, pp 110–112
2. Anyah RO, Semazzi FHM (2004) Simulation of the sensitivity of Lake Victoria basin climate to lake surface temperatures. Theor Appl Climatol 79:55–69
3. Anyah RO, Semazzi FHM (2006) NCAR-AGCM ensemble simulations of the variability of the Greater Horn of Africa climate. Special Issue, Theor Appl Climatol 86:39–62
4. Anyah RO, Semazzi FHM, Xie L (2006) Simulated physical mechanisms associated with climate variability over Lake Victoria Basin in East Africa. Mon Wea Rev 134:3588–3609
5. Anyah RO, Semazzi FHM (2007) Variability of East African rainfall based on multi-year RegCM3 model simulations. Int J Climatol 27:357–371
6. Anyah RO, Semazzi FHM (2009) Idealized simulation of hydrodynamic characteristics of Lake Victoria that potentially modulate regional climate. Int J Climatol 29(7):971–981
7. Asnani GC (1993) Tropical meterology Vol. 1 and Vol. 2 Indian Institute of Tropical Meteorology Pashan: Pune, 1012pp
8. AAwange J, Anyah R, Agola N, Forootan E, Omondi P (2013) Potential impacts of climate and environmental change on the stored water of Lake Victoria Basin and economic implications. Water Res Res 49(12):8160–8173. https://doi.org/10.1002/2013WR014350
9. Awange JL, Saleem A, Sukhadiya RO, Ouma YO, Hu K (2019) Physical dynamics of Lake Victoria over the past 34 years (1984–2018): is the lake dying? Sci Total Env 658:199–218
10. Awange JL, Palancz B, Völgyesi L (2020) Hybrid imaging and visualization. Employing machine learning with mathematica - Python. Springer Nature International, Berlin
11. Awange JL, Forootan E, Kuhn M, Kusche J, Heck B (2014) Water storage changes and climate variability within the Nile Basin between 2002 and 2011. Adv Water Res 73:1–25. https://doi.org/10.1016/j.advwatres.2014.06.010
12. Awange JL, Hu KX, Khaki M (2019) The newly merged satellite remotely sensed, gauge and reanalysis-based multi-source weighted-ensemble precipitation: evaluation over Australia and Africa (1981–2016). Sci Total Env 670:448–465
13. Awange JL, Ferreira VG, Forootan E, Khandu Andam-Akorful SA, Agutu NO, He XF (2016) Uncertainties in remotely sensed precipitation data over Africa. Int J Climatol 36(1):303–323. https://doi.org/10.1002/joc.4346
14. Awange J, Forootan E, Kusche J, Kiema JKB, Omondi P, Heck B, Fleming K, Ohanya S, Gonçalves RM (2013) Understanding the decline of water storage across the Ramser-lake Naivasha using satellite-based methods. Adv Water Res 60:7–23. https://doi.org/10.1016/j.advwatres.2013.07.002
15. Awange JL, Sharifi M, Ogonda G, Wickert J, Grafarend E, Omulo M (2008) The falling Lake Victoria water levels: GRACE, TRIMM and CHAMP satellite analysis of the lake basin. Water Res Manag 22:775–796. https://doi.org/10.1007/s11269-007-9191-y

16. Awange JL, Ogallo L, Kwang-Ho B, Were P, Omondi P, Omute P, Omulo M (2008) Falling Lake Victoria water levels: is climate a contribution factor? Clim Change. https://doi.org/10.1007/s10584-008-9409-x
17. Awange JL, Aluoch J, Ogallo LA, Omulo M, Omondi P (2007) Frequency and severity of drought in the lake Victoria region (Kenya) and its effect on food security. Clim Res 33:135–142
18. Awange JL, Ong'ang'a O (2006) Lake Victoria: ecology resource and environment. Springer, Berlin, p 354
19. Bates B, Kundzewicz ZW, Wu S, Palutikof J (2008) Climate change and water. IPCC Technical Paper VI. ISBN: 978-92-9169-123-4. http://www.ipcc.ch/pdf/technical-papers/climate-change-water-en.pdf
20. Canter MJ, Ndegwa SN (2002) Environmental scarcity and conflict: a contrary case from Lake Victoria. Global Env Polit 2(3):40–62
21. Dahle C, Flechtner F, Gruber C, König R, Michalak G, Neumayer K-H (2012) GFZ GRACE Level-2 processing standards document for Level-2 product release 0005. Scientific Technical Report STR12/02-Data, GFZ, Potsdam, Germany
22. East African Community Secretariat (2004) The vision and strategy framework for management and development of Lake Victoria Basin. Lake Victoria Development Programme, Arusha, Tanzania
23. Fenoglio-Marc L, Kusche J, Becker M (2006) Mass variation in the mediterranean sea from GRACE and its validation by altimetry, steric and hydrologic fields. Geophys Res Lett 33(19). https://doi.org/10.1029/2006GL026851.
24. Fenoglio-Marc L, Rietbroek R, Grayek S, Becker M, Kusche J, Stanev E (2012) Water mass variation in the Mediterranean and Black Sea. J Geodyn 59–60:168–182. https://doi.org/10.1016/j.jog.2012.04.001
25. Fleming K, Awange JL (2013) Comparing the version 7 TRMM 3B43 monthly precipitation product with the TRMM 3B43 version 6/6A and BoM datasets for Australia, Australian Meteorological and Oceanographic Journal (in press)
26. Hecky RE, Mugidde R, Ramlal PS, Talbot MR, Kling GW (2010) Multiple stressors cause rapid ecosystem change in Lake Victoria. Freshwater Biol 55:19–42. https://doi.org/10.1111/j.1365-2427.2009.02374.x
27. Geheb K, Crean K (2003) Community-level access and control in the management of Lake Victoria's fisheries. J Env Manag 67(2):99–106
28. Hecht JE, Vincent K, Kahata J (2011) Costs imposed by climate change in three Ecoregions of East Africa. Study prepared for USAID East Africa, Office of Regional Economic Growth and Integration by International Resource Group, Washing D.C
29. Huffmann G, BolvinD (2012) TRMM and other data precipitation data set documentation, ftp://precip.gsfc.nasa.gov/pub/trmmdoc/3B42_3B43_doc.pdf. Accessed 25 sep 2012
30. Indeje M, Semazzi FHM, Ogallo LJ (2000) ENSO signals in East African rainfall seasons. Int J Climatol 20:19–46
31. Johnson J (2009) Climate change and fishery sustainability in Lake Victoria. In: African Journal of Tropical Hydrobiology and Fisheries. Lake Victoria Stakeholders Conference, Kampala, 27–30 October 2008
32. Kanangire C (2012) Transboundary water resources management in Lake Victoria Basin. Presentation at the Infrastructure Consortium for Africa, First Meeting, March 8-8, 20112, Frankfurt, Germany
33. Kayombo S, Jorgensen SE (2006) Lake Victoria: experience and lessons learned brief. In: International Lake Environment Committee, Lake Basin Management Initiative, Kusatsu, Japan. www.worldlakes.org/uploads/Victoria_7.15.04.pdf
34. Klees R, Zapreeva EA, Winsemius HC, Savenije HHG (2007) The bias in GRACE estimates of continental water storage variations. Hydrol Earth Syst Sci Discuss 11:1227–1241

35. Kull D (2006) Connections between recent water level drops in Lake Victoria, dam operations and drought. http://www.irn.org/programs/nile/pdf/060208vic.pdf
36. Kummerow C, Simpson J, Thiele O, Barnes W, Chang A, Stocker E, Adler R, Hou A, Kakar R, Wntz F, Aschroft P, Kozu T, Hing Y, Okamoto K, Iguchi T, Kuroiwa H, Im E, Haddad Z, Huffman G, Ferrier B, Olson W, Zipser E, Smith E, Wilheit T, North G, Krishnamurti T, Nakamura K (2000) The status of the Tropical Rainfall Measuring Mission (TRMM) after two years in orbit. J Appl Meteorol 39(12):1965–1982. https://doi.org/10.1175/1520-0450(2001)040<1965:TSOTTR>2.0.CO;2
37. Kusche J (2007) Approximate decorrelation and non-isotropic smoothing of time-variable GRACE-type gravity field models. J Geodesy 81(11):733–749. https://doi.org/10.1007/s00190-007-0143-3
38. Kusche J, Schmidt R, Petrovic S, Rietbroek R (2009) Decorrelated GRACE time-variable gravity solutions by GFZ, and their validation using a hydrological model. J Geodesy 83:903–913. https://doi.org/10.1007/s00190-009-0308-3
39. Lejju JB (2012) The influence of climate change and human-induced environmental degradation on Lake Victoria. African Books Collective. 104p. ISBN:9994455672
40. LVBC (2007) Regional transboundary diagnostic analysis of the lake Victoria Basin. Lake Victor basin Commission, Kisumu, Kenya
41. LVBC (2011) Analysis of trade in Lake Victoria ports and Basin. Lake Victoria Basin commission, African Center of Technology Studies (ACTS). 48pp. ISBN 9966-41-155-0
42. Mitchell TD, Carter TR, Jones PD, Hulme M, New M (2003) A comprehensive set of high-resolution grids of monthly climate for Europe and the globe: the observed record (1901–2000) and 16 scenarios (2001–2100). J Clim, Submitted (August 2003)
43. Mngube MF (2011) Lake Victoria a new frontier for development in East Africa. In: 2nd East African Community Symposium, Lake Victoria Basin Commision, April 29, 20011, Arusha, Tanzania
44. Mogaka H, Gichere S, Davis, R and Hirji R (2005) Climate variability and water resources degradation in Kenya: improving water resources development and management. World Bank Working Paper No. 69, Washington D.C, December 2005
45. Mutenyo IB (2009) Impacts of irrigation and hydroelectric power developments on the victoria nile in Uganda. PhD Thesis. School of Applied Sciences, Cranfield University, United Kingdom
46. Mwiturubani AD (2010) Climate change and access to water resources in the Lake Victoria basin. In: Climate change and natural resources conflicts in Africa, Institute for Security Studies (ISS), Pretoria, South Africa, 17 p
47. Nicholson S, Some B, McCollum J, Nelkin E, Klotter D, Berte Y, Diallo B, Gaye I, Kpabeba G, Ndiaye O, Noukpozounkou J, Tanu A, Thiam A, Toure AA, Traore A (2003) Validation of TRMM and other rainfall estimates with a high-density gauge dataset for West Africa: Part II: validation of TRMM rainfall products. J Appl Meteorol 42(10):1355–1368. https://doi.org/10.1175/1520-0450(2003)042<1355:VOT
48. Nicholson SE (1999) Historical and modern fluctuations of lakes Tanganyika and Rukwa and their relationship to rainfall variability. Clim Change 41:53–71
49. Nicholson SE (1998) Historical fluctuations of Lake Victoria and other lakes in the Northern Rift Valley of East Africa. In: Lehman JT (ed) Environmental change and response in East African lakes. Kluwer, Dordrecht, pp 7–35
50. Nikulin G, Jones C, Samuelsson P, Giorgi F, Sylla MB, Asrar G, Büchner M, Cerezo-Mota R, Christensen OB, Dequè M, Ferna'ndez J, Hänsler A, van Meijgaard E, Sushama L (2012) Precipitation climatology in an ensemble of CORDEX-Africa regional climate simulations, pp 0894–8755. https://doi.org/10.1175/JCLI-D-11-00375.1
51. Ntiba MJ, Kudoja WM, Mukasa CT (2001) Management issues in the Lake Victoria watershed. Lakes Reser Res & Manag 6:211–216
52. Owor M, Taylor RG, Tindimugaya C, Mwesigwa D (2009) Rainfall intensity and groundwater recharge: empirical evidence from the Upper Nile Basin. Env Res Lett 4:035009. https://doi.org/10.1088/1748-9326/4/3/035009

53. Omondi P, Awange J, Ogallo Ininda J, Forootan E (2013) The influence of low frequency sea surface temperature modes on delineated decadal rainfall zones in Eastern Africa region. Adv Water Res 54:161–180
54. Omondi P, Awange J, Ogallo LA, Okoola RA, Forootan E (2012) Decadal rainfall variability modes in observed rainfall records over East Africa and their relations to historical sea surface temperature changes. J Hydrol 464–465:140–156. https://doi.org/10.1016/j.jhydrol.2012.07.003
55. Palmer MA, Reidy Liermann C, Nilsson C, Flörke M, Alcamo J, Lake PS, Bonds N (2008) Climate change and the world's river basins: anticipating management options. Front Ecol Env 6:81–89
56. Peterson TC, Easterling DR, Karl TR, Groisman P, Nicholls N, Plummer N, Torok S, Auer I, Bohm R, Gullett D, Vincent L, Heino R, Tuomenvirta H, Mestre O, Szentimrey T, Salinger J, Førland EJ, Hanssen-Bauer I, Alexandersson H, Jones P, Parker D (1998) Homogeneity adjustments of in-situ atmospheric climate data: a review. Int J Climatol 18:1493–1517
57. Preisendorfer R (1988) Principal component analysis in meteorology and oceanography. Elsevier, Amsterdam
58. PPA (2007) Bujagali II - Economic and financial evaluation study final report. Power Planning Associates Ltd. http://siteresources.worldbank.rg/EXTBUJHYDPOWPRO/Resources/BujagaliEconFnclEvalStudyMainText.pdf
59. Riebeek H (2006) Lake Victoria-s falling waters. http://earthobservatory.nasa.gov/Study/Victoria/printall.php
60. Rummel R, Balmino G, Johannessen J, Visser P, Woodworth P (2002) Dedicated gravity field missions - principles and aims. J Geodyn 33:3–20
61. Saji NH, Goswami BN, Vinayachandran PN, Yamagata T (1999) A dipole mode in the tropical Indian Ocean. Nature 401:360–363
62. Schreck CJ, Semazzi FHM (2004) Variability of the recent climate of eastern Africa. Int J Climatol 24:681–701
63. Song Y, Semazzi FHM, Xie L, Ogallo LJ (2004) A coupled regional climate model for LVB of East Africa. Int J Climatol 24:57–75
64. Sun L-Q, Semazzi FHM, Giorgi F, Ogallo L (1999) Application of the NCAR regional climate model to Eastern Africa, 2. Simulation of internannual variability of short rains. J Geophys Res -Atm. 104:6549–6562
65. Sutcliffe JV, Parks YP (1999) An outline of the Nile Basin. The Hydrology of the Nile. IAHS Special Publication no. 5, 1999
66. Sutcliffe J, Petersen G (2007) Lake Victoria: derivation of a corrected natural water level series. Hydrol Sci J Sci Hydrol 52(6):1316–1321
67. Swenson S, Wahr J (2009) Monitoring the water balance of Lake Victoria, East Africa, from space. J Hydrol 370(1–4):163–176. https://doi.org/10.1016/j.jhydrol.2009.03.008
68. Sewagudde S (2009) Lake Victoria-s water budget and the potential effects of climate change in the 21st Century. Afr J Trop Hydrobiol Fish 12:22–30
69. Tapley B, Bettadpur S, Ries J, Thompson P, Watkins M (2004a) GRACE measurements of mass variability in the Earth system. Science 305:503–505. https://doi.org/10.1126/science.1099192
70. Tapley B, Bettadpur S, Watkins M, Reigber C (2004) The gravity recovery and climate experiment: mission overview and early results. Geophys Res Lett 31(L09607). https://doi.org/10.1029/2004GL019920
71. Wahr J, Molenaar M, Bryan F (1998) Time variability of the Earth's gravity field: hydrological and oceanic effects and their possible detection using GRACE. J Geophys Res 103(B12):30205–30229. https://doi.org/10.1029/98JB02844

72. Wandiga SO (2006) Climate change induced vulnerability to Malaria and Cholera in the Lake victoria region. A final report submitted to assessments of impacts and adaptations to climate change, Project No. AF 91, The International START Secretariat, Washington D.C
73. Werth S, Güntner A, Schmidt R, Kusche J (2009) Evaluation of GRACE filter tools from a hydrological perspective. Geophys J Int 179:1499–1515. https://doi.org/10.1111/j.1365-246X.2009.04355.x
74. Yin X, Nicholson SE (1998) The water balance of Lake Victoria. Hydrol Sci - J Sci Hydrol 43(5):789–811

Chapter 12
Lake Victoria Basin: Droughts and Food Security

"For the Greater Horn of Africa (GHA) region, rather than responding successfully to the frequent recurrent droughts that afflict the region, the communities are invariably devastated by famine crisis, instabilities in national economies and political tensions. For example, the Ethiopian "biblical" famines of 1973-74 and 1984-85 left about 200,000 and 400,000, people dead, respectively, with the former disaster resulting in the overthrow of Emperor Haile Selassie. The latter contributed to the end of the Marxist regime of Mengistu Haile Mariam."

—*Mpelesoka et al.* [21]

12.1 Summary

Given the significance of extreme climate (droughts and floods) on Lake Victoria's waters, using monthly and seasonal precipitation data for the period 1961–1999, this Chapter, establishes drought frequency and severity in relation to food security in the Lake Victoria region in Kenya. Use is made of percentiles together with time series analysis for a 40 yr period, with the lower 25% of the quartile designated as the threshold values that indicate drought years. The percentiles and Drought Severity Index (DSI) are then used to ascertain the physical conditions and severity of the drought years. The 1980s and 1990s are found to be drier decades (the 1980s had more drought events than any other decade considered herein), with lower negative anomalies in comparison to the 1960s and 1970s, and the number of drought events and their severity have been increasing in recent years (especially during the 1990s). In general, drought affects the crop planting seasons of March to May (MAM) and September to November (SON), and leads to increased food imports in the years following the drought. Incidentally, MAM and SON are the main rainy seasons (with MAM being the main rainy season) that replenishes the lake. If the present

J. Awange, *Lake Victoria Monitored from Space*,
https://doi.org/10.1007/978-3-030-60551-3_12

trends persist, the Lake Victoria region will face not only more severe drought events, but also significant reductions in its food security [9].

12.2 Introductory Remarks

As already discussed in Chaps. 4 and 9, droughts have contributed to declines noticed in Lake Victoria water levels, which has had a number of social and economic effects, e.g., affecting power production, with resulting economic losses [23]. Combined with other existing factors, these droughts contribute to a worsening food situation in the region where a majority of the population are poor and living below poverty line, see e.g., [2]. Food insecurity (i.e. a lack of basic food to provide enough energy for day to day activity) in this region could lead to a catastrophe [27], since many of the local communities living around Lake Victoria are among the poorest [7, 11] and most food insecure in the region, due to high population densities [2, 15, 28], widespread poverty, recurrent droughts, crop failures, high mortality rates (from tuberculosis, malaria, and HIV/AIDS), and environmental degradation (contamination, pollution, land/forest degradation, biodiversity degradation, introduction of exotic species; [2]). Shipton [26] reviewed the famine and food shortage situation in Africa and focused specifically on those who suffer and the action they take, information which may be vital to the lake region.

Food security is dependent on social, economic as well as climatic factors. Climatic extremes, particularly droughts and floods, affect the state of food security in Africa [18, 25]. Climatic extremes are normal climatic occurrences and their impacts on society vary with respect to vulnerability and the success or failure of coping mechanisms at both the household and national level. Absence of adaptation measures to climate variability results in negative impacts on food security, health, and ultimately, life. An example is the 1999–2001 drought that affected most parts of Kenya, including some areas that normally receive high rainfall. At its peak, 4.5 million people in Kenya lost their livelihoods and ability to cope, and were subsequently entirely dependent on relief food provided jointly by the government and donors [13]. Adger [1] discussed adaptation to climate change related phenomena in developing countries while [12] described food assistance programs. In addressing the Brazilian drought, Awange et al. [8] proposed the conceipt of living with droughts. The impacts of such extremes (in the context of climate change) on Lake Victoria Basin (LVB) has been assessed in [10, 21, 24], while Agutu et al. [3] proposed remotely sensed approach to monitoring agricultural droughts in the region to improve food security.

In the LVB, floods are more frequent than drought and have had adverse impacts. The climate within the region is highly variable with a higher natural susceptibility to frequent climate extremes than neighbouring regions. In the past, the majority of interventions in the region have been in flood management. However, drought, which is a slow and progressive extreme climatic event, has until recently been the least anticipated extreme in this region filled with wetlands. As a result, its impacts are more adverse than those of floods because of the inadequacy of existing coping mechanisms.

Although drought is a natural, recurring phenomenon which cannot be controlled [4, 14, 16, 17], prediction of its frequency, severity and probability, e.g., [8, 10, 20, 21] is essential for farmers in that it helps in planning and management. In the absence of such information, food security is threatened. Lack of such information in Uganda, for instance, exacerbated the impact of drought on maize yield and led Tumwesigye [29] to characterize drought patterns to help in the development of drought resistant maize cultivars. This Chapter focuses on the severity and frequency of drought events in the Kenyan part of the LVB, with the hope of providing mitigation and/or adaptation measures necessary for key stakeholders to improve drought management.

12.3 Planting Seasons in Lake Victoria Basin-LVB

The study of [9] was centred on the Kenyan part of LVB, with data taken from Kisumu station, 0.1 °S, 34.75 °E (Fig. 1.3), which is approximately 1500 m above mean sea level. Lake Victoria Basin has 4 seasons: the hot dry season (December to February, DJF); the major rainy season (March to May, MAM); the short dry season (June to August, JJA); and the short rainy season (September to November, SON) [9, 11]. Kisumu has an annual average rainfall of 1340 mm, and experiences relatively high rainfall in the region. For instance, [5] found that both annual rainfall and rainfall during the long rainy season were greater at Kisumu than at any of the 8 stations in Kenya that they analysed. The highest rainfall in the LVB occurs in the north-western parts and gradually decreases towards the southeast (see Chap. 11). The average annual rainfall for the LVB is 1424 mm, and varies between 891 mm in parts of Mara catchments to a maximum of 2168 mm in the middle of Yala Basin [6]. The principal occupation in the LVB is subsistence agriculture (e.g. maize, sorghum, beans, cassava), and planting takes place largely during the long and short rainy seasons (February to March and September to October, respectively). This contrasts with the rest of Kenya, where planting generally occurs in March to April and October to November [5]. Recently, however, this pattern has changed. During the 2005 long dry spell, for example, the planting season started as late as April–May in most parts of LVB (Kenya), and in 2006 it started in late March in most places.

As a response to these changing patterns, farmers have intensified the planting of drought resistant crops (e.g. sweet potatoes and cassavas). Cassava crops are, however, in decline, as consumer preferences shift towards maize and rice; poor cassava yields are also attributed to destruction by moles [19]. This further exacerbates food insecurity. Maize is the preferred crop, and (together with legumes) is often planted in both rainy seasons, due to the fact that one yield is often insufficient to sustain farmers throughout the year. In addition to the problem of drought, poor soil fertility resulting from a lack of crop rotation has also contributed to serious food insecurity in the region. To address these emerging threats of drought and soil infertility, research is being done on drought-resistant varieties of maize.

Table 12.1 Seasonal rainfall anomalies for the Kisumu meteorological station

YEAR	DJF	MAM	JJA	SON
1961	–	−1.8	2.6	6.2
1962	4.3	3.2	−0.2	0.7
1963	1.7	1.4	−1.0	−2.1
1964	1.4	0.5	1.1	−1.1
1965	−0.5	1.7	−3.0	2.1
1966	1.2	0.2	0.8	−2.3
1967	−3.1	−1.1	−3.7	4.0
1968	0.1	1.1	1.6	1.2
1969	3.3	−1.8	−1.8	−0.9
1970	1.5	−0.3	−2.1	−2.1
1971	−2.0	2.2	−0.8	−0.9
1972	−0.3	−0.6	2.5	1.9
1973	0.7	−1.0	−2.6	−2.1
1974	−3.1	−0.3	1.8	−0.8
1975	−2.1	0.9	0.0	−1.4
1976	−0.2	−1.5	0.2	1.0
1977	−0.1	2.5	1.4	−1.3
1978	2.3	1.4	1.7	0.8
1979	1.9	−0.3	−0.7	−0.8
1980	−0.5	−0.3	−1.0	−1.4
1981	−2.0	0.4	0.8	−2.9
1982	−1.2	−3.7	2.7	5.0
1983	−1.9	−1.7	−0.4	−0.1
1984	−1.5	−3.1	3.2	0.9
1985	−0.1	2.0	0.5	−2.6
1986	−0.6	1.4	−0.6	−0.5
1987	0.1	−1.7	−0.2	1.5
1988	0.0	1.1	−0.5	0.1
1989	0.2	0.2	−1.9	0.2
1990	0.9	−1.0	−0.3	−2.5
1991	−0.2	0.8	−2.2	0.4
1992	−2.3	−1.8	4.0	0.4
1993	1.6	−0.4	0.3	−2.9
1994	−2.2	3.5	−0.4	0.1
1995	−0.7	0.9	−0.9	1.2
1996	0.0	1.5	−0.2	1.6
1997	−1.9	−0.3	−1.1	−1.4
1998	6.7	−1.4	−1.0	−2.1
1999	−1.5	1.7	2.3	−0.2

Table 12.2 Food imports (t) in Kenya (Central Bureau of Statistics, Statistical Abstract, 2003. Extract on Net Imports, Principle Articles, Table 52b)

YEAR	Maize (tons)	Wheat (tons)	Rice (tons)
1961	102130	13004	4037
1962	25590	52470	4128
1963	36	8523	4594
1964	264	6	672
1965	81452	5999	2096
1966	143444	24965	2021
1967	73	4512	207
1968	39	–	2995
1969	268	45	245
1970	14334	–	1149
1971	29077	13000	10203
1972	104	64821	2150
1973	82	77083	23
1974	7281	3744	1502
1975	3578	2917	4
1976	32	50	10001
1977	32	33	24
1978	80	90888	11
1979	18	21152	241
1980	323873	48462	1239
1981	77394	49239	4573
1982	89056	139326	11880
1983	–	81946	44768
1984	405443	149906	507
1985	125454	143793	562
1986	700	115281	61745
1987	–	217857	39129
1988	–	75578	10000
1989	2	123535	30006
1990	–	322632	27983
1991	–	242612	61163
1992	414929	100808	58920
1993	12874	314410	37150
1994	650387	353076	93519

(continued)

Table 12.2 (continued)

YEAR	Maize (tons)	Wheat (tons)	Rice (tons)
1995	11965	249134	28177
1996	6759	486917	39159
1997	1101105	388138	62435
1998	368761	478865	62893
1999	75520	578543	53358
2000	409416	636045	105803
2001	308606	617542	172246
2002	229288	515179	148705

In the study of Awange et al. [9], the data used consisted of:

1. Rainfall data (1961–1999). Kisumu station was selected because of its position in relation to other stations in the region (see Fig. 1.3). There are a number of other meteorological stations in the LVB, but all are found in the same (Lake Victoria) climatological zone. Awange et al. [9] therefore used one station to represent the region in order to avoid duplication of results. Kisumu is very close to Lake Victoria, and thus can sufficiently account for both the catchment and over-lake rainfall. Rainfall anomaly data are presented in Table 12.1.
2. Food import data (1961–2002). Secondary data on food imports was obtained from abstracts provided by the Central Bureau of Statistics (CBS) (available at the Kenyan Ministry of Agriculture's Library) and are presented in Table 12.2. Maize being the staple food in the region, the quantities of maize imports from 1961–2002 were used to represent overall food imports. In Kenya, food is not normally imported for specific regions affected by droughts; rather, the affected region triggers imports of food to the national reserve. When drought is severe in the LVB, for example, the Government checks the national reserve for surplus. If the food stored in the national reserve begins to dwindle, and the other regions are not in a position to supply the national reserve, then an order is made to import food.

12.4 Drought Analysis

For easier analysis and comparison purposes, both rainfall and food data were standardized in [9]. The daily records data were averaged to get the monthly mean

$$X_m = \frac{\sum d}{n}, \tag{12.1}$$

where d is the daily measurements and n the number of days in the given month. The monthly means were averaged for the 3 months in a season to obtain the seasonal mean. To get the annual means, the monthly means for all the 12 months of the year were averaged.

12.4.1 Determination of Drought Years

Once the rainfall mean was computed in (12.1), the "percent of normal" (quartile) was calculated by dividing actual precipitation by normal precipitation—typically considered to be a 30 yr mean—and multiplying by 100% (Eq. 12.3). Usually time scales range from a single month, or a group of months representing a particular season, to an annual or water year. Normal precipitation for a specific location was considered to be 100%. The normal precipitation N was calculated by

$$N = \frac{\sum X_A}{30}, \tag{12.2}$$

i.e. 30 yr mean of a season, where X_A is the annual mean. Percent of normal P_n was then given by

$$P_n = \frac{A_p}{N} \times 100\%, \tag{12.3}$$

where A_p is the actual precipitation. The Drought Severity Index (DSI) was then determined by considering as "dry" all those observations <25% (first quartile) of the ranked historical records, while those which were >75% (fourth quartile) were considered as "wet".

12.4.2 Standardization of Data

In order to compare the data, the rainfall and food data were standardized using

$$Z = \frac{X_t - X_{30}}{\sigma}, \tag{12.4}$$

where Z is the standardized value, X_t is the observed value at a particular time, X_{30} the 30-year mean for the parameter and σ is the standard deviation. This gave the anomalies that were used for plotting the time series charts (see Table 12.1).

12.4.3 Data Analysis Methods

Anomalies of the parameters were used in the plotting for comparison between rainfall and food variables from the standardized values of Z in Eq. 12.4. Time series and relational analysis were then performed. The data were sorted in descending order; 25% of the normal and below represented the "lower threshold". The threshold value was then plotted in the time series, identifying severe and extreme drought as the 2 main categories (Table 12.3). This method was used to determine the frequency, severity and duration of droughts in LVB.

12.5 Drought Years and Food Security

12.5.1 Drought Years

Figure 12.1 indicates that the period from 1960 to 1970 was wetter with higher positive anomalies. The period 1970–1980 shows a mixed spell of wet and dry years, while 1983–1997 shows a largely dry spell with lower negative anomalies. Table 12.4 gives drought years obtained from observing the lower threshold values in Fig. 12.2, which were calculated as 1.2, 1.05, 0.98 and 1.38 for DJF, MAM, JJA and SON, respectively. Figure 12.2 and Table 12.5 show the severity and duration of drought years. The worst drought in terms of severity was 1980–1985, and drought events have been increasing in recent years. Because of the El Ñino rains of 1998, the 1990s were not particularly dry, but significant droughts did occur in 1990 and in 1997. A general pattern of drought emerges:

- During the hot dry season (DJF), severe drought can be expected approximately every 7–8 years.
- During the long rainy season (MAM), severe drought can be expected every 5–8 years.
- During the dry season (JJA) there is no clear cycle of drought events.
- During the short rainy season (SON), severe drought can be expected every 3–4 years.

12.5.2 Drought in Relation to Food Security

In the LVB, the long and short rainy seasons (i.e. MAM and SON, respectively) are critical to food security. Figure 12.3 shows the relationship between the various seasonal droughts and food imports. For all the seasons considered, there is a significant rise in food imports in the years 1966, 1970, 1980, 1984, 1992, 1994 and 1997. These years followed the drought years identified in Tables 12.4 and 12.5, with a lag of 1 yr.

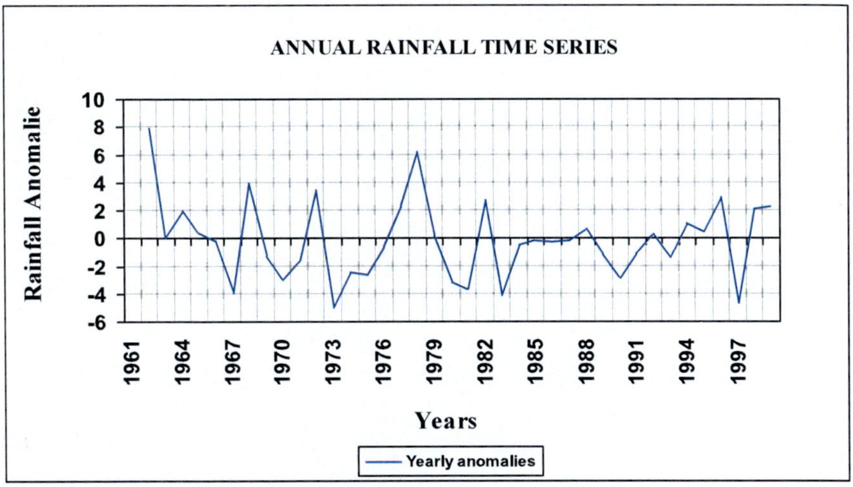

Fig. 12.1 Time series of annual rainfall anomalies at Kisumu during 1961–1999. *Source* Awange et al. [9]

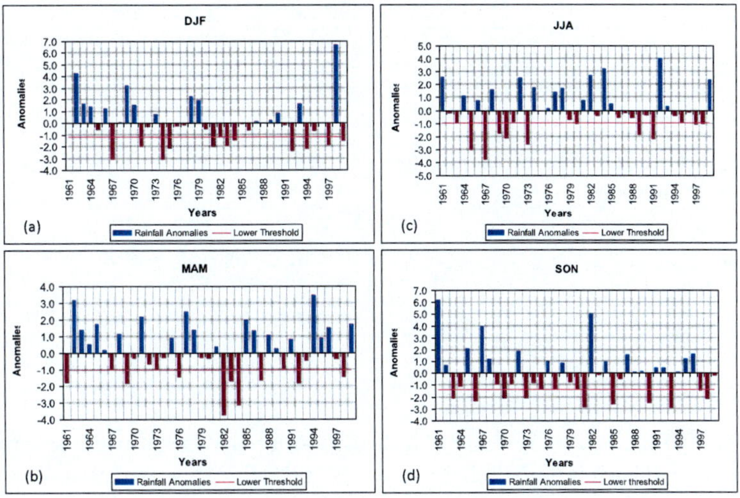

Fig. 12.2 Seasonal droughts from 1961–1999 (see Sect. 12.3 for season abbreviations). *Source* Awange et al. [9]

Table 12.3 Percentage range for dry and wet spells

Percentage rainfall relative to mean	Physical conditions
>175	Extremely wet
>150	Very wet
125	Near normal
100	Normal
75	Near normal
<50	Severe drought
<25	Extreme drought

Table 12.4 Seasonal drought years

Seasons	Drought years
DJF	1967, 1971, 1974, 1975, 1981, 1982, 1983, 1984, 1992, 1994, 1997, 1999
MAM	1961, 1967, 1969, 1976, 1982, 1983, 1984, 1987, 1992, 1998
JJA	1963, 1965, 1967, 1969, 1970, 1973, 1980, 1989, 1991, 1995, 1997, 1998
SON	1963, 1966, 1970, 1973, 1975, 1977, 1980, 1981, 1985, 1990, 1993, 1997, 1998

Table 12.5 Periods of drought and their intensity. Extreme: rainfall <25% of normal; severe: <50% to 25% of normal

Drought	Year
Extreme	1973–1975, 1981–1985, 1990–1993, 1996–1997
Severe	1963, 1967, 1970–1971, 1980

From the drought profile in Table 12.6, a relationship between drought severity and populations affected by food relief was established (Fig. 12.4), which indicates that as drought severity increases, population receiving food relief also increases, and it also shows that after a significant drop of drought severity from 1971 to 1975, the population receiving food relief during this time also decreased.

The information on drought patterns in the LVB is vital to farmers' drought management planning, as well as to that of the Ministry of Agriculture and the National Disaster Management body.

The variability of rain influences not only drought but also food imports (Fig. 12.4), and droughts in the LVB probably contribute to national food imports. This can be analysed using MAM and SON data. Figure 12.3b, d shows wet periods for both seasons, suggesting that the increase of the food imported in 1966 could have been

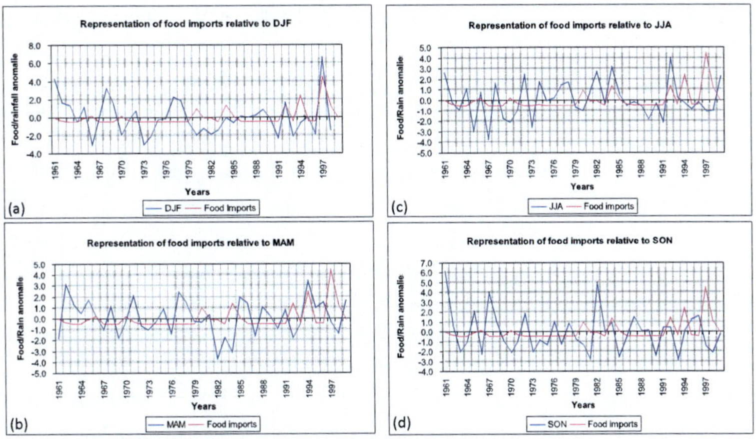

Fig. 12.3 Time series of food imports relative to the 4 seasons during 1961–1999. *Source* Awange et al. [9]

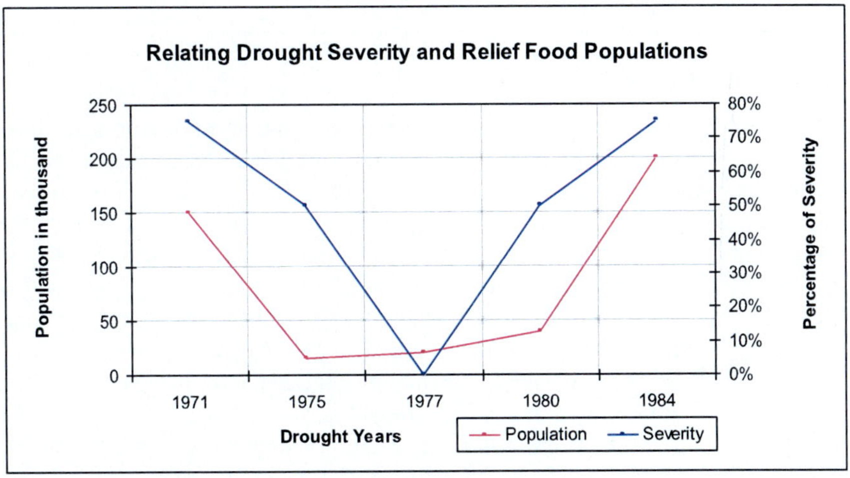

Fig. 12.4 Relationship between drought severity and food-insecure population in Kenya. *Source* Awange et al. [9]

Table 12.6 Drought profile in relation to numbers of people receiving food relief in the Lake Victoria region of Kenya (all droughts were geographically widespread). *Source* Kenya Government Disaster Preparedness Paper

Year	Drought	No. of people
1983–1984	Extreme	200,000
1980	Severe	40,000
1975	Severe	16,000
1971	Extreme	150,000

triggered by droughts from other regions, with the Lake Victoria region unable to make up shortfalls in the national basket.

In 1969 and 1979, MAM and SON were dry (negative anomalies, Figs. 12.3 and 12.4). This could have affected the planting of maize during February–March and August–September, leading to a food shortage that triggered the food imports of 1970 and 1980. The extreme drought for Kenya as a whole during 1981–1985 (Table 12.5) severely affected the planting season of MAM, and triggered the food imports of 1984. The 1992 increase in food imports could be attributed to droughts in other parts of the country, since the rainfall anomalies of 1991 indicate a wet season in the LVB for MAM and SON. Similar to the El Ñino year of 1997, rainfall in 1994 was above normal, which could have led to crop failure due to floods. The food imports of 1994 therefore could not be attributed to drought.

The LVB is dependent on national food reserves in years of drought (Fig. 12.4), while also exerting an influence over these reserves through its fluctuating level of food production. Droughts, therefore, contribute to food insecurity, not only for the LVB, but for the entire country. This information should therefore help the planners and managers intensify agricultural activities in the region for better crop yield to ensure food security in the basin.

For the LVB, farmers, policy makers and other stake holders should consider the following courses of action.

1. Development of crop varieties that are higher yielding, nutritious, drought resistant, early-maturing, and disease and pest tolerant. For instance, the PIONEER and PANNAR maize types are capable of maturing within 3 months, even with low rainfall. The setback is that the crop is favoured by weevils, and significant losses could occur through lack of proper treatment (this is a particular danger where yields are large, since farmers may fail to treat the entire crop in an attempt to reduce treatment costs). Information on such types of seed needs to be disseminated to farmers and the seeds made accessible and affordable to them.
2. The need to educate the people on the drought patterns ascertained in [9, 21], and the need to embrace drought resistant crops. For example, cassava—a favourite crop from the 1960s to the 1980s—is rapidly losing favour among farmers, because of its relatively poor performance in drought conditions and because of shifts in consumer preference.
3. Research and development of improved technologies in agriculture; for example, irrigation, such as has been achieved by the Yala Swamp project (despite doubts that were initially expressed concerning its environmental consequences—mostly on wetlands—see [11].
4. Studies to establish the stored water potential of LVB (see Sect. 12.6.1 below). This will assist in further studies on hydrological droughts (i.e., reduced stream flow, inflow into the lake and wetlands, see e.g., [30, 31]) within the region.

12.6 Space Sensing of Droughts in Greater Horn of Africa

The GHA, a semi-arid region, is made up of Burundi, Djibouti, Ethiopia, Eritrea, Kenya, Rwanda, Somalia, South Sudan, Sudan, Tanzania and Uganda. Long rains over equatorial GHA region mostly occur during March–May (MAM, Fig. 12.5b) while short rains occur in October–December (OND, Fig. 12.5d) corresponding to the migration of the intertropical convergence zone (ITCZ) from south to north and vice versa [10]. Furthermore, Ethiopia, South Sudan, Sudan, and parts of Uganda experience a single rainy season from June–September (JJAS) Fig. 12.5c).

Apart from its seasonal differences, rainfall variability over the GHA is closely associated with the large-scale regional and global circulations such as ENSO, fluctuation of the Indian Ocean and Atlantic Ocean SST, and moisture fluxes over the Congo region, e.g., [10]. Mean temperature patterns over GHA follow the annual rainfall pattern. Space techniques tend to remote sense larger spatial areas and have been employed to sense drought occurrences, frequencies, intensity and areal increments in Greater Horn of Africa (GHA), which includes lake Victoria. The subsections that follow present synopsis of some of the studies that have been undertaken.

12.6.1 Hydro-meteorological Drought Patterns

Spatio-temporal patterns of hydrological droughts over the GHA were explored by Awange et al. [10] based on total water storage (TWS) changes derived from time-variable gravity field solutions of Gravity Recovery And Climate Experiment (GRACE, 2002–2014) discussed in Chap. 5, together with those simulated by Modern Retrospective Analysis for Research Application (MERRA, 1980–2014). These hydrological extremes were related by Awange et al. [10] to meteorological drought events estimated from observed monthly precipitation products of Global Precipitation Climatology Center (GPCC, 1979–2010) and Tropical Rainfall Measuring Mission (TRMM, 1998–2014). The major focus of the work of Awange et al. [10] was on the application of spatial Independent Component Analysis (sICA) to extract distinguished regions with similar rainfall and TWS with similar overall trend and seasonality. The work highlighted:

- Identification of 4 statistically independent regions facing hydro-meteorological droughts.
- Hydrological and meteorological droughts are highly correlated over tropic regions.
- Inconsistencies were found between MERRA and GRACE TWS over semi-arid regions.
- IOD indicated bigger inter-annual impact on hydro-meteorological drought than ENSO.
- Prolonged (39 months) meteorological drought was found over Sudan-western Ethiopia.

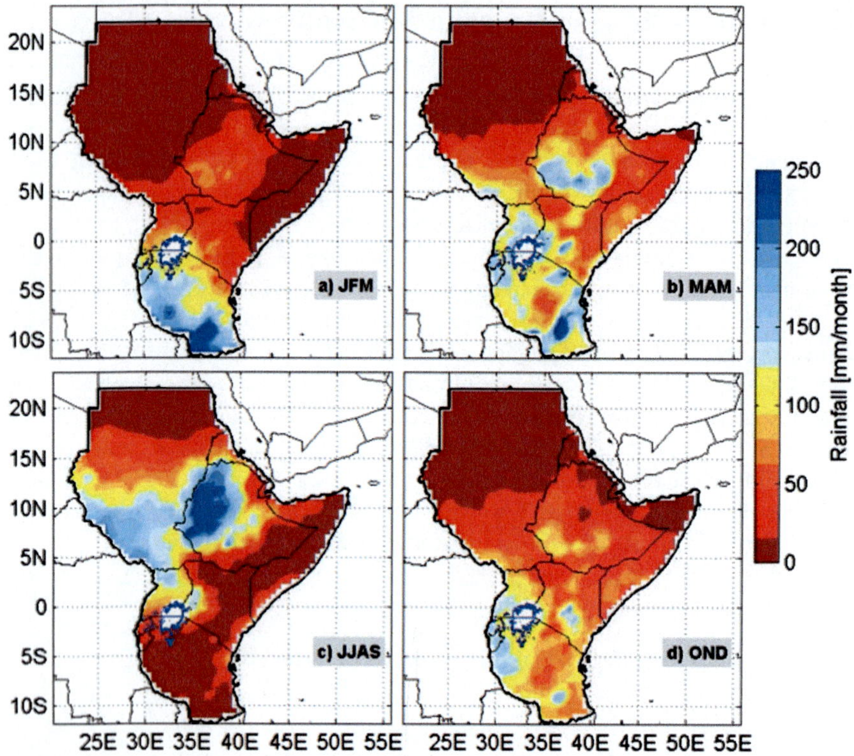

Fig. 12.5 Spatial variability of seasonal rainfall over GHA from 1979–2010 based on GPCC v6 data: **a** January–March (JFM), **b** March–May (MAM), **c** June–September (JJAS), and **d** October–December (OND). Note that the signals over Lake Victoria has been masked out to avoid spurious trends due to varying lake levels. *Source* Awange et al. [10]

Rainfall and TWS were used to estimate Standard Precipitation Indices (SPIs) and Total Storage Deficit Indices (TSDIs), respectively that were employed to characterize frequency and intensity of hydro-meteorological droughts over GHA. Significant positive (negative) changes in monthly rainfall over Ethiopia (Sudan) between 2002 and 2010 leading to a significant increase in TWS over the central GHA region were noted in both MERRA and GRACE TWS (2002–2014). However, these trends were completely reversed in the long-term (1980–2010) records of rainfall (GPCC) and TWS (MERRA). The four independent hydrological sub-regions extracted based on the sICA (i.e., Lake Victoria Basin, Ethiopia- Sudanese border, South Sudan, and Tanzania), see Figs. 12.6 and 12.7, indicated fairly distinct temporal patterns that matched reasonably well between precipitation and TWS changes.

While meteorological droughts were found to be consistent with most previous studies in all sub-regions, their impacts are clearly observed in the TWS changes resulting in multiple years of extreme hydrological droughts. Correlations between SPI and TSDI were found to be significant over Lake Victoria Basin, South Sudan, and

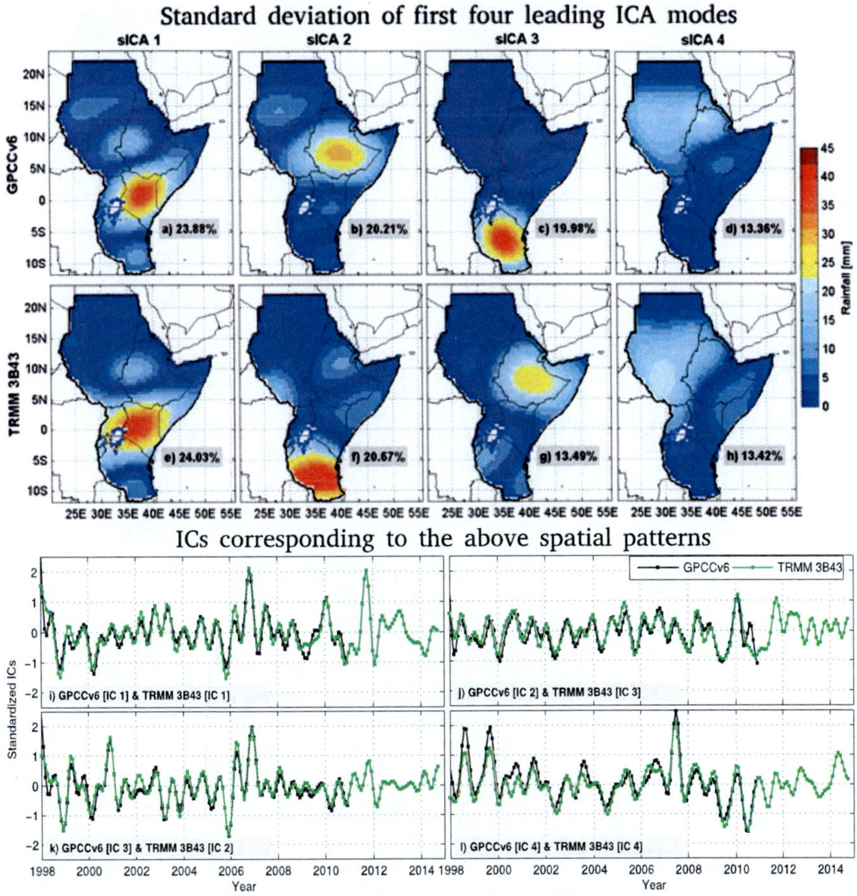

Fig. 12.6 The first four dominant modes of rainfall variability over the GHA region based on **a–d** GPCC (1979–2010), and **e–h** TRMM (1998–2014). The corresponding temporal patterns (or ICs) are shown in the bottom panels **(i–l)**, where a low-pass filter was applied to enhance their interpretation. **i** Temporal patterns corresponding to a & e, **j** b & g, **k** c & f, and **l** d & h. *Source* Awange et al. [10]

Tanzania. The low correlations between SPI and TSDI over Ethiopia are likely related to inconsistency between TWS and precipitation signals. Further, Awange et al. [10] found that hydrological droughts in these regions were significantly associated with Indian Ocean Dipole (IOD) events while El Niño Southern Oscillation (ENSO) played a secondary role. For more details, see Awange et al. [10].

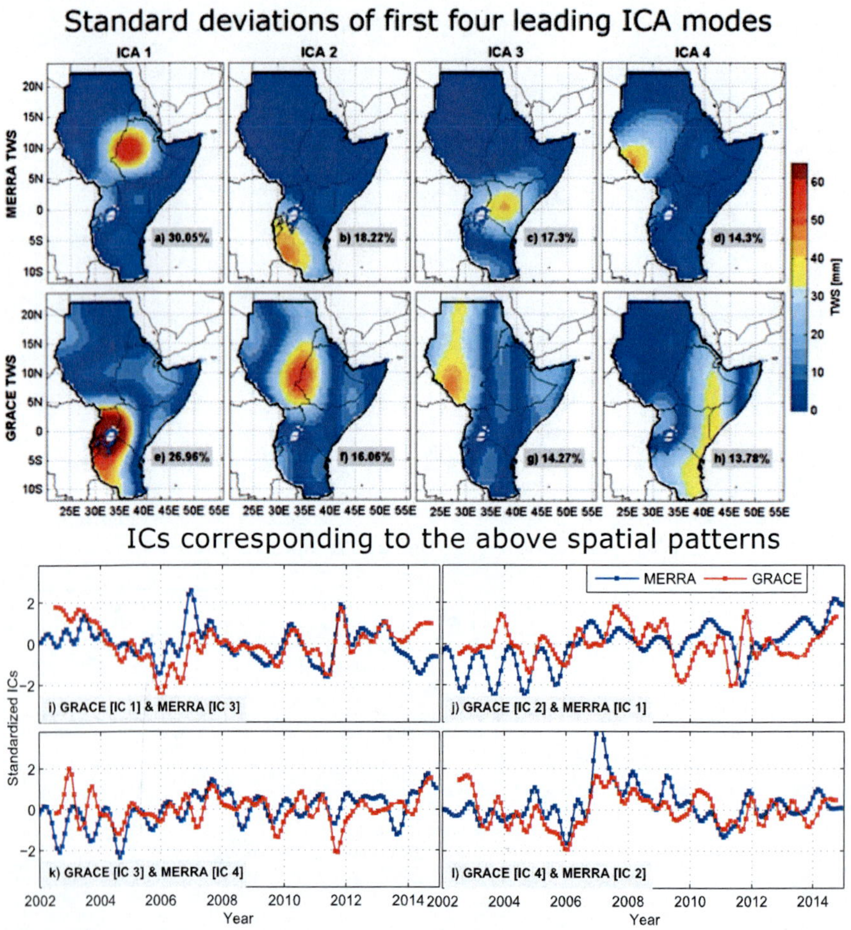

Fig. 12.7 The first four dominant modes of TWS variability over the GHA region based on **a–d** MERRA (1980–2014) and **e–h** GRACE data (2002–2014). The corresponding temporal patterns (or ICs) are shown in the bottom panels (**e–h**), where a 6-month low-pass filter was applied to enhance their interpretation. **i** Temporal patterns corresponding to c & e, **j** a & f, **k** d & g, and **l** b & h. *Source* Awange et al. [10]

12.6.2 Characterizing Agricultural Drought

Heavy reliance of East Africa (EA) on rain-fed agriculture makes it vulnerable to drought-induced famine. Yet, most research on EA drought focuses on meteorological aspects with little attention paid on agricultural drought impacts. Agricultural droughts are defined by Awange [30, 31] as soil water deficiency, plant water stress, and reduced biomass and yield. The inadequacy of in-situ rainfall data across EA has also hampered detailed agricultural drought impact analysis. Recently, however,

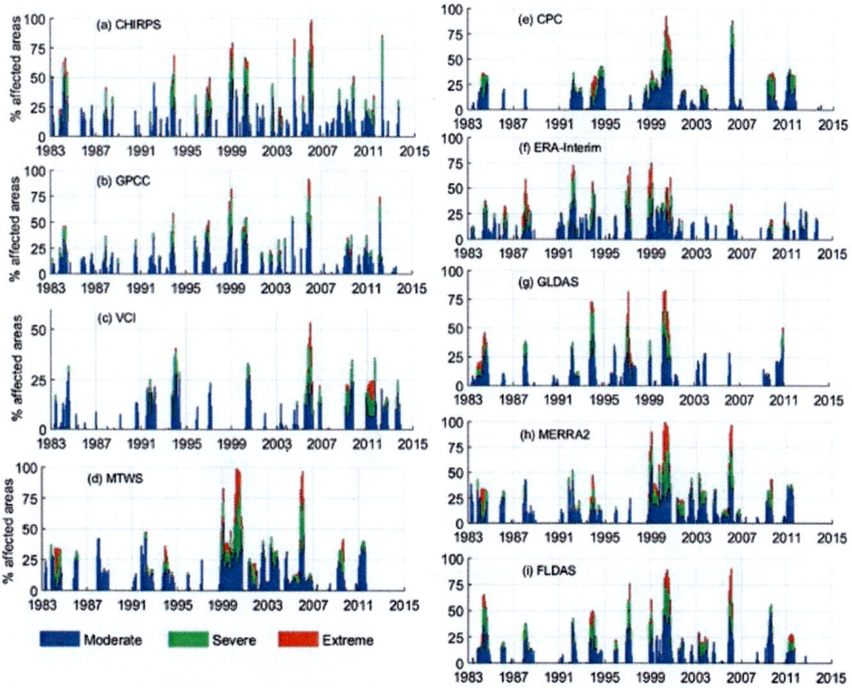

Fig. 12.8 Percentage of area affected by various drought intensities during the period 1983–2013. Percentage areas are computed by integrating the regional spatial and temporal patterns then determining percentage of pixels under each drought category (see Agutu et al. [3]). The rainfall products have the highest percentage areas under drought followed closely by soil moisture products and finally the lowest percentage areas are by VCI. In addition to the soil moisture products having different percentage areas under drought, CPC is consistent with MERRA-2, GLDAS is consistent with ERA-interim while FLDAS is in between. *Source* Agutu et al. [3]

there has been increased data availability from remote sensing (rainfall, vegetation condition index—VCI, terrestrial water storage—TWS), reanalysis (soil moisture and TWS), and land surface models (soil moisture).

To support agricultural drought monitoring from diverse indicators, it is imperative to identify and provide information on the most effective agricultural drought indicator or a combination of indicators for the EA region [3]. Therefore, Agutu et al. [3] employed these products to characterise EA droughts between 1983 and 2013 in terms of severity, duration, and spatial extent. Furthermore, the capability of these products to capture agricultural drought impacts was assessed using maize and wheat production data. Their results showed that while all products were similar in drought characterisation in dry areas, the similarity of CHIRPS and GPCC extended over the whole EA. CHIRPS and GPCC also identified the highest proportion of areas under drought followed closely by soil moisture products whereas VCI had the least coverage. Drought onset was marked first by a decline/lack of rainfall,

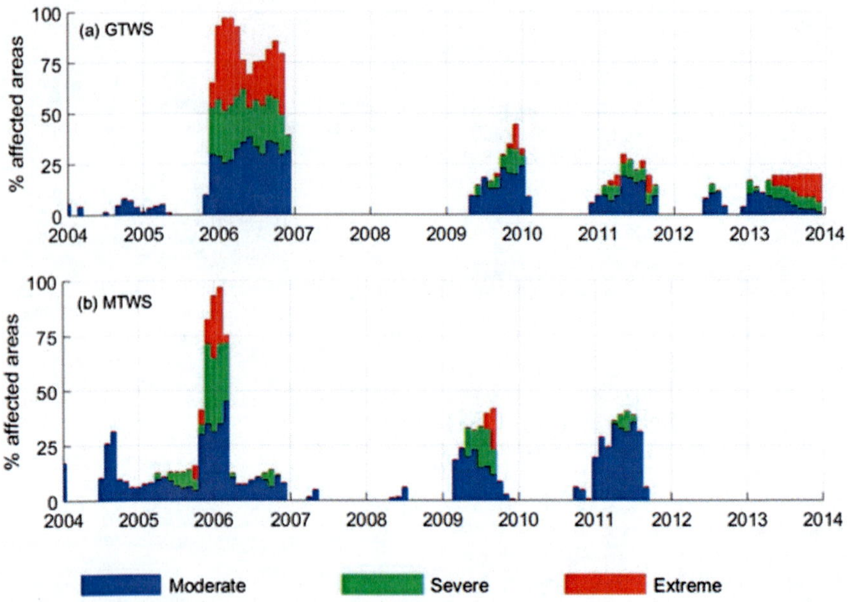

Fig. 12.9 Comparison of performance between GTWS and MTWS in terms of percentage of areas affected by various drought categories. Percentage areas are computed as in Fig. 12.8. They have consistent performance, with GTWS having a lag in drought detection probably due to groundwater that is lacking in MTWS. *Source* Agutu et al. [3]

followed by VCI/soil moisture, and then TWS. VCI indicated drought lag at 0–4 months following rainfall while soil moisture and TWS products had variable lags vis-à-vis rainfall. GLDAS mischaracterized the 2005–2006 drought vis-à-vis other soil moisture products. Based on the annual crop production variabilities explained, Agutu [3] identified CHIRPS, GPCC, FLDAS, and VCI as suitable for agricultural drought monitoring/characterization in the East African region for the study period (1983–2013). Finally, GLDAS explained the lowest percentages of the Kenyan and Ugandan annual crop production variances. In general, Agutu et al. [3]:

- characterized drought and assessed utility of drought indicators in East Africa (EA), which includes the Lake Victoria Basin,
- ordered drought detection as first rainfall, VCI/moisture then TWS,
- VCI was found to have 0–4 months of lag in detecting drought cycle with respect to rainfall,
- ranked the suitability of agricultural drought indicators for EA as: CHIRP, GPCC, FLDAS, and VCI,
- found that the land surface models' products generally performed inconsistently over EA.

The assessment of the products are presented in Figs. 12.8, 12.9 and 12.10, and their correlation with maize and wheat presented in Fig. 12.11. These findings are

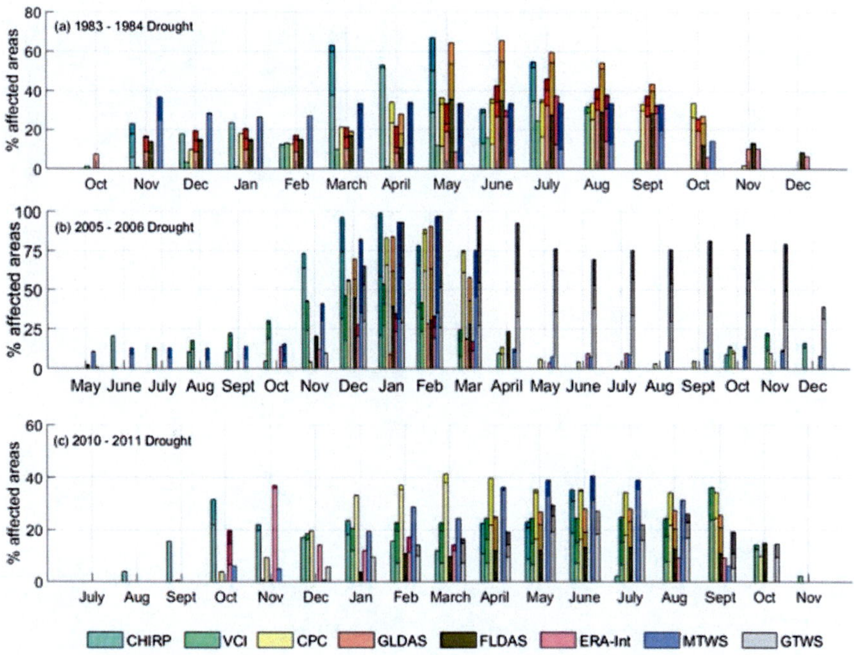

Fig. 12.10 Percentage of areas affected by various drought intensities during the 1983–1984, 2005–2006, and 2010–2011 drought episodes. Each bar has up to 3 colour grades (gradients) representing from bottom moderate, severe, and extreme droughts at the top. Percentage areas are computed as in Fig. 12.8 but only for the duration of drought. VCI has a lag of about 2–3 months in identifying the drought cycle in relation to CHIRPS. The rest of the products have inconsistent lags in relation to rainfall across the three drought episodes. *Source* Agutu et al. [3]

important for the gauge data deficient EA region as they provide alternatives for monitoring agricultural drought.

12.6.3 Influence of Coupled Ocean-Atmosphere Phenomena

Drought-like humanitarian crises in the Greater Horn of Africa (GHA) are increasing despite recent progress in drought monitoring and prediction efforts. Notwithstanding these efforts, there remain challenges stemming from uncertainty in drought prediction, and the inflexibility and limited buffering capacity of the recurrent impacted systems. The complexity of the interactions of ENSO, IOD, IPO and NAO, arguably remains the main source of uncertainty in drought prediction. To develop practical drought risk parameters that potentially can guide investment strategies and risk-informed planning, Mpelesoka et al. [21] quantified drought characteristics that underpin drought impacts management and highlighted the following:

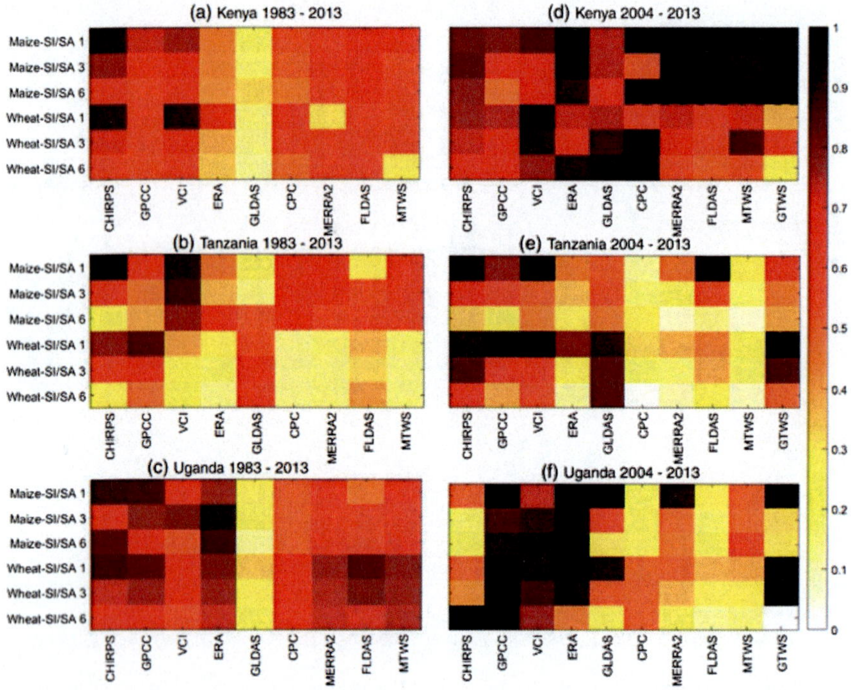

Fig. 12.11 Proportions of variability in national annual crop (maize and wheat) production (R^2) explained by various drought indices for Kenya, Tanzania, and Uganda. CHIRPS, GPCC, and VCI consistently explains relatively higher variability in crop production, while the soil moisture products have inconsistent performance (a, b, and c). The figures d, e, and f should be interpreted with care as the datasets used to fit the models are only 10 years long. Both SI and SA are computed at annual scales. (The y axis indicates the crop (maize/wheat), Standardized index SI (for a, b, and c), and Standardized Anomaly SA (for d, e and f) while 1,3 and 6 indicate the standardization time scales for the indicators on the x axis). *Source* Agutu et al. [3]

- Zonality in drivers of drought occurrence over the GHA.
- Probabilities of drought-year occurrences range from 10 to 40%.
- Most droughts last 14 and 24 months for the 12- and 24-month timescales.
- Drought areal-extent decadal trends range from 0.6%, Tanzania to 3.7%, Ethiopia.

Drought characteristics that include probability of drought-year occurrences, durations, areal-extent and their trends over 11 decades (1903–2012) were derived from the Standardized Precipitation Index (SPI). Transient probability of drought-year occurrences, modelled on Beta distribution, across the region ranged from 10 to 40%, although most fall within 20–30%. For more than half of the drought events, durations of up to 4, 7, 14 and 24 months for the 3-, 6-, 12- and 24-month timescales were evident, while 1 out of 10 events persisted for up to 18 months for the short timescales, and up to 36 months or more for the long timescales. Apparently, only drought areal-extent showed statistically significant trends of up to 3%, 1%, 3.7%,

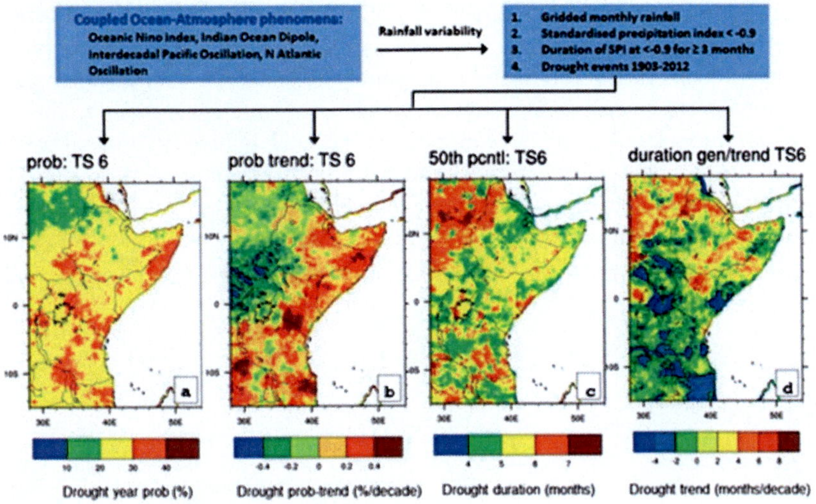

Fig. 12.12 Graphical abstract of Mpelesoka et al. [21]

2.4%, 0.7%, 0.3% and 0.6% per decade over Sudan, Eritrea, Ethiopia, Somalia, Kenya, Uganda and Tanzania, respectively. Since there was no evidence of significant changes in drought characteristics, the peculiarity of drought-like crises in the GHA was attributed by [21] (at least in part) to unaccounted for systematic rainfall reduction. This highlighted the importance of distinguishing drought impacts from those associated with new levels of aridity. In principle drought is a temporary phenomenon while aridity is permanent, a difference that managers and decision-makers should be more aware [21]. For details on the study, refer to Mpelesoka et al. [21], while a summarized abstract of the work is presented in Fig. 12.12.

12.7 Concluding Remarks

Lake Victoria is a riparian resource shared by 3 East African countries: Kenya, Uganda and Tanzania. It serves as a source of water for irrigation, transport, domestic and livestock uses and a source of livelihood for over 42 million people living around the lake. Its fish (tilapia and Nile perch) are exported the world over. Its role as an indicator of environmental and climate change on long term scales has been documented in Nicholson [22]. The vital role of this lake and its surroundings as a source of food has been documented e.g. in Awange and Ong'ng'a [11]. The lake and its environment is under threat from environmental pollution and more recently from declining water levels.

Increased precipitation variability in the LVB has led to an increase in drought events in the recent past. Drought in the region is inversely related to food security (though the relationship is an indirect one), and if the current drought trend persists, it is likely to have further negative effects. In the past, negative effects of drought may have stemmed from an inability to anticipate and cope with drought events. However, severe climate extremes were not as frequent as in recent decades, and since the current trend of rainfall variability is expected to increase, drought severity is likely to be of greater magnitude, and with much greater extremes. The effect of future droughts on food security are thus likely to be severe, and the task of recovering from them costly, if communities in this region do not anticipate trends of increased drought severity, and if they fail to adopt coping strategies. Drought analysis is therefore vital for proper food management.

References

1. Adger WN, Huq S, Brown K, Conway D, Hulme M (2003) Adaptation to climate change in the developing world. Prog Dev Stud 3:179–195
2. Agola NO and Awange JL (2014) Globalized poverty and environment. 21st Century challenges and innovative solutions. Springer, Berlin
3. Agutu NO, Awange JL, Zerihun A, Ndehedehe CE, Kuhn M, Fukuda Y (2017) Assessing multi-satellite remote sensing, reanalysis, and land surface models' products in characterizing agricultural drought in East Africa. Remote Sens Env 194:287–302. https://doi.org/10.1016/j.rse.2017.03.041
4. Akong JJ, Downing TE (1988) Small holder vulnerability and response to drought. In: Parry ML, Carter TR, Konjin NT (eds) The impact of climate variations on agriculture, vol 2. Assessments in semi-arid regions. Kluwer, Dordrecht, pp 221–247
5. Amissah-Arthur A, Jagtap S, Rosenzweig C (2002) Spatiotemporal effects of El Niño events on rainfall and maize yield in Kenya. Int J Climatol 22:1849–1860
6. APFM (Associate Programme on Flood Management) (2004) Strategy for flood management for Lake Victoria Basin, Kenya. www.apfm.info/pdf/strategy_kenya_e.pdf
7. Aseto O, Ong'ang'a O, Awange JL (2003) Poverty reduction: a challange of Lake Victoria Basin (Kenya). African Herald Publishing House, Kendu Bay
8. Awange JL, Mpelasoka F, Goncalves R (2016) When every drop counts: analysis of droughts in Brazil for the 1901–2013 period. Sci Total Env 566–567:1472–88. https://doi.org/10.1016/j.scitotenv.2016.06.031
9. Awange JL, Aluoch J, Ogallo L, Omulo M, Omondi P (2007) An assessment of frequency and severity of drought in the Lake Victoria region (Kenya) and its impact on food security. Clim Res 33:135–142. https://doi.org/10.3354/cr033135
10. Awange JL, Khandu Forootan E, Schumacher M, Heck B (2016) Exploring hydro-meteorological drought patterns over the Greater Horn of Africa (1979–2014) using remote sensing and reanalysis products. Adv Water Res 94:45–59. https://doi.org/10.1016/j.advwatres.2016.04.005
11. Awange JL, Ong'ang'a O (2006) Lake Victoria: ecology, resources and environment. Springer, Heidelberg
12. Barrett CB (2002) Food security and food assistance programs. In: Gardner B, Rausser G (eds) Handbook of agricultural economics, vol 2. Elsevier Science, Amsterdam, pp 2103–2190
13. DMCN (Drought Monitoring Centre Nairobi) (2002) Factoring of weather and climate information and products into disaster management policy. A contribution to strategies for disaster reduction in Kenya. UNDP, Government of Kenya, and WMO, Nairobi
14. Downing TE, Kangethe W, Karnaw C (1989) Coping with drought in Kenya: national and local strategies. Lynne Rienner, Boulder CO

15. Ehrlich PR, Ehrlich AH, Daily GC (1993) Food security, population and environment. Pop Dev Rev 19:1–32
16. Garcia R, Escudero J (1982) The constant catastrophe: malnutrition, famines and drought. Drought and man series, vol. 2. IFIAS Project. Pergamon Press, New York
17. Glantz HM (1987) Drought and hunger in Africa: denying famine a future. Cambridge University Press, Cambridge
18. Kadomura H (1994) Climate changes, drought, desertification and land degradation in the Sudano-Sahelian region: a historic geographical perspective. In: Kadomura H (ed) Savannization process in tropical Africa. Tokyo Metropolitan University, II, Country briefs, pp 203–228
19. Kiptot E, Franzel S, Hebinck P, Richards P (2006) Sharing seed and knowledge: farmer to farmer dissemination of agroforestry technologies in western Kenya. Agrofor Syst 68:167–179
20. Mpelasoka F, Awange JL, Goncalves RM (2018) Accounting for dynamics of mean precipitation in drought projections: a case study of Brazil for the 2050 and 2070 periods. Sci Total Env. https://doi.org/10.1016/j.scitotenv.2017.10.032
21. Mpelasoka F, Awange JL, Zerihun A (2018) Influence of coupled ocean-atmosphere phenomena on the Greater Horn of Africa droughts and their implications. Sci Total Env 610–611, 691–702. https://doi.org/10.1016/j.scitotenv.2017.08.109
22. Nicholson SE (1998) Historical fluctuations of Lake Victoria and other lakes in the Northern Rift Valley of East Africa. In: Lehman JT (ed) Environmental change and response in East African lakes. Kluwer, Dordrecht, pp 7–35
23. Otieno HO, Awange JL (2006) Energy resources in East Africa. Springer, Heidelberg
24. Phoon SY, Shamseldin AY, Vairavamoorthy K (2004) Assessing impacts of climate change on Lake Victoria Basin, Africa: people-centred approaches to water and environmental sanitation. 30th Water Engineering and Development Centre (WEDC) Int Conf. Vientiane, Lao PDR, pp 392–397
25. Scoones L, Chibudu C, Chikura S, Jeranyama P, 8 others, (1996) Hazards and opportunities: farming livelihoods in dryland Africa. Zed Books, London, Lessons from Zimbabwe
26. Shipton P (1990) African famines and food insecurity: anthropological perspectives. Annu Rev Anthropol 19:353–394
27. Skees JR (2000) A role for capital markets in natural disasters: a piece of the food security puzzle. Food Pol 25:365–378
28. Swaminathan MS (2001) Food security and sustainable development. Curr Sci 81:948–954
29. Tumwesigye EK, Musiitwa F (2004) Characterizing drought patterns for appropriate development and transfer of drought resistant maize cultivars in Uganda. In: Fiesen DK, Palmer AFE (eds) Integrated approaches to higher maize productivity in the new millennium: Proceedings of the 7th Eastern and Southern Africa regional maize conference, 5–11 February 2002, Nairobi, Kenya. CIMMYT (International Maize and Wheat Improvement Center) and KARI (Kenya Agricultural Research Institute), Nairobi, pp 260–262
30. Awange JL (2012) Environmental monitoring using GNSS. Global navigation satellite system, Springer, Berlin
31. Awange JL (2018) GNSS environmental sensing. Revolutionizing environmental monitoring, Springer, Berlin

Chapter 13
Early Warning System: Is NDVI the Answer?

13.1 Summary

Normalized Difference Vegetation Index (NDVI), which is a measure of vegetation vigour, and lake water levels respond variably to *precipitation* and its *deficiency*. For a given lake catchment, NDVI may have the ability to depict localized natural variability in water levels in response to weather patterns. This information may be used to decipher natural from unnatural variations of a given lake's surface.

This chapter, based on the work of [42], evaluates the potential of using NDVI and its associated derivatives VCI (vegetation condition index), SVI (standardised vegetation index), AINDVI (annually integrated NDVI), green vegetation function (F_g), and NDVIA (NDVI anomaly) to depict Lake Victoria's water levels. Thirty years of monthly mean water levels and a portion of the Global Inventory Modelling and Mapping Studies (GIMMS) AVHRR (Advanced Very High Resolution Radiometer) NDVI datasets are used. Their aggregate data structures and temporal co-variabilities are analysed using GIS/spatial analysis tools.

Locally, NDVI is found to be more sensitive to drought (i.e., responded more strongly to reduced precipitation) than to water levels. It shows a good ability to *depict water levels one-month in advance, especially in moderate to low precipitation years. SVI and SWL (standardized water levels) used in association with AINDVI and AMWLA (annual mean water levels anomaly) readily identified high precipitation years*, which are also when NDVI has a low ability to depict water levels. NDVI also appears to be able to highlight unnatural variations in water levels. Omute et al. [42] proposed an *iterative approach* for the better use of NDVI, which may be useful in developing an early warning mechanisms for the management of Lake Victoria and other lakes with similar characteristics.

J. Awange, *Lake Victoria Monitored from Space*,
https://doi.org/10.1007/978-3-030-60551-3_13

13.2 Introductory Remarks

As discussed in the previous chapters, Lake Victoria, the world's second largest fresh water lake, is refilled largely (80%) by direct rainfall, e.g., [3, 6]. Its annual weather patterns and water use are major factors that significantly affect its water level [32]. The extent to which these affect the water level is still debatable, with Kull [32], see also Chap. 4 suggesting that 55% of the fall in water level occurred in the period 2002–2006 because of excessive release through the Owen Falls and Kiira hydro power dams as opposed to 45% due to drought. In 2002–2006, however, the lake receded at an alarming rate [4, 32] bringing to focus the need for the constant monitoring of its water levels.

Lake Victoria is a shared resource with immense importance in sustaining livelihoods (i.e., agricultural, fisheries, hydropower generation, and freshwater, see e.g., Awange and Ong'ang'a [6], not only within its locality (East Africa), but also as far as the Nile River stretches, see, e.g., [7, 55, 56]. In light of its immense importance, a multi-faceted approach for monitoring its surface dynamics is necessary to provide information that would assist water resources managers and policy makers in making decisions that would enhance the sustainable use of the lake's waters.

Existing water-level monitoring technologies, such as *tide gauges* and *radar altimetry*, are limited to the empirical evaluation of water-surface changes. Tide gauges, for example, are limited in their spatial coverage and their in-situ use requires solving the problem of accessibility to both the observation sites as well as the data, which are often only available to government authorities [17]. To overcome these setbacks, satellite altimetry methods, see e.g., [11, 13], which are semi-realtime and less precise than the gauges, are being adopted, e.g., [16]. The tide gauges, however, do not provide any other relevant proxy data, such as that provided by *satellite imagery*. Imagery can, for example, provide a time series of various measures of vegetation vigour, such as the main interest of this chapter, the Normalized Difference Vegetation Index (NDVI; Tucker [50]) for the extended evaluation of a lake's surface dynamics in relation to its catchment's response to weather and climate patterns.

In an environment where there is an increasing uncertainty in the prospect for adequate rainfall (see e.g., Chap. 12 and [38]), the use of NDVI should be viewed as an attempt to provide input to developing enhanced water management systems that incorporate other observable such as drought indices, to adequately forecast the dynamics of a lake's surface. Drought, a condition of precipitation deficiency, has a direct negative impact on all useable water sources [35]. NDVI variables should quantify and/or decipher a lake's surface dynamics as affected by local weather/drought patterns from anthropogenic water use, while also accounting for weather patterns in a given lake's catchment, see e.g., [36, 47].

NDVI has traditionally been used for vegetation cover modelling, see, e.g., [45] and was extended, for example, to epidemiological studies where, in the work of Hay [28], it was used to predict malaria seasons in Kenya. NDVI has also been shown to provide a good overview of the prevailing plant water stress as a function of the prevailing weather conditions [24, 27]. This relationship is already being exploited by the Famine Early Warning Systems Network (FEWS [21]) to monitor crops and range lands in semi arid sub-Saharan Africa. It is also a known covariate with other environmental variables, such as surface cover and rainfall, especially in semi arid regions [40, 41]. In the study of Chen [18], the potential use of NDVI to identify critical soil-loss prone areas in Xiangxi watershed, the river basin nearest to the Three Gorges Dam was highlighted.

In tropical Africa, natural vegetation is generally very sensitive to precipitation. Camberlin [14] established that, even with varying mean annual rainfall amounts, for a third of tropical Africa, inter-annual NDVI correlates with corresponding inter-annual rainfall. A correlation of $r > 0.5$ has been observed in semi-arid regions, with generally lower correlations in areas with mean annual rainfall above 600 mm. However, such correlations are region dependent. Significant correlations have been observed even in areas with more than 600 mm mean annual rainfall [14], e.g., the Lake Victoria Basin (LVB). Thus, depending on the location and the spatial extent of the study area, varying correlation patterns should be expected due to differing precipitation patterns.

Nicholson [39] used the Global Inventory Modelling and Mapping Studies (GIMMS) NDVI dataset (1982–1985) to compare vegetation response to rainfall in East Africa and the Sahel regions and found that in these regions, annually integrated NDVI (AINDVI) closely varied spatially with mean annual rainfall, especially at two-months time lag. The time lag is important because vegetation does not directly respond to rainfall, but rather to soil moisture, which is a multi-month integral of rainfall [39]. However, NDVI saturates at rainfall thresholds of above 1000 mm due to water logging and below 200 mm where it readily depicts bare soil. Furthermore, Anyamba [2] suggest the possibility of seasonally predicting NDVI using observed rainfall for Africa. This would require the use of numerous active ground-based weather stations [53]. Use of NDVI is exemplified in the work of [30] who applied NDVI derived from MODIS imagery to investigate duckweed blooms and other floating vegetation in Lake Maracaibo, Venezuela. Propastin [46] applied NDVI to model the link between wetland and biomass conditions in the Ili River delta and changes in the Backhash Lake water levels, Kazakhstan, and showed that NDVI displayed significant intra- and inter-annual variability strongly correlated with the Ili River discharge and Backhash Lake's water levels. For Lake Victoria Basin (LVB), NDVI derived from MODIS has recently been applied by to monitor vegetation changes over the basin and identified the hotspot regions where such changes occur. Details of this study are discussed in Chap. 14.

In this chapter, the application of NDVI and its derivatives, namely SVI (standardised vegetation index), AINDVI (annually integrated NDVI), green vegetation function (F_g), and NDVIA (NDVI anomaly) as proxy datasets for Lake Victoria's natural water level variations and therefore their use for monitoring lake water surface levels is assessed. The assessment is based on the assumption that for Lake Victoria, of all possible natural factors, weather patterns are the major natural factor governing its natural water level trends [5, 32, 48]. This is compounded by water releases for hydroelectric power generation, which have been proposed as having largely contributed to falling water levels [4, 32]. Thus, the possibility of using NDVI to discern such anthropogenic fall in water level from that due to drought is also assessed.

Broadly speaking, this chapter aims at identifying and evaluating the possibilities of using catchment NDVI as a proxy variable for depicting, as well as monitoring, lake surface water levels. It seeks to understand the general relationship between the Lake Victoria catchment NDVI and its water levels and therefore the applicability of its aggregated NDVI in depicting natural water level patterns. It is assumed that any functional contributions of the catchment to the water balance in terms of surface runoff, drainage, land use or otherwise can only be aggregate. Therefore, for each of the NDVI images, the focus is on the mean NDVI. To identify and understand such possibilities and associated patterns, spatial-temporal and quantitative variabilities of Lake Victoria's catchment (Advanced Very High Resolution Radiometer (AVHRR) NDVI 1987–2003) and the observed water levels were used.

13.3 Use of NDVI to Study Lake Victoria

According to the United States Department of Agriculture (USDA [52]), Lake Victoria's water level is known to be extremely sensitive to moderate changes in rainfall over the lake and its large catchment [3]. This is heightened by the fact that it is typically shallow. Thus, its recharge capacity is very dependent upon its large surface area of approximately 69,295 km^2 [3] and the tributary flows from its catchment, which extends to an area of over 184,000 km^2 (Fig. 13.1).

Land use in the catchment is largely agricultural (Fig. 13.2). It is therefore prone to low vegetation cover, especially in the dry seasons, while also having a mixture of dense forest covers. Diverse land cover types exist, ranging from bare ground, rain-fed and irrigated crops, to urban areas (Fig. 13.2). On the Ugandan side of the catchment, the most dominant vegetation cover is rain-fed shrub crops followed by generally open shrub land. In Kenya, the land cover is mostly rain-fed herbaceous crops, similar to Tanzania, except that in the latter, the crops are often punctuated by variable mixtures of savannah shrubs. A detailed description of the lake basin can be found, e.g., in Awange and Ong'ang'a [6].

13.3.1 NDVI Data Sources

The data used includes daily rainfall observations from the Entebbe meteorological station (1990–2005). This station is chosen as being representative of the other stations within the basin due to the fact that they are within the same climatic zone, see e.g., [5]. The primary datasets are the GIMMS AVHRR NDVI (1987–2003), and thirty years of monthly mean water levels (1976–2005) from the Jinja tide gauge station in Uganda. The NDVI used is a spatial and temporal subset of an improved continuous AVHRR calibrated global dataset processed under the GIMMS project.

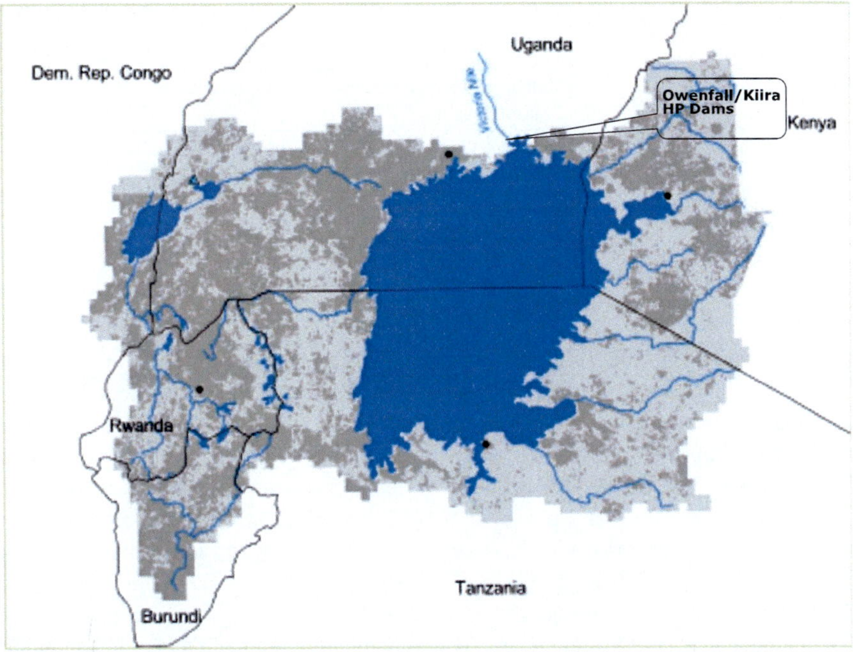

Fig. 13.1 The Lake Victoria Catchment Basin with the Owen Falls hydroelectric dam depicting the position of the Jinja tide gauge station. *Source* [54]

Fig. 13.2 Examples of land use (i.e., cattle farming) and vegetation cover (i.e., shrubs) in the Ugandan part of Lake Victoria catchment area. *Source* [42]

The AVHRR NDVI data processing and quality is comprehensively discussed by Pinzon [43, 44]. It has been critiqued by some authors, e.g., Tucker [51] cites [19, 25] who question the non-surface uncertainty in AVHRR NDVI data. However, Tucker [51] argues that their findings are mainly a result of inappropriate use of the data.

The water levels are observations relative to an established height datum of 1122.86 m above mean sea level, while the lake's maximum mean level is 1134.5 m.

Table 13.1 Data source used

Data	Source	Period
Daily rainfall	Entebbe Meteorological Station, Uganda	1990–2005
AVHRR NDVI	GIMMS, University of Maryland	1987–2003
Monthly mean water levels	Jinja tide gauge (Uganda) and	1976–2005
Satellite altimetry	NASA Physical Oceanography DAAC (PODAAC)	1993–2002

Observations of water level fluctuations provided by altimetry missions are of major importance in monitoring water reservoirs, as they provide independent information in addition to the traditional in-situ datasets [13, 20]. Satellite altimetry water levels from Topex/Poseidon (T/P) and its follow-on, Jason-1, were incorporated for the period 1993–2009. T/P was the first successful repeat-orbit altimetry mission and operated in several phases in its original orbit until September 2002. The spacecraft was in an orbit with an altitude of about 1300 km and provided repeat measurements over a given geographical position at approximately 10-day intervals. Technical details of the spacecraft can be found in [23] and the references therein. During the ten years of its operation, T/P provided 369 cycles of precise and accurate observations of water-level variations with a near global coverage. According to the pre-launched report of Benada [12], the sea-level measurements of T/P had a precision of 2.4 cm and an accuracy of 14 cm for typical oceanic conditions, with small geographically correlated errors. Since its launch in December 7th 2001, Jason-1 has provided sea and lake surface height observations with the same global coverage and orbital repeatability of 10-days (see Chap. 5 for detailed discussion of satellite altimetry mission).

In order to obtain the lake level heights (LLH) of Lake Victoria during the ten years of T/P's satellite operation (1993–2002), data (Merged Geophysical Data Record (MGDR)) provided by the NASA Physical Oceanography DAAC (PODAAC) at the Jet Propulsion Laboratory, California Institute of Technology was used. To cover the remaining period of the study, the Jason-1 Interim Geophysical Data Record (IGDR) datasets that are available via the PODAAC website was examined. These satellite altimetry data provided independent datasets fulfilling the same task as the Jinja tide gauge, while offering the possibility of comparison (and controlling outliers) with the tide gauge dataset, see Fig. 5.5. Table 13.1 summarizes the data source used.

13.4 Exploitation of NDVI Data

NDVI is the ratio of the reflectance "difference" between the near-infrared (NIR) and visible (R=red) bands and the "sum" of the reflectance of these two bands, e.g., Tucker [49] expressed as

$$NDVI = (NIR - R)/(NIR + R). \tag{13.1}$$

It is primarily a biophysical measure of vegetation vigour that directly assesses the level of photosynthesis rather than the extend of canopies [27], and is a spatially continuous variable. It can be derived from imagery provided by any satellite sensor that scans in the visible (red) and near infra-red bands, and is expressed as a normalized ratio ranging from -1 to $+1$. NDVI was primarily developed to improve the sensitivity of the simple ratio $SR = NIR/R$ to vegetation structure and cover. The NDVI is sensitive to soil reflectance in sparse vegetation, however, it saturates faster in dense vegetation. In dense cover, the NIR signal increases while the red signal stagnates [33]. On the other hand, the SR saturates slowly and has a good ability to differentiate soil from vegetation [9].

The impact of drought is directly observable in the five major useable water sources; soil moisture, groundwater, snow pack, stream flow, and reservoir storage [35]. NDVI depicts vegetation vigour, which is an ancillary reflection of precipitation variability, especially in semi arid regions. It therefore depicts *plant water stress*, which is a direct function of drought. This dynamic characteristic of NDVI makes it very useful as a drought metric, specifically for agricultural drought impact monitoring [8], for example by identifying when the available amount of water falls below that required by plants [37].

In practice, NDVI as a drought measure is normally used in its linear transforms, e.g., the green vegetation fraction Fg, as suggested by Gutman and Ignatov [26], which is the normalized NDVI given as [15]

$$F_g = (NDVI_i - NDVI_{min})/(NDVI_{max} - NDVI_{min}), \tag{13.2}$$

where $NDVI_i$ is the NDVI for the individual pixel i from Eq. (13.1), $NDVI_{min}$ is the NDVI of bare soil and $NDVI_{max}$ is the NDVI of full vegetation cover. $NDVI_{min}$ and $NDVI_{max}$ are constant global values. This F_g value is used to monitor the spatial dynamics of greenness, especially in dense vegetation cover [33]. Gutman and Ignatov [26] suggested values of 0.04 and 0.52, while Carlson and Ripley [15] suggested 0.2 and 0.7 as the typical minimum and maximum values, respectively. Furthermore, the F_g value has been extended to model the fractional cover through $F_c = F_g^2$, that is the proportion of the ground area covered by the projection of plant leaves and stems [27].

The use of F_g is synonymous with the vegetation condition index (VCI) of Kogan [31], given by

$$VCI_{ijk} = (NDVI_{ijk} - NDVI_{i,min})/(NDVI_{i,max} - NDVI_{i,min}), \tag{13.3}$$

except that the VCI uses local minimum and maximum NDVI values. The VCI specifically assesses the temporal variability of the NDVI signal while minimizing the first-order effects of spatially variable environmental factors like climate, soil and, vegetation types and topography [31]. In Eq. (13.3), the $NDVI_{i,min}$ and $NDVI_{i,max}$ are the multi-year minimum and maximum NDVI values, respectively, for NDVI image pixel i, while j and k represent a given month and year, respectively. NDVI is also commonly used as the standardized vegetation index (SVI), first proposed by Liu and Negron-Juarez [34], i.e.,

$$SVI_{ijk} = (NDVI_{ijk} - NDVI_{ij,min})/\sigma NDVI_{ij}, \tag{13.4}$$

where $\sigma NDVI_{ij}$ is the standard deviation of NDVI for pixel i in month j. This is statistically a Z-score of observed NDVI based on multi-year mean NDVI values and their standard deviations on a chosen accumulation time scale. Bayarjargal et al. [10] suggest that this time scale can be weekly, bi-weekly, monthly or seasonal as appropriate, depending upon the required precision and data availability. This is synonymous with the standard precipitation index of McKee et al. [35]. Also in use is the NDVI anomaly (NDVIA), the departure of NDVI from its long-term (multi-year) mean for a chosen accumulation period, expressed as [1]

$$NDVIA_{ijk} = NDVI_{ij,mean} - NDVI_{ijk}, \tag{13.5}$$

where $NDVI_{ijk}$ is the monthly NDVI for pixel i in month j for year k, $NDVI_{ij,mean}$ is the multi-year average NDVI for pixel i in month j.

13.4.1 Processing of the NDVI Data

The GIMMS NDVI images have the pixel values (NDVI values) factored by 10000 to enhance visualization without loss of information, therefore the values range between $-10000(-1)$ to 10000 (1). For the Lake Victoria Basin (LVB), pixels with NDVI values of less than zero were identified to be wetlands, marsh lands, beaches, or muddy waters [42]. Such land cover is quite common within the lake catchment area, since transition zones between the lake and dry land exist. These bad/low pixels were all masked out, including open water (-10000), as well as missing values (-2006). This was necessary because any water to water-logged land cover will tend to have saturated constant NDVI values whose pixel values will distort the mean. The analyses were restricted to the LVB using a Catchments Area Mask (CAM), which was digitized from the general outline of the catchment (Fig. 13.1) and projected onto the NDVI image coordinate systems (Albers Conical Equal Area). Its area outside the lake, but within the catchment, was rasterized at a matching 8 km resolution. GIS map algebra tools were used (Fig. 13.3) for masking out the bad pixels from each of the NDVI bi-monthly *images* (X). Furthermore, using GIS overlay tools, the resulting *Images* (Y) were further area masked by the CAM to obtain the required NDVI images CAM(Y). Since the catchment lies within the equatorial belt and is therefore sufficiently vegetated, the masking inherently also removes insufficiently vegetated pixels, thereby improving the accuracy of NDVI modelling, see e.g., [29].

Within a given month of any year, there were *bi-monthly* images with approximately the same NDVI statistics, and one could choose either image. However, a

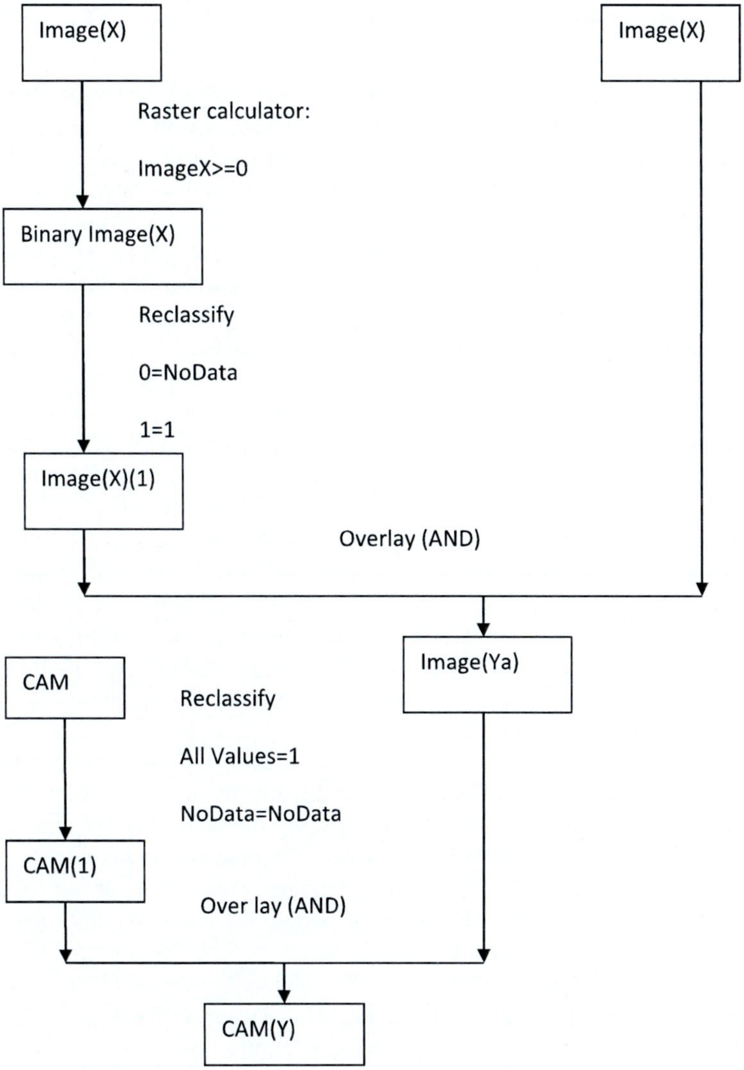

Fig. 13.3 Processing structure chart for the GIMMS NDVI images. *Source* [42]

monthly mean NDVI value was computed from the two images to formulate the primary NDVI values used (Fig. 13.3). This was done for all studied months. Entire catchment mean NDVI values rather than the individual pixel values were considered. Even then, within each of the images, the distributions of NDVI values are approximately normal, thus suggesting heterogeneous vegetation cover in the catchment, despite the low spatial resolution (i.e., monthly).

13.4.2 Analysis of the NDVI Data

The global and internal structures of the datasets were explored and found to be useful for designing the NDVI data derivation algorithms, and in developing assumptions used in the data analysis. Various NDVI/water level co-variability trends were assessed. NDVI as a predictor variable was analysed both in its raw form (Eq. 13.1) and its transformed forms (drought metrics, Eqs. 13.3–13.5). Correlation analyses were then performed to identify the extent of the association between the water levels and NDVI. Correlation coefficient trends were also analysed for the identification of the possible uses and weaknesses of NDVI in the depiction of natural water variability. Visualization of the quantitative and spatial trends of NDVI, and a refinement of the identified relationships was achieved using GIS map algebra and overlay tools.

The data were analysed on the basis of a calendar rather than a vegetative year, which starts the month following that one with the minimum NDVI. This is because for the Lake Victoria Basin (LVB), the driest months can be any of the months in the two dry seasons, yet a minimum NDVI only occurs in August in any of the years (see Fig. 13.5b). For seasonal analysis, the data were classified over three monthly intervals with the onset of the first dry season being December of the preceding year, and ending in February. This is based on the known local pattern of rainfall in the LVB as indicated in Fig. 13.5a. The seasons are typically: Dec-Feb (dry), Mar-May (very wet), Jun-Aug (dry) and Sep-Nov (wet). This pattern has been used in many studies, for example, an analysis of drought severity by Awange et al. [5] and in assesing the physical dynamics of the lake [3].

Exploratory Data Analysis: The water levels tend to be normally distributed. To determine years that require further analysis, the severity in the variability of water levels over annual time scales was depicted using the percentage of normal (PN) (Eq. 13.8). The monthly mean levels (X_M) were averaged to give annual mean levels (X_A) for each of the years (Eq. 13.6). These were then used to compute the long-term annual mean ($X_{30} = 11.38$) for a thirty year period (Eq. 13.7). The PN values from Eq. 13.8 provide the ranking of the years in terms of water level drops (Fig. 13.4). The normal water level is considered to be X_{30} with a PN value of 100%. Above this are years of greater than normal water levels, while below are those reflecting water stress. Annual and long-term means are given as [4, 5]

$$X_A = \sum X_M/12, \tag{13.6}$$

and

$$X_{30} = \sum X_A/N, \tag{13.7}$$

where N is the number of years considered, in this case 30 years (1976–2005). Using Eqs. 13.6 and 13.7, one obtains the percentage of normal (PN) as [4, 5]

$$PN = \frac{X_A}{X_{30}} \times 100\%, \tag{13.8}$$

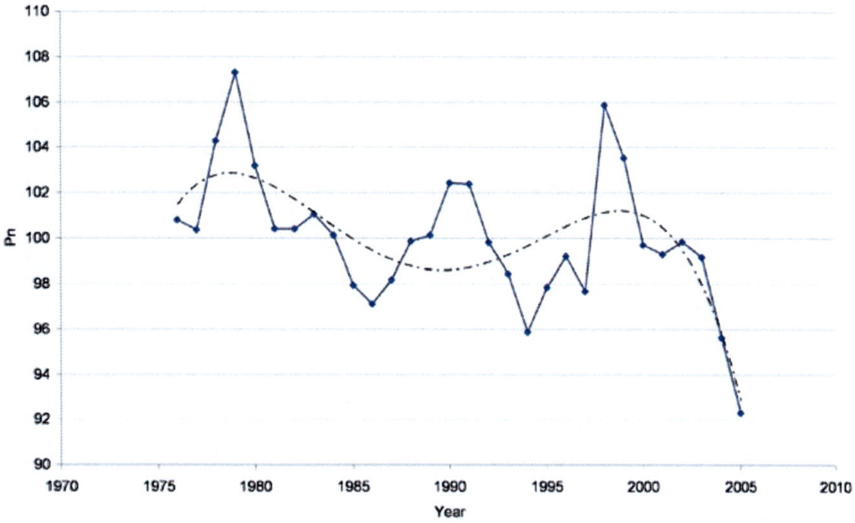

Fig. 13.4 Water level departures from normal (100%), 1976–2005 (see Eq. 13.8). The dashed-dotted lines show a polynomial fit to the data. In general, a declining trend is noticed. *Source* [42]

It can be seen from Fig. 13.4 that the water levels were above or near normal for most years and consistently so between 1976 and 1984. However, they are almost consistently below the normal between 1992–2005, except for 1998 and 1999. In the East African region, the El Ñino-Southern Oscillation (ENSO) rains are known to have started in late October 1997, but intensified during most of 1998. In early 1999, only moderately high rainfall was experienced as an extended effect of the ENSO rains. Thus, the observed high water levels in 1999 appear to be a result of the higher precipitation in 1998. It is then seen that the water level sharply fell between 2002 and 2005.

The annual variation of these variables for a sample of years (two high precipitation years 1991, 1999, one normal 2003, and one low 1994) were compared using their mean monthly values as shown in Fig. 13.5a, b, c. To compare the water level variability with NDVI, PN values for NDVI were compared with corresponding PN values for water levels for the common years 1987–2003 (Fig. 13.5d).

Figure 13.5a, b suggest a seasonal response of NDVI to precipitation. Using the monthly means, the correlation coefficients (r) between NDVI and precipitation for 1991, 1994 and 2003 with no time lag are 0.39, 0.78 and 0.19, respectively. With rainfall proceeding the NDVI by one-month (i.e., one-month NDVI lag), on average, higher values of r were obtained, with r equal to 0.45, 0.55 and 0.37, respectively. However, with a two-months lag, this association is almost non existent, with r values of 0.27, 0.07 and 0.03, respectively. It can also be seen that the 2003 precipitation is much lower relative to 1991, 1994 and 1999 (Fig. 13.5a). The significant drop in r values for 1994 when considering time lags is due to the relatively much higher rainfall resulting in an NDVI lag that does not affect the correlation values. This leads

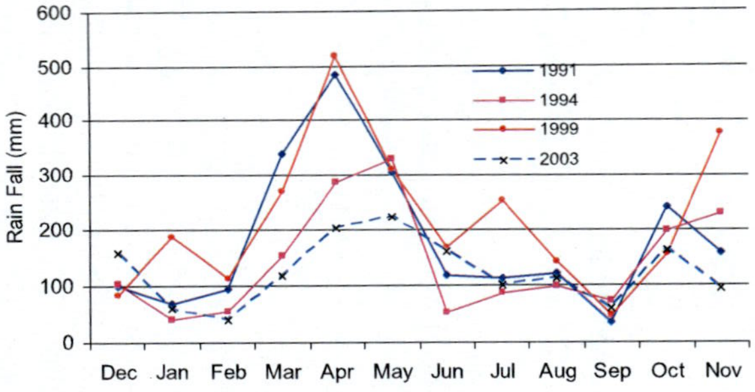

(a) Variations in Monthly Mean R/F.

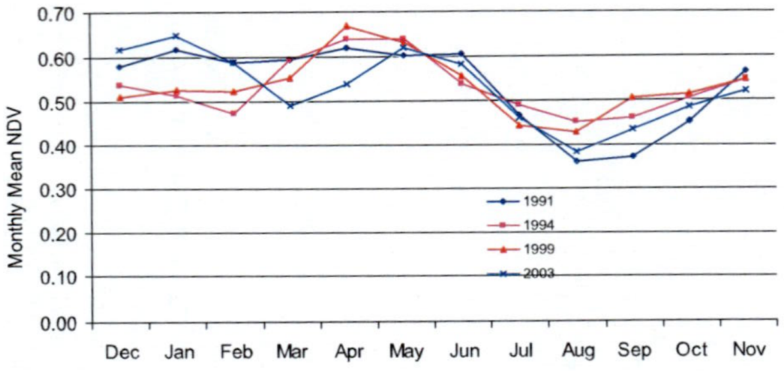

(b) Variations in Monthly Mean NDVI.

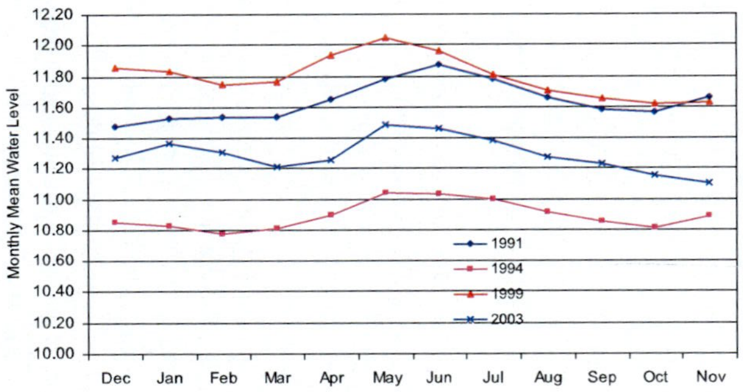

(c) Variation in Monthly Mean Water Level (m).

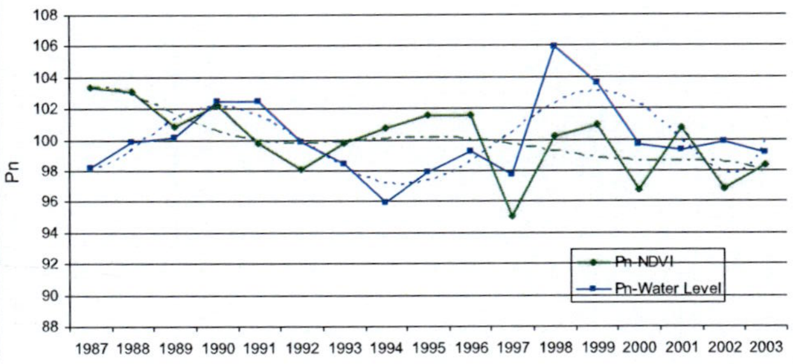

(d) WL and NDVI Departures from Normal, 1987-2003.

Fig. 13.5 Variation of monthly means; **a** rainfall **b** NDVI **c** water levels from the tide gauge data, and **d**, a plot of water levels and NDVI departures from normal (PN) for the period 1987–2003. *Source* [42] to the proposal that considering time lags is only appropriate for moderate drought years.

Hence, as expected, there appears to be a good localized association between rainfall and NDVI in the Lake Victoria catchment when one considers a one-month time lag, i.e., when rainfall of the preceding month is compared with the NDVI of the current month. There is also the possibility of a direct correspondence between the maximum rainfall and maximum NDVI occurrences in any of the years. However, there is a month difference between an occurrence of minimum rainfall (Sept) and minimum NDVI (Aug). This may be because NDVI is very sensitive to any slight reduction in rainfall.

Figure 13.5c shows that within a given year, the lake's monthly mean water levels vary slightly, as indicated by the amplitudes. This is in contrast to the monthly mean NDVI, which shows comparatively larger variations in its small values (Fig. 13.5b). The smaller gradients in Fig. 13.5c compared to the sharper ones in Fig. 13.5b are a further indication of the sensitivity of NDVI to precipitation, as opposed to the delayed response of water levels to surface run off from the catchment. The water levels are maximum in May-June following the March-April-May rainfall season, e.g., [5], whereas maximum rainfall and NDVI are in April-May. Again, there is no direct correspondence between the minimum water level, which is generally between Oct-Dec, and the minimum rainfall or NDVI (Aug-Sept). Overall, these observations seem to indicate that within a year, NDVI responds faster to weather changes than water levels.

The comparison of PN values computed from Eq. 13.8 (Fig. 13.5d) shows that over annual time scales, NDVI may be less sensitive to weather patterns than water levels, whose PN trend is more sinusoidal than NDVI. Their sinusoidal curves appear to be similar in configuration but with opposite gradients. However, there is almost no association ($r = 0.11$) between the annual mean NDVI and annual mean water levels. Considering annual mean water levels lagging annual mean NDVI by one year, there is still a poor correlation ($\mathrm{r} = -0.26$), which increases to -0.29 for a two year lag. With NDVI lagging the water levels, there is only a very small association ($r = -0.23$).

Co-variability trend analyses were then performed, firstly using NDVI in its primary 'raw' form and water levels; and then using the NDVI derived drought metrics. The metrics are the standardized or normalized NDVI (Eq. 13.4). Data anomalies and temporal differences were also used. The annually integrated NDVI (AINDVI) trend (i.e., the sum of mean monthly NDVI in a given year) was derived for use as a proxy indicator of precipitation patterns. An attempt was made to refine the sensitivity of the NDVI/water level covariability trend based on a suggested threshold AINDVI.

AINDVI Image Analysis: To interpret the AINDVI spatial trend, sample AINDVI images for the years 1988, 1992, 1996, 1997, 1998 and 2003, which are typically representative of the inter-annual precipitation patterns were extracted. Figure 13.6a shows the distribution of NDVI for a sample high precipitation year (1988) and a drought year (1997). Just like the primary source NDVI images, the distribution of AINDVI within the pixels is typically normal, reaffirming the heterogeneous vegetation at 8 km spatial resolution. It appears to emphasize the fact that due to the

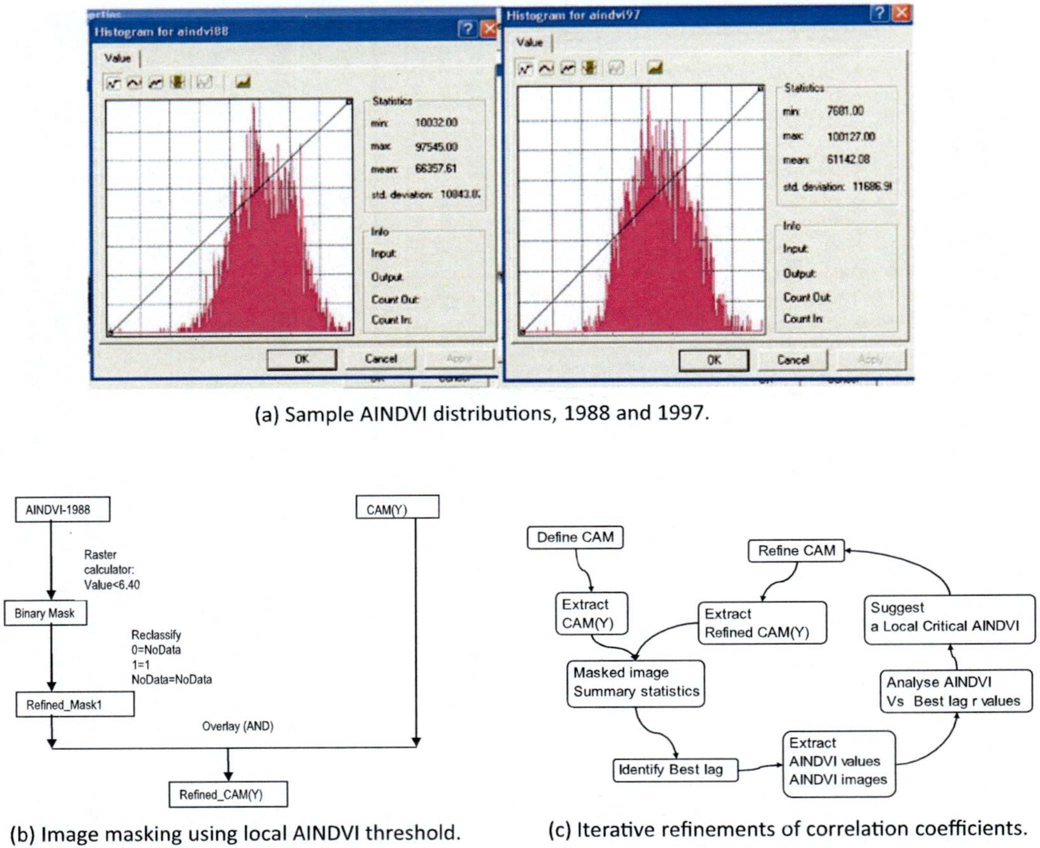

Fig. 13.6 **a** Sample AINDVI distributions, 1988 (high precipitation year) and 1997 (drought year), **b** image masking using local AINDVI threshold, and **c** iterative refinements of correlation coefficients. *Source* [42]

low spatial resolution of the GIMMS NDVI, no large distinct areas of homogenous vegetation cover that would affect the study results can be identified.

Iterative process: Improving the NDVI /water level sensitivity using the AINDVI Spatial Pattern: The North-Western and North-Eastern parts of the lake catchment were found to have higher AINDVI. Considering that the water level/NDVI relationship is very sensitive to small changes in AINDVI (precipitation), while AINDVI still tends to have an upper limit for the case of high correlations, the NDVI images were further masked. The local upper threshold AINDVI was obtained by comparing the local AINDVI variability curve with the observed NDVI /water level r values. To improve on the association, image areas with AINDVI higher than the upper threshold, here set to 6.4, were masked out to provide refined CAM(Y) images (Fig. 13.6b). With the exception of 1991, 1996 and 1998 (Fig. 13.7a), the local upper threshold AINDVI for obtaining good correlations ($r > 0.5$) is about 6.46 (see Fig. 13.7c). This further masking was based on the 1988 AINDVI image which had the highest AINDVI in the dataset (see Fig. 13.7c). It provided a refined CAM as the catchment area with AINDVI pixel values less than 6.4 and masked the NDVI pixels that had a high probability of having values greater than the upper threshold. Repetitive masking of higher outlier AINDVI pixels was carried out as outlined in Fig. 13.6c.

13.5 Analysis Employing Variants of NDVI

Raw NDVI and water level covariability: The strength of association between water level and raw NDVI variability was assessed based on temporal lags. Zero, one and two-months lags were used with NDVI lagging the water levels (Table 13.2, Fig. 13.7a). For a given year, the one-month lag, for example, means that the mean NDVI of December was associated with the mean water level of the following January. Within a year, there is generally a low association between the monthly mean NDVI and the corresponding monthly mean water levels. However, with NDVI lagging the water levels by one or two-months, higher correlations were obtained, except for the years 1988 and 1997 for the two-months lag. The year 1988 does not indicate any association between its NDVI and water level at all, while for 1997, contradictory results are shown with very low association for the two-months lag, but good associations for the zero and one-month lags are realized. However, there were two extreme weather patterns in 1997, with drought occurring throughout much of the year, with exceptionally high ENSO rainfall from October onwards [22]. The observed low correlations in 1998 appear to be indicative of the peak of heavy ENSO rains (Table 13.2).

The reverse was also tested, i.e., water levels lagging NDVI by one month (Fig. 13.7b). The hypothesis was that current rainfall can only be reflected in NDVI as cumulated moisture over time, but it should be directly observable in water levels as a summation of direct rainfall over the lake surface and surface run off. This may be false because relatively low correlations (Fig. 13.7b) were obtained, especially when compared to those in Fig. 13.7a.

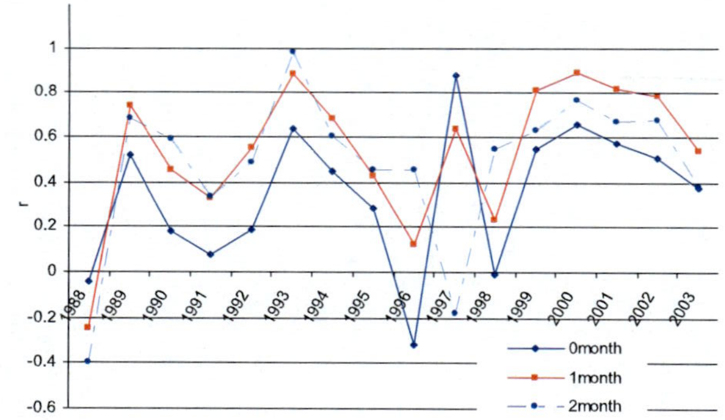

(a) NDVI lagging water levels, zero, one and two months lags.

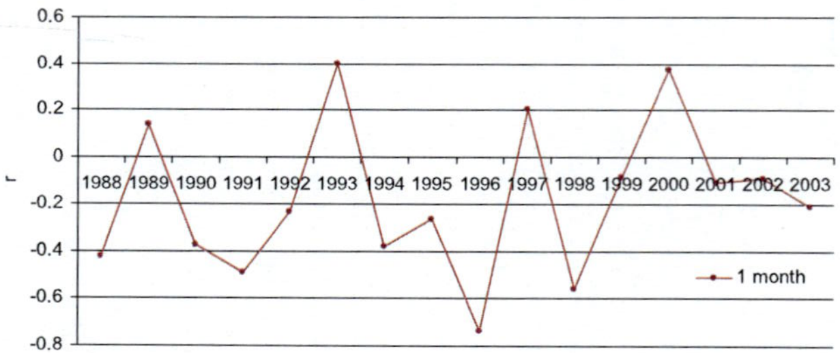

(b) Water levels lagging NDVI, one month lag.

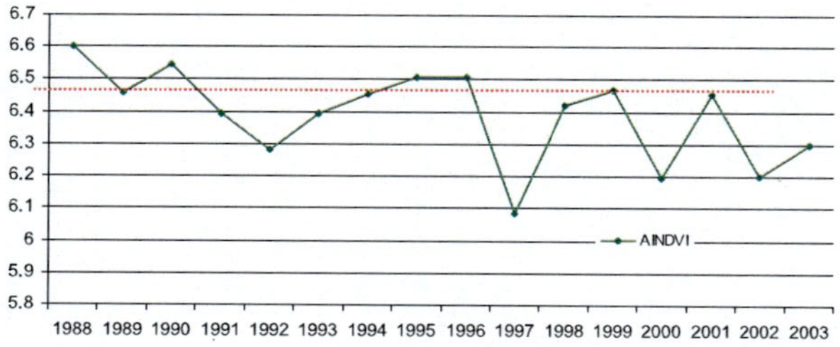

(c) Interannual Variation of AINDVI in the catchment.

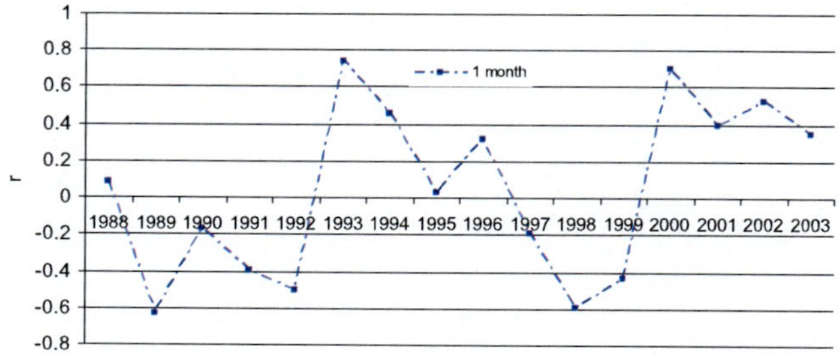

(d) SVI lagging standardised WL, one month Lag.

Fig. 13.7 Correlation Variability; **a** NDVI lagging water levels by zero, one and two-months, **b** water levels lagging NDVI by one-month, **c** interannual variation of AINDVI in the catchment (with the exception of 1991, 1996 and 1998 (Fig. 13.7a), the local upper threshold AINDVI for obtaining good correlations ($r > 0.5$) is about 6.46), and **d**, SVI lagging standardised WL (water levels), one-month Lag. *Source* [42]

Table 13.2 Correlations between NDVI and water levels for the lags of zero, one, and two months. *Source* [42]

Year	0-lag	1-month lag	2-months lag
1987	−0.08	–	–
1988	−0.05	−0.25	−0.40
1989	0.52	0.74	0.68
1990	0.18	0.45	0.59
1991	0.07	0.33	0.33
1992	0.18	0.55	0.48
1993	0.63	0.88	0.98
1994	0.44	0.68	0.60
1995	0.28	0.42	0.45
1996	−0.32	0.12	0.45
1997	0.88	0.63	−0.19
1998	−0.01	0.23	0.54
1999	0.54	0.81	0.62
2000	0.65	0.89	0.76
2001	0.57	0.81	0.66
2002	0.50	0.78	0.67
2003	0.37	0.54	0.38

On average, when the NDVI of the preceding month is associated with water levels of the current month (one-month lag), better correlations were obtained (average $r = 0.54$) than for a two-months lag (average $r = 0.49$). In addition, long-term monthly means of NDVI and water levels across the study years were calculated. Their correlation analysis appeared to confirm that one-month lag ($r = 0.81$) is better than no lag ($r = 0.47$) or two-months lag ($r = 0.73$) for modelling the predictability of water levels from NDVI.

Use of AINDVI variability trend: To explain the observed variations in Fig. 13.7a, the annually integrated NDVI (AINDVI) was computed as the sum of the monthly mean NDVI for each of the years (Fig. 13.7c). Figure 13.7a, c indicate that with the exception of 1997 (two-months lag), at low AINDVI, the r values tend to be higher. When the AINDVI was analysed against the r values of Fig. 13.7a, good negative correlations were obtained for zero month ($r = -0.59$) and one-month lags ($r = -0.50$), but poor correlation ($r = -0.01$) for the two-months' lag. The later is possibly diluted by the outlier character of 1997. This observation equally applies when Fig. 13.7b is compared to Fig. 13.7c.

The range of variability of AINDVI is quite small (6.1–6.6) as indicated in Fig. 13.7c, however, this small variation still produces marked differences in the NDVI/ water level association observed in Fig. 13.7a. This means that the use of NDVI in the depiction of water levels for a given year is very sensitive to the AINDVI and therefore to the annual precipitation. The good positive correlations at lower precipitation (low AINDVI, e.g., Fig. 13.7a, c) indicate that both the water levels and

NDVI are not only sensitive to low precipitation, but they respond to it more than to excessive rainfall. Even then, NDVI responds to a reduction in precipitation much faster than the water levels. This explains the better correlations at the one-month to two-months' lags.

For higher annual precipitation, there are weaker associations between NDVI and water levels. A case in point is 1998, owing to the ENSO rains. Although, this was the case in 1998, one would expect that there was much more rainfall in 1988, 1989, 1991, 1995, and 1996 as inferred from their higher AINDVI. However, NDVI saturates at higher precipitation, as seen in 1998. High NDVI is associated with sufficient soil moisture content rather than total water in the system which, when in excess, causes water logging and subsequently stagnation of photosynthetic activities. Therefore, these years appear to have had optimum moisture rather than excessive rainfall. In addition, the commissioning of the Kiira dam does not appear to be the cause of the poor associations in 1996, but rather it is the relatively high rainfall that was just enough to lead to higher AINDVI values than in 1998.

One would also expect poorer correlations in 1999 than in 1998 as suggested by the higher AINDVI in 1999, but the opposite is observed. This is because the ENSO rains peaked in 1998, and continued only to the beginning of 1999. Therefore, there was no NDVI saturation in 1999, but rather the moisture delivered at the start of the year and the moderate rainfall thereafter appears to have sustained the vegetation vigour throughout most of the year. Hence, the higher AINDVI values in 1999 (Fig. 13.7c) than in 1998, while still producing higher r values (Fig. 13.7a).

Generally, poor correlations were obtained when water levels were lagged behind NDVI in the analyses (Fig. 13.7b). This enhances the argument that NDVI responds more sensitively to weather patterns than do water levels. In addition, NDVI responds to cumulative moisture (rainfall) whereas the Lake Victoria water level only instantaneously responds to rainfall, with a time delay in its response to runoff. If the cumulative effect of rainfall in the specific days were to be reflected in the monthly mean water levels, the water levels would have good correlations with the NDVI of the next month(s) (perhaps at even no lag), and also at relatively higher, but non-NDVI saturating amounts of rainfall (e.g., 1996). This in essence would agree with the observations of [39], that rainfall (and therefore water levels) should precede NDVI for better associations. These arguments seem to explain the outlier characteristics of 1988 and 1996, which show no saturation in NDVI (Fig. 13.7c), but also display low correlations (Fig. 13.7a).

The sensitivity NDVI/water level association to the AINDVI spatial pattern: The AINDVI refined CAM was used on a sample of CAM(Y) NDVI images for 1988, 1991, 1996 and 1998, which displays a correlation or AINDVI greater than or equal to the upper threshold AINDVI of 6.4 (e.g., 1991). The results are shown in Table 13.3. In essence, to achieve better models for depicting lake water levels from catchment NDVI, we applied the proposed iterative approach (Fig. 13.6c).

The variability in the NDVI/water level relationship is well explained by the AINDVI, which is sensitive to changes in precipitation (AINDVI). It was found that masking out NDVI pixels with a higher than the threshold AINDVI value improved the ability of NDVI to depict the water levels, especially in drought conditions. When

Table 13.3 Refined correlations following the suggested iterative process (Fig. 13.6c). *Source* [42]

	AINDVI	Refined AINDVI	% Change AINDVI	r	Refined r	% Change r
1988	6.6	5.62	15	−0.25	−0.34	36
1991	6.39	5.62	12	0.33	0.32	3
1996	6.51	5.54	15	0.12	0.2	67
1998	6.42	5.62	12	0.23	0.37	61

NDVI pixels with higher than the suggested threshold AINDVI were masked out, the suggested iterative process in Fig. 13.6c improved the correlations by more than 60% for 1996 and 1998 (Table 13.3). For example, for 1996 there is a 15% reduction in AINDVI leading to a 67% improvement in the correlation coefficient. However, there is no proportionality in the changes. In 1991 and 1998, there was a 12% reduction in the AINDVI, but the change in r for 1991 is almost negligible, with a 3% reduction compared to a 61% increase for 1998. This appears to be because the 1991 AINDVI of 6.39 is only slightly below the 6.4. Even then, the results for 1991 seem to confirm the existence of a localized threshold AINDVI.

One month was identified as the most appropriate time frame for lagging NDVI behind water levels, as shown in Fig. 13.7a. Therefore, the following analyses are restricted to further understanding the applicability of the one-month lag using NDVI drought metrics.

Using the standardised vegetation index (SVI): The SVI (Fig. 13.7d) was computed using Eq. 13.4, assuming a normal distribution of these variables. The standardisation yields the SVI when the input is NDVI. The correlations between SVI and water levels were very low and yet the fluctuation was high. SVI was not found to be a useful direct predictor of water levels. However, its correlation with the standardized water levels (SWL) shown in Fig. 13.7d to some extent follows the same trend as that of Fig. 13.7a for the period 1992–2003. In Fig. 13.7d, the 1998 coefficient indicated in Fig. 13.7a is attenuated relative to those of 1988 and 1996 that have been amplified. This suggests that the SVI and SWL relationship has the ability to reveal higher than normal rainfall events. For example, in comparison to Fig. 13.7a, 1998 and 1999 in Fig. 13.7d are both attenuated proportionately to excess rainfall. The ENSO rains that peaked in 1998 continued to the early months of 1999. Therefore, whereas 1988 and 1996 had lower correlations compared to 1998 and 1999, they did not have higher than normal rainfall, which is a requirement for their optimum NDVI. This seems to support the argument for their relatively higher AINDVI (Fig. 13.6b). Similarly, abnormal rainfall patterns, and therefore NDVI patterns, appear to have occurred in the years 1989–1992 and 1997. In Fig. 13.7a, these have good positive correlations, but negative correlations in Fig. 13.7d. This does not happen for the other years, that similarly show positive correlations in Fig. 13.7a.

Vegetation condition index (VCI): The aggregated VCI was computed using Eq. 13.9, which is synonymous with Eq. 13.3, except that monthly mean NDVI was used instead of the pixel values, and the multi-year minimum and maximum NDVI were identified as 0 and 0.999, respectively. These values signify the existence of both bare soils and extremely vigorous vegetation cover types in the catchment. For the Lake Victoria catchment, the VCI is equal to the NDVI (Eq. 13.9). Thus, the VCI analyses, results, and interpretations are not different from those of raw NDVI. The aggregated VCI is computed as

$$VCI = NDVI/0.999. \tag{13.9}$$

Since the functioning of the VCI is similar to that of the green vegetation fraction (F_g) in Eq. (13.2), its global minimum and maximum NDVI values were used to compute two sets of F_g using 0.04 and 0.52 (Gutman and Ignatov, [26]) and 0.2 and 0.7 (Carlson and Ripley [15]), respectively. However, more than 50% of the mean monthly NDVI values are greater than 0.52. Thus, for the Gutman and Ignatov [26] values, more than 50% of the obtained F_g values were greater than unity. Within the context of the fractional cover $Fc = F_g^2$, these values were unrealistic. However, the Carlson and Ripley [15] values fitted the NDVI dataset and provided a correlation pattern shown in Fig. 13.8a, which is similar to Fig. 13.5d. Thus, like the VCI, the analyses and use of the F_g give results that are also exactly the same as those of the raw NDVI, with only a scaling effect on the NDVI values.

Use of NDVI Anomaly (NDVIA): The NDVI Anomaly (NDVIA) in Eq. (13.5) was used firstly with the raw water levels and then with similarly computed water level anomalies. In both cases, the NDVIA lagged the water levels by one-month. The NDVIA has a weak ability to depict the water levels as indicated by its generally poor and very noisy trend in Fig. 13.8b. The analysis of both anomalies resulted in a pattern (Fig. 13.8c) similar to that found using the SVI (Fig. 13.7d) and therefore other relatable patterns (Fig. 13.7a). Like the SVI, the NDVIA equally has the ability to attenuate the NDVI/water level co-variability as affected by higher or lower than expected rainfall events (Fig. 13.8c).

In addition, the annual mean water levels anomaly (AMWLA) and annual mean NDVIA (AMNDVIA) were computed as the means of their respective monthly anomalies for each of the study years. Their variabilities are shown in Fig. 13.8d, e respectively. Unlike the AMWLA (Fig. 13.8d), the variability in AMNDVIA, shown in Fig. 13.8e, does not explain any of the NDVI and water level trends.

Combined use of anomalies, SVI/SWL co-variability and AINDVI: The AMWLA inter-annual variability (Fig. 13.8d) generally followed the correlation variability in Fig. 13.7a, especially from 1991 onwards, with some discrepancies, namely that it only peaked in 1994, whereas in Fig. 13.7a, the peaks were in 1993 and 2000. Even then, it correctly does so in the years 1991–1997, but falls short of the same in the other years, for example, 1998–2003, that are very much attenuated. Like the SVI, the anomaly trend appears to have the ability to distinguish years with very anomalous water levels, despite a strong correlation with NDVI. For example, 1998 is a known ENSO year and it is very well depicted as being more anomalous than 1996, which on the other hand shows a lower correlation in Fig. 13.7a than 1998.

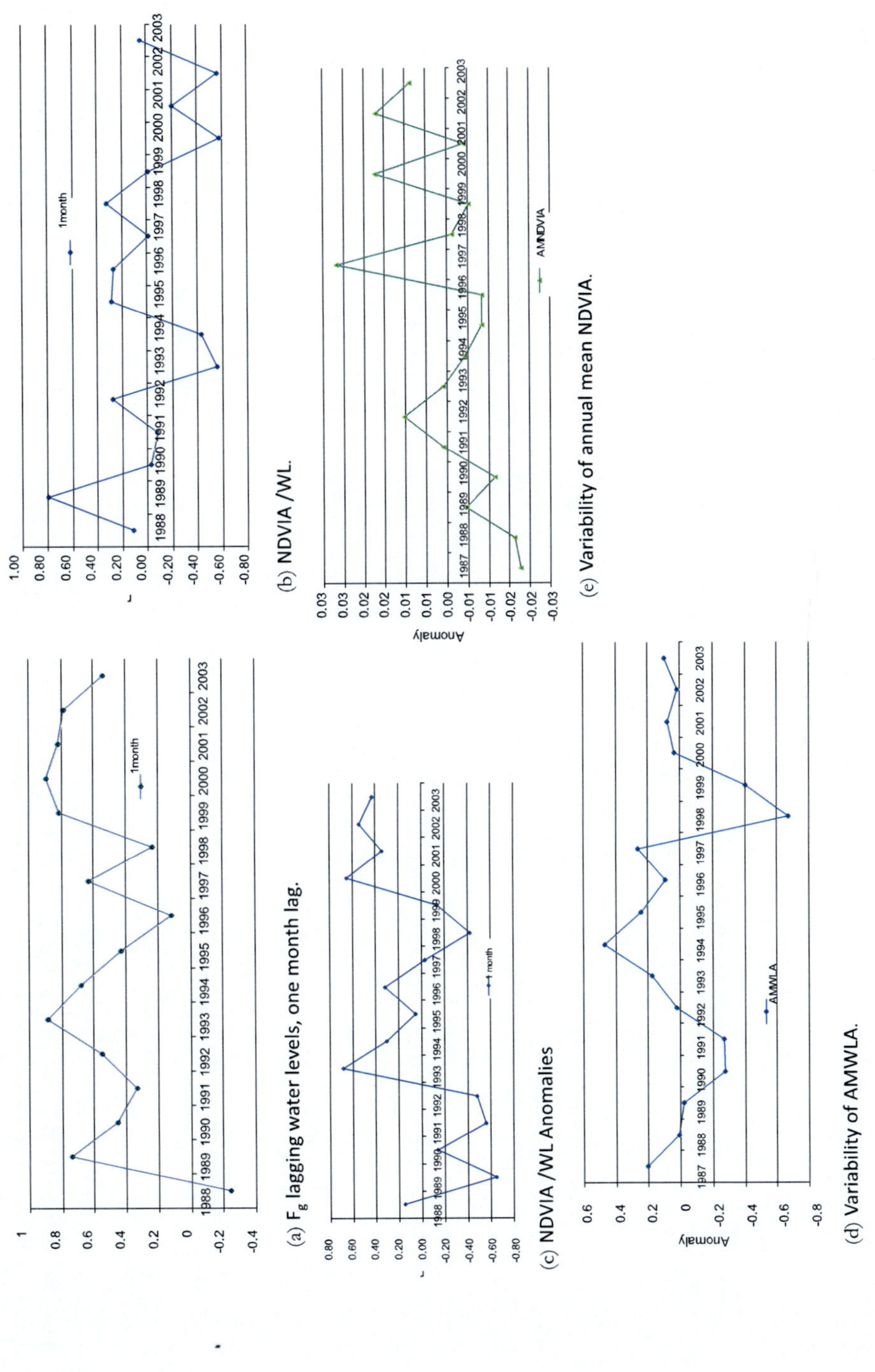

Fig. 13.8 Correlation variability; **a** F_g lagging water levels, one-month lag, **b** NDVIA /WL (water levels), **c** NDVIA /WL Anomalies, **d** variability of AMWLA, and **e**, variability of annual mean NDVIA. *Source* [42]

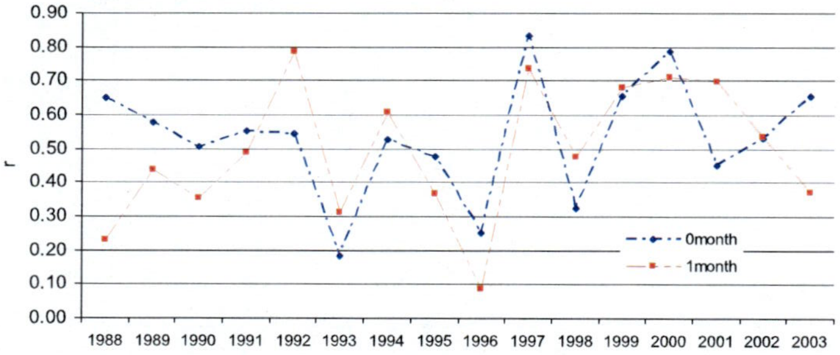

(a) NDVI differences lagging WL differences, zero and one month lag.

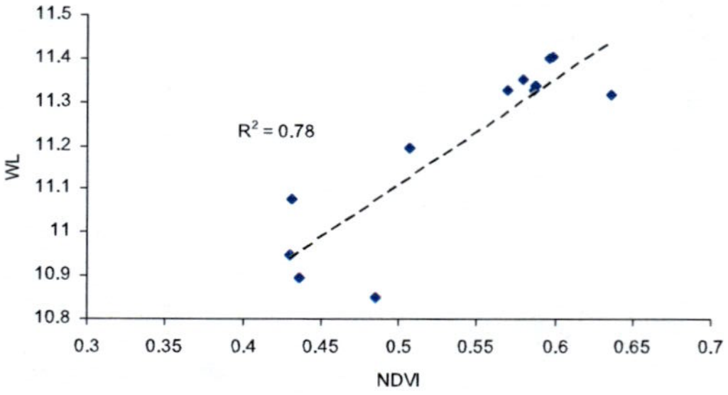

(b) NDVI lagging water levels, one month lag, 1993.

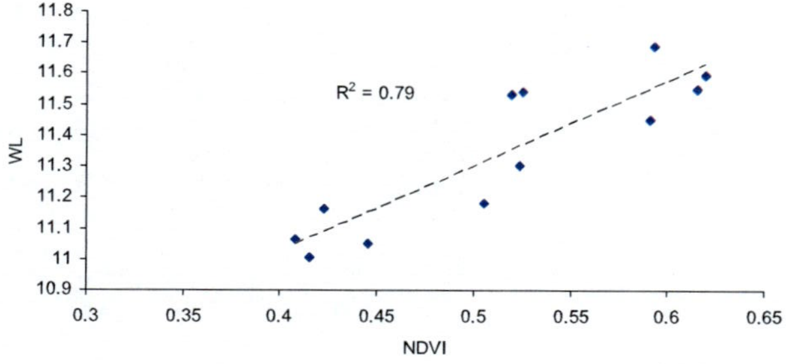

(c) NDVI lagging water levels, one month lag, 2000.

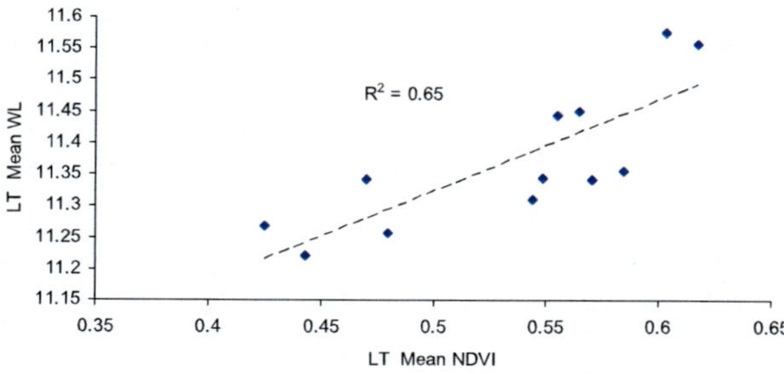

(d) Long term NDVI lagging long term water levels, one month lag.

Fig. 13.9 Correlation Variability; **a** NDVI differences lagging WL differences at zero and one-month lag, **b** NDVI lagging water levels at one-month lag for 1993 **c** NDVI lagging water levels at one-month lag for 2000, and **d**, long term NDVI lagging long-term water levels with a one-month lag. *Source* [42]

The trend also clearly isolates 1989, 1991, 1998, and 1999 as having higher than normal water levels. Thus the hypothesis of depicting higher than normal rainfall events using the SVI/SWL trend indicated for 1989–1992 and 1997–1999 as earlier suggested may be valid. A typical example is 1997, where there was drought for most of the year [22], with lower than normal water levels (Figs. 13.4 and 13.8d) and the lowest AINDVI (Fig. 13.7c). However, Fig. 13.7d does not indicate it as being a very dry year. This is because of the short occurrence of extremely heavy ENSO rainfall from November 1997. Thus, this mixture of severe drought and short heavy rains explains why it is less attenuated in Fig. 13.7d than 1989–1992, 1998 and 1999, which were much wetter as inferred from their higher AINDVI values (Fig. 13.7c).

The importance of the 1997 trend can be stressed by comparing it with 1994. The AINDVI for 1997 is much lower (Fig. 13.7c) and has on average higher water levels (Fig. 13.4) with a lower positive anomaly (Fig. 13.8d), again a result of the short but heavy ENSO rains that contributed to its higher average water level compared to 1994. However, there was long dry spell in 1997 that severely reduced its vegetation vigour (Fig. 13.7c). This appears to confirm the fact that NDVI is more sensitive to drought than are the water levels. If the precipitation in 1997 was sufficiently low throughout the year, as it appears to have been in 1994, which has a higher AINDVI yet lower water levels, the 1997 AINDVI would not be very anomalous in contrast to its water levels (Fig. 13.5d). This confirms the existence of a mixture of long severe drought and short extremely heavy rainfall during 1997.

These observations point to the use of the SVI/SWL association in conjunction with the AINDVI and the AMWLA in isolating years of higher than expected rainfall, regardless of there being no saturation in NDVI for a given year, e.g., 1989–1992, and for cases where a year may in fact be classified as a drought year, e.g., 1997. This also emphasizes the fact that both the water levels and the NDVI are more sensitive to drought than excessive rainfall. This is explained by the good association that is indicated for 1997 (Fig. 13.7a) in contrast to some of the wetter years, e.g., 1990–1992, 1998 and 1999. These indicate a lower association between the NDVI and water levels, therefore, any effort towards a practical use of NDVI/water level correlations may be fruitless.

It also seems that the anomaly trend has the ability to highly attenuate years of possible anthropogenic influence on water level variation (2000- onwards). This hypothesis may be true because the years 1992–1997, which are also known to depart from normal (Fig. 13.8d), try to maintain their configuration of Fig. 13.7a and yet the years from 2000 are highly attenuated in Fig. 13.8d.

Use of differenced image NDVI: Inter-monthly NDVI differences in water levels and in NDVI were analysed (Fig. 13.9). The differences in NDVI and in water levels from month to month were computed for each of the years (note, the January differences were computed as differences from December of the preceding year). Their correlations were computed at zero and one-month lags. These are, on average, reasonably high, but noisy. They appear to mimic their corresponding patterns in Fig. 13.7a, especially for the later years, i.e., 1996 onwards (Fig. 13.9a).

As shown in Fig. 13.9a, both lags are very noisy, but on average provide almost the same good correlations. Their average r values are 0.53 and 0.49 for the zero and one-month lag, respectively. Thus, the observed patterns in 1996 and 1998 can still be explained in terms of higher rainfall as depicted by the AINDVI (Fig. 13.7c).

The use of one-month lag in Fig. 13.9a is well explained by the AINDVI ($r = -0.59$). This is unlike the zero lag which, despite having a higher average r value, has a correlation with AINDVI of −0.46. However, these apparent higher associations may only be so because 1988 attained positive correlations of 0.65 and 0.23 for the zero and one-month lags, respectively (Fig. 13.9a). Despite the 1988 ($r = -0.25$) value, generally, the co-variability of the raw NDVI and water levels at one-month lag shown in Fig. 13.7a (average $r = 0.54$) is still better than from the differences and is still well explained by the AINDVI ($r = -0.50$).

Example of regression fits: Regression fits for NDVI lagging water levels using examples of moderate precipitation years, 1993 and 2000, show a good linear trend. Similarly, long-term monthly means of the two datasets computed for the full study period show that up to 80% (see the coefficient of determination R^2 values) of the association between the NDVI/water levels can be explained by simple linear fits (Fig. 13.9b, c, d).

13.6 Concluding Remarks

This Chapter used aggregated NDVI values and demonstrated that for the Lake Victoria Basin (LVB) [42]:

1. NDVI can be used in depicting the expected natural trends of its water levels.
2. NDVI can identify suspicious, possibly anthropogenic variations by isolating drought episodes. This is especially so when considering a *one-month temporal* lag of water level with respect to NDVI. It appears to be mostly applicable in normal rainfall and drought years, during which time NDVI is more sensitive to low precipitation than the water levels are.
3. The use of NDVI can be *iteratively* improved upon by considering the catchment specific threshold annually integrated NDVI (AINDVI). As a drought metric, NDVI governs its suggested use by acting as a surrogate for identifying high rainfall years when it cannot be appropriately used. It can then be used to decipher abnormal precipitation trends by the use of the water level and NDVI anomalies, namely the AINDVI, SVI and AMWLA trends. This iterative approach is thus proposed.

These conclusions may be useful for developing early warning mechanisms for Lake Victoria water management. Indeed, one should forsake using NDVI in years when the predicted AINDVI is greater than the suggested local threshold. These are years with plenty of rainfall, hence there would be no stress on the lake's waters, and yet, as explained, the NDVI /water level relationships do not generally hold. These mechanisms may function well in advance, given that there are possibilities of forecasting NDVI and therefore AINDVI. However, further analysis is recommended using high resolution imagery and undertaking analyses using pixel-specific NDVI, rather than the aggregate catchment NDVI.

References

1. Anyamba A, Tucker CJ, Eastman JR (2001) NDVI anomaly patterns over Africa during the 1997/98 ENSO warm event. Int J Remote Sens 22(10):1847–1859. https://doi.org/10.1080/01431160010029156
2. Anyamba A, Tucker CJ, Mahoney R (2002) From El nino to La nina: vegetation response patterns over East and Southern Africa during the 1997–2000 period. J Clim 15(21):3096–3103. https://doi.org/10.1175/1520-0442(2002)015<3096:FENOTL>2.0.CO;2
3. Awange JL, Saleem A, Sukhadiya RO, Ouma YO, Hu K (2019) Physical dynamics of Lake Victoria over the past 34 years (1984–2018): is the lake dying? Sci Total Environ 658:199–218
4. Awange JL, Ogalo L, Bae K-H, Were P, Omondi P, Omute P, Omulo M (2008) Falling Lake Victoria levels: is climate a contributing factor? Clim Chang 89:281–297. https://doi.org/10.1007/s10584-008-9409-x
5. Awange JL, Aluoch J, Ogallo LA, Omulo M, Omondi P (2007) Frequency and severity of drought in the Lake Victoria region, Kenya and its effects on food security. Clim Res 33(2):135–142
6. Awange JL, Ong'ang'a O (2006) Lake Victoria: ecology resource and 571 environment. Springer, Berlin
7. Awange JL, Forootan E, Kuhn M, Kusche J, Heck B (2014) Water storage changes and climate variability within the Nile basin between 2002 and 2011. Adv Water Resour 73:1–25. https://doi.org/10.1016/j.advwatres.2014.06.010
8. Agutu NO, Awange JL, Zerihun A, Ndehedehe CE, Kuhn M, Fukuda Y (2017) Assessing multi-satellite remote sensing, reanalysis, and land surface models' products in characterizing agricultural drought in East Africa. Remote Sens Environ 194:287–302. https://doi.org/10.1016/j.rse.2017.03.041
9. Baret F, Guyot G (1991) Potentials and limits of vegetation indices for LAI and APAR assessment. Remote Sens Environ 35(2/3):161–173. https://doi.org/10.1016/0034-4257(91)90009-U
10. Bayarjargal Y, Karnieli A, Bayasgalan M, Khudulmur S, Gandush C, Tucker CJ (2006) A comparative study of NOAA-AVHRR derived drought indices using change vector analysis. Remote Sens Environ 105(1):9–22
11. Becker M, LLovel W, Cazenave A, Güntner A, Crétaux J-F (2010) Recent hydrological behavior of the East African great lakes region inferred from GRACE, satellite altimetry and rainfall observations. Comptes Rendus Geosci 342:223–233. https://doi.org/10.1016/j.rse.2006.06.003
12. Benada JR (1997) TOPEX/POSEIDON user's handbook. Technical report, Jet Propulsion Laboratory, California Institute of Technology Generation B (MGDR-B) Version 2.0
13. Birkett CM (1995) The contribution of TOPEX/POSEIDON to the global monitoring of climatically sensitive lakes. J Geophys Res Ocean 100(12):25179–25204. https://doi.org/10.1029/95JC02125
14. Camberlin P, Martiny N, Philippon N, Richard Y (2007) Determinants of the interannual relationships between remote sensed photosynthetic activity and rainfall in tropical Africa. Remote Sens Environ 106(2):199–216. https://doi.org/10.1016/j.rse.2006.08.009
15. Carlson TN, Ripley DA (1997) On the relation between NDVI, fractional vegetation cover, and leaf area index. Remote Sens Environ 62(3):241–252. https://doi.org/10.1016/S0034-4257(97)00104-1

16. Charon B, Reynolds C (2005) Anew remote sensing tool for water resources management. http://www.pecad.fas.usda.gov/cropexplorer/global_reservoir/index.cfm. Accessed 21 Mar 2006
17. Charon B, Brad D (2004) Anew remote sensing tool for water resource management. Earth Obs Mag 13(6). http://www.eomonline.com/Common/Archives/2004octnov/. Accessed 29 Apr 2006
18. Chen L, Qian X, Shi Y (2011) Critical area identification of potential soil loss in a typical watershed of the three Gorges reservoir region. Water Resour Manag 25(13):3445–3463. https://doi.org/10.1007/s11269-011-9864-4
19. Chilar J, Chen JM, Li Z, Huang F, Latifovic R, Dixon R (1998) Can interannual land surface signal be discerned in composite AVHRR data? J Geophys Res 103:23163–23172. https://doi.org/10.1029/98JD00050
20. Cretaux JF, Birkett C (2006) Lake studies from satellite radar altimetry. Comptes Rendus Geosci 338(14/15):1098–1112. https://doi.org/10.1016/j.physletb.2003.10.071
21. FEWS NET (Famine Early Warning Systems Network) (2007) Africa data dissemination service. http://earlywarning.cr.usgs.gov/adds/index.php. Accessed 1 Mar 2008
22. FAO (Food and Agriculture Organization) (1998) Food supply situation and crop prospects in Sub Saharan Africa: Africa report, no. 1/98. FAO/GIEW, Rome, Italy. http://www.fao.org/DOCREP/004/W8261E/w8261e03.htm. Accessed 28 Feb 2008
23. Fu L-L, Christensen E, Yamarone C Jr, Lefebvre M, Escudier M, Menard P, Dorrer Y (1994) TOPEX/POSEIDON mission overview. http://hdl.handle.net/2014/34628
24. Gontia NK, Tiwari KN (2010) Estimation of crop coefficient and evapotranspiration of wheat (Triticum aestivum) in an irrigation command using remote sensing and GIS. Water Resour Manag 24(7):1399–1414. https://doi.org/10.1007/s11269-009-9505-3
25. Gutman GG (1999) On the use of long-term global data of land reflectances and vegetation indices derived from the advanced very high resolution radiometer. J Geophys Res 104:6241–6255. https://doi.org/10.1029/1998JD200106
26. Gutman G, Ignatov A (1998) The derivation of the green vegetation fraction from NOAA/AVHRR data for use in numerical weather prediction models. Int J Remote Sens 19(8):1533–1543. https://doi.org/10.1080/014311698215333
27. Hatfield JL, Prueger JH, Kustas WP (2004) Remote sensing of dryland crops. In: Ustin S (ed) Remote sensing for natural resources and environmental monitoring: manual of remote sensing, vol 4, 3rd edn. Wiley, New Jersey, pp 531–568
28. Hay SI, Snow RW, Rogers DJ (1998) Predicting malaria seasons in Kenya using multi temporal meteorological satellite sensor data. Trans R Soc Trop Med Hyg 92(1):12–20
29. Kastens JH, Kastens TL, Kastens DLA, Price KP, Martinko EA, Lee R (2005) Image masking for crop yield forecasting using AVHRR NDVI time series imagery. Remote Sens Environ 99(3):341–356. https://doi.org/10.1016/j.rse.2005.09.010
30. Kiage LM, Walker ND (2009) Using NDVI from MODIS to monitor duckweed bloom in Lake Maracaibo, Venezuela. Water Resour Manag 23:1125–1135. https://doi.org/10.1007/s11269-008-9318-9
31. Kogan FN (1997) Global drought watch from space. Bull Am Meteorol Soc 78(4):621–636. https://doi.org/10.1175/1520-0477(1997)078<0621:GDWFS>2.0.CO;2
32. Kull D (2006) Connection between recent water level drops in Lake Victoria, dam operations and drought. Technical report, International River Network. http://www.irn.org/programs/nile/
33. Liang S (2004) Quantitative remote sensing of land surfaces. Wiley, New Jersey
34. Liu WT, Negrón Juárez RI (2001) ENSO drought onset prediction in northeast Brazil using NDVI. Int J Remote Sens 22(17):3483–3501. https://doi.org/10.1080/01431160010006430
35. McKee TB, Doesken NJ, Kleist J (1993) The relationship of drought frequency and duration to time scales. In: Preprints, 8th conference on applied climatology, January 17–22, 1993, Anaheim, CA. http://scholar.google.com/scholar?
36. Mendicino G, Versace P (2007) Integrated drought watch system: a case study in Southern Italy. Water Resour Manag 21(8):1409–1428. https://doi.org/10.1007/s11269-006-9091-6

37. McVicar TR, Jupp DLB (1998) The current and potential operational uses of remote sensing to aid decisions on drought exceptional circumstances in Australia: a review. Agric Syst 57(3):399–468. https://doi.org/10.1016/S0308-521X(98)00026-2
38. Mpelasoka F, Awange JL, Zerihun A (2018) Influence of coupled ocean-atmosphere phenomena on the Greater Horn of Africa droughts and their implications. Sci Total Environ 610–611:691–702. https://doi.org/10.1016/j.scitotenv.2017.08.109
39. Nicholson SE, Davenport ML, Malo AR (1990) A comparison of the vegetation response to rainfall in the Sahel and East Africa, using normalized difference vegetation index from NOAA AVHRR. Clim Chang 17(2/3):209–241
40. Nicholson SE (2001) Application of remote sensing to climatic and environmental studies in arid and semi-arid lands. In: IEEE international geoscience and remote sensing symposium, 2001, July 9–13, 2001, Sydney, NSW: IGARSS'01, vol 3, pp 985–987. https://doi.org/10.1109/IGARSS.2001.976722
41. Nicholson SE (2001b) Climatic and environmental change in Africa during the last two centuries. Clim Res 17(2):123–144. https://doi.org/10.3354/cr017123
42. Omute P, Corner R, Awange JL (2012) The use of NDVI and its derivatives for monitoring Lake Victoria's water level and drought conditions. Water Resour Manag 26:1591–1613. https://doi.org/10.1007/s11269-011-9974-z
43. Pinzon J, Brown ME, Tucker CJ (2004) Satellite time series correction of orbital drift artifacts using empirical mode decomposition. In: Huang N (ed) Hilbert-Huang transform: introduction and applications, Chap. 10, Part II. Applications. World Scientific Publishing, Singapore
44. Pinzon J, Brown ME (2007) GIMMS NDVI data description, global inventory mapping and modelling studies. http://gimms.gsfc.nasa.gov/ndvi/ndvie/GIMMSdocumentation_NDVIe.pdf. Accessed 28 Feb 2007
45. Privette JL, Fowler C, Wick GA, Baldwin D, Emery WJ (1995) Effects of orbital drift on advanced very high resolution radiometer products: normalized difference vegetation index and sea surface temperature. Remote Sens Environ 53(3):164–171
46. Propastin PA (2008) Application of remote sensing simple model for monitoring Balkhash Lake water levels and Ili River discharges: application of remote sensing. Lakes Reserv: Res Manag 13:77–81. https://doi.org/10.1111/j.1440-1770.2007.00354.x
47. Subash N, Mohan HSR (2011) A simple rationally integrated drought indicator for rice-wheat productivity. Water Resour Manag 25(10):2425–2447. https://doi.org/10.1007/s11269-011-9817-y
48. Swenson S, Wahr J (2009) Monitoring the water balance of Lake Victoria, East Africa, from space. J Hydrol 370(1–4):163–176. https://doi.org/10.1016/j.jhydrol.2009.03.008
49. Tucker CJ (1980) Remote sensing of leaf water content in the near infrared. Remote Sens Environ 10(1):23–32. https://doi.org/10.1016/0034-4257(80)90096-6
50. Tucker CJ (1979) Red and photographic infrared linear combinations for monitoring vegetation. Remote Sens Environ 8(2):127–150. https://doi.org/10.1016/0034-4257(79)90013-0
51. Tucker CJ, Pinzon JE, Brown ME, Slayback D, Pak EW, Mahoney R, Vermote E, El Saleous N (2005) An extended AVHRR 8-km NDVI dataset compatible with MODIS and SPOT vegetation NDVI data. Int J Remote Sens 26(20):4485–4498. https://doi.org/10.1080/01431160500168686
52. USDA/FAS (United States Department of Agriculture, Foreign Agricultural Services) (2007) Global reservoir monitor. http://www.pecad.fas.usda.gov/cropexplorer/global_reservoir/index.cfm. Accessed 21 Mar 2008
53. Verdin J, Funk C, Senay G, Choularton R (2005) Climate science and early warning. Philos Trans R Soc B 360(1463):2155–2168. https://doi.org/10.1098/rstb.2005.175410.1175/1520-0442(2002)
54. World Resources Institute (2006) Earth trends: the environmental information portal [Image]. http://earthtrends.wri.org/text/water-resources/map-300.html. Accessed 12 May 2008
55. Yates DN, Strzepek KM (1998) Modelling the Nile basin under climate change. J Hydrol Eng 3(2):98–108. https://doi.org/10.1061/(ASCE)1084-0699(1998)3:2(98)
56. Awange J (2021) Nile Waters, weighed from space. Springer International Publishers

Chapter 14
Vegetation Variability "Hotspots" (2003–2018)

14.1 Summary

Lake Victoria's surface area has recently been shown to have shrunk by 0.3% compared to its 1984 value, a decline that has been associated with climatic as well as anthropogenic factors. Climatic factors include, e.g., reduced rainfall, which impacts not only on the lake's water level but also on the basin's vegetation that forms the lake's catchment. Understanding the locations of vegetation changes and the driving forces of such changes, therefore, is of most critical importance to major stakeholders regarding environmental management, policies and planning. For Lake Victoria Basin (LVB; Kenya, Uganda, Tanzania, Rwanda and Burundi), human development and climatic variability/change have subjected the region to significant changes in its vegetation characteristics whose spatio-temporal patterns are, however, not well understood. To understand this variability in vegetation for the period 2003–2018, this chapter presents the work of Morgan et al. [11], which employs the use of remotely sensed MODIS (Moderate Resolution Imaging Spectroradiometer), CHIRPS (Climate Hazards Group InfraRed Precipitation with station data) precipitation data, Google Earth Pro imagery, Gravity Recovery and Climate Experiment (GRACE)-based Mascon's total water storage (TWS) products and the statistical PCA (Principal Component Analysis). The chapter aims at determining (i) "significant hotspots", i.e., vegetation areas within the LVB largely impacted, and (ii), the extent of which anthropogenic and climatic variability have contributed to the "hotspots" formation (see Fig. 14.1). The results indicate a total of 8 hotspots; 5 in Uganda and 1 each in Kenya, Tanzania and Rwanda. Google Earth Pro imagery of all the hotspots show the changes in anthropogenic processes as the primary driver for the long-term changes in vegetation characteristics. Conversely, the analysis of PCA and Mascon's TWS concluded that only the Tanzanian hotspot may have been driven somewhat by climate variability. Climate variability is understood to be the driver of short-term vegetation changes while the long-term effects are driven primarily by human influence.

J. Awange, *Lake Victoria Monitored from Space*,
https://doi.org/10.1007/978-3-030-60551-3_14

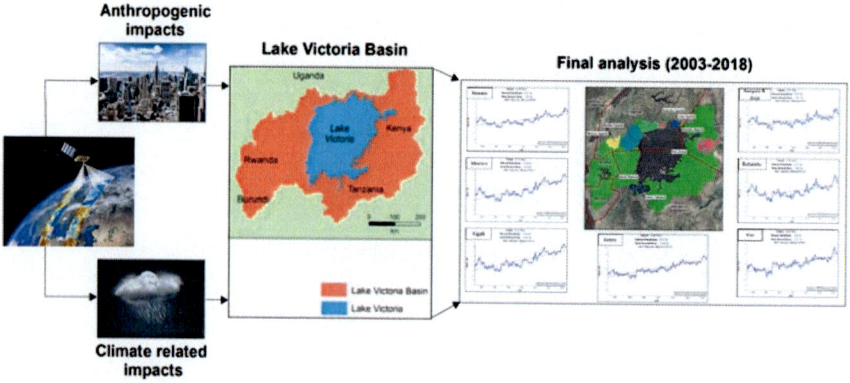

Fig. 14.1 Graphic abstract summarizing the chapter's work. *Source* [11]

14.2 Vegetation and Hydrological Characteristics

Studies have shown that alterations to the hydrological characteristics of freshwater areas correlate with changes in shoreline vegetation, e.g., [12, 26, 36, 38]. Decreases in water-levels can allow vegetation types and other plant species the freedom to regenerate [23, 54]. Conversely, anthropogenic processes that act to clear vegetation for urbanisation and development can indirectly lead to decreased water-levels of freshwater bodies that lie in proximity to where such stressors occur [49, 54]. Such water bodies provide important socio-economic services for the regions in which they are located, hence why they are sourced for extraction. For example, [27, 49, 54] investigate the extent of decline of flora species macrophyte around streams that are in close proximity to cultivated and urbanised regions and conclude that there are significant overall decrease of species richness for each of the streams they surveyed. Yet the converse is also true; anthropogenic activities such as use of fertilizers and industrial wastes also indirectly fuel increase in microphytes as demonstrated by [31] who studied macrophytes' abundance changes in eutrophicated tropical reservoir of Salto Grande in Brazil.

For the Lake Victoria Basin (LVB; 31°39'–34°53' E and 0°20'–3 °N) that houses the second largest freshwater lake in the world [6, 35] employed medium-resolution remotely sensed data of Landsat (5, 7 and 8) and Sentinel-2 images for the period 1984–2018 and indicated that the surface area of the lake had shrunk significantly in recent decades at a rate of 5.97 km^2/year (from manual digitisation) and 5.27 km^2/year (from Modification of Normalized Difference Water Index (MNDWI)). Four areas that were identified as "hotspots", which had major influences in the overall reduction of Lake Victoria's surface area were Birinzi, Uganda—33.9 km^2; Mwanza Gulf, Tanzania—21.5 km^2; Emin Pasha Gulf, Tanzania—14.6 km^2; and Winam Gulf, Kenya—12.8 km^2 [6].

To understand this decline, monitoring of its vegetational changes is vital and this predisposes it to remote sensing techniques due to its sheer size. This is because in-situ recordings over such a large area is impractical and extremely susceptible to inaccuracy, whilst photogrammetric processes would prove to be far too expensive when conducted yearly [6]. Even with this realization, there is little information pertaining to the changes in areal extent of LVB's vegetation coverage and their triggers. Studies that have been conducted have focussed largely on variance of NDVI values in the region (e.g., [41]) but have not equated the extent of which this contributes to changes in vegetation coverage. For instance, a 2014 study conducted in the Winam Gulf (Kenyan section of LVB) utilised coarse resolution 300 m MERIS (Medium Resolution Imaging Spectrometer) and fine resolution 30 m Landsat-7 imagery to analyse the NDVI of aquatic vegetation in the area, and performed accuracy assessments for each Normalised Difference Vegetation Index (NDVI) extraction method [14]. With the gradual decrease in the surface area of Lake Victoria [6], there is reason to suggest that this would be a driving force for changes in the characteristics of the surrounding vegetation within the catchment.

Factors that have been known to influence vegetation variability include both climatic and anthropogenic. For example, an NDVI variability study in the USA from 1982–1992 confirmed with a 99% confidence that El Ñino Southern Oscillation (ENSO) was the primary driver for influencing NDVI interannual variability [32]. A South African study [46] found that seasonal rainfall changes were primary NDVI drivers in areas where the annual rainfall was between 300 and 900 mm, or if the total rainfall between the rainy and dry seasons were profound. Within LVB, [37, 41] showed the influence of climate on vegetation by looking at the relationship between vegetation and rainfall in the region. Furthermore, climate variability through global teleconnections such as ENSO and Indian Ocean Dipole (IOD) on the one hand, and seasonal trends on the other hand, are also known influencers of short-term variability of vegetation in the LVB region [1, 6, 15, 41–43, 53, 55].

This chapter extends the work of [6], which analysed the physical dynamics of Lake Victoria by studying vegetation changes in its entire basin (i.e., LVB), which forms its catchment. The difference between the two studies is that [6] aimed at looking at changes on the physical surface of the lake and the triggers therein from 1984 to 2018. The current chapter employs high temporal-resolution MODIS (Moderate Resolution Imaging Spectroradiometer), CHIRPS (Climate Hazards Group InfraRed Precipitation with Station) precipitation, Google Earth Pro imagery and Gravity Recovery and Climate Experiment (GRACE)-based Mascon's total water storage products for the 2003–2018 period, and the statistical PCA (principal component analysis), to determine (i) significant "hotspots", i.e., the main differences vegetation areas, which are negatively impacted within LVB, and (ii), the extent of which anthropogenic and climatic variability have contributed to the "hotspots" formation. This chapter seeks to understand vegetational changes within LVB, which as catchment, is associated with the Lake's physical surface changes. Investigating the actual causality however, is out of the scope of the current chapter.

The remainder of the chapter is organised as follows: In Sect. 14.3, study area, the data and methods are presented. Section 14.4 covers results and discussion before concluding in Sect. 14.5.

14.3 Data and Methods

14.3.1 Lake Victoria Basin: Background

Lake Victoria, the world's largest tropical lake and second-largest freshwater lake is located in Eastern Africa (31°39'–34°53' E and 0°20'–3 °N), with its shoreline covering three countries, Kenya (6%), Uganda (45%) and Tanzania (49%) [10, 35]. Its basin area (Fig. 2.2) is almost three times the size of the lake and extends over the three East African countries together with Rwanda and Burundi. The basin refers to the network of rivers and streams that transport water and nutrients into the lake and is about 193000 km^2 [9, 10, 35]. It therefore facilitates the livelihood and wellbeing of over 40 million people across 5 countries at a density of over 300 people per km^2, with this population expected to triple by the year 2050 [6, 7, 40]. Water levels for Lake Victoria are primarily dependent on rainfall (approx. 80% of total recharge) [6, 10, 30]. The remaining 20% is the result of discharge from within the catchment area [9, 10].

14.3.2 Data

Atmospherically corrected satellite images and hydroclimate data are employed to analyse the degree of which climate variability and anthropogenic activities has impacted on vegetation changes in the LVB. The data used for the investigation is elaborated upon below, and summarized in Table 14.1.

Table 14.1 A summary of data employed in this chapter. Source [11]

Description	Sensor	Time-line	Purpose	Data source
NDVI	MODIS	Dec. 2003–2018	Vegetation changes	https://earthexplorer.usgs.gov/
Google Earth Pro	Landsat	Dec. 2003–2016	Anthropogenic impacts	Application-based
GSFC Mascons	GRACE	January 2003–July 2016	Water altimetry changes	https://ccar.colorado.edu/grace/gsfc.html
Infrared rainfall data	CHIRPS	1984–2018	Impacts of climate change	http://chg.geog.ucsb.edu/data/chirps/

14.3.2.1 MODIS

MODIS NDVI data products (eMODIS NDVI V6) are available for direct download from the United State Geological Survey (USGS) Earth Explorer database (https://earthexplorer.usgs.gov/). The eMODIS products includes either 7- or 10-day composite datasets with it's native highest spatial resolution of 250 × 250 m pixels size. The eMODIS NDVI V6 products provide NDVI datasets over entire world.

In this chapter, the 10-day intervals products are download from the USGS website for the period 2003–2018 at three years interval, i.e., 2003, 2006, 2009, 2012, 2015, and 2018. The month of December is selected as it comes after the short-rainy season of September–November, therefore potentially making it more suitable for identifying NDVI changes [6]. A total of 18 images are downloaded for December, i.e., 3 images for each year, with the temporal duration of the images being 10 or 11 days, i.e., 1st–10th, 11th–20th and 21st–31st for the evaluated month (December). The imagery itself have coarse spatial resolution that encompasses the entire area of the African continent. These products are regarded as high temporal resolution, therefore, might be a suitable products for identifying areas of spatio-temporal changes in vegetation coverage [13]. The focus of this chapter is on the vegetation of the catchment area for the entire LVB, therefore, the areal extent of interest is the lakes's surrounding areas (land area with vegetation cover) resulting in output masking the lake's surface. A subset image is created for 2018 MODIS NDVI data using this shapefile of the LVB. All the rest of MODIS imagery are clipped using 2018 subset image. During the sub-setting of the images, a snap to pixel technique is used to make sure every pixel for all years represent the same geographical location for capturing genuine changes in NDVI values for the evaluated years.

14.3.2.2 Google Earth Pro

Google Earth Pro (GEP) is one of the online platform that uses base images with the option that provides time series image analysis through image slider tool [48]. This platform provides high to medium resolution satellite images, for instance IKONOS and Landsat, respectively, over the world with variations from one region to another. For this chapter, the freely available historical Landsat images from GEP are used to visually analyse further (Sect. 14.4.4 and Fig. 14.7) the identified NDVI hotspots where vegetation changes occur (see Fig. 14.6). The triggers for these changes could be factors that include anthropogenic activities, e.g., urbanisation, forest and agriculture areas clearing. Furthermore, such visual analysis aid in showing the extent of anthropogenic contributions to vegetation changes.

14.3.2.3 GSFC Mascons

Mass-concentration (mascon) grids are monitored by the Gravity Recovery and Climate Experiment (GRACE) dual-satellite. The GSFC mascon estimation is processed by a general approach that it models the best-fit trend and annual time-variable gravity signals [34, 47]. It is worth mentioning that although the GSFC solution is comprised of 41,168, 1 arc-degree mascon cells, the original spatial resolution within a region is still 300 km. This means the time series of GSFC for each mascon within the same constraint region (e.g., basins) is highly correlated to the near mascons. In this chapter, GSFC mascon is used to monitor the equivalent water changes for every month, and compared to rainfall and vegetation changes.

14.3.2.4 CHIRPS Rainfall

Climate Hazards Group InfraRed Precipitation with Station data (CHIRPS) is a rainfall product that was developed to support the United States Agency for International Development Famine Early Warning Systems Network (FEWS NET) [20]. It merges satellite and in-situ observations to perform high spatial ($0.05^\circ \times 0.05^\circ$) resolution and monthly temporal resolution, e.g., National Oceanic and Atmospheric Administration's (NOAA's) Rainfall Estimate [33], African Rainfall Climatology [39], CHPclim dataset [21] and gauge products [20]. Generally, climate change is inferred from climatology data that spans more than 30 years. Given that CHIRPS data are available for 34 years, undertaking a PCA analysis on it can indicate the impacts of climate change through the analysis of its time series. When this is compared with the epochs of vegetation, one can infer on the impacts of climate change/variability on vegetation. The data is available on https://earlywarning.usgs.gov/fews/datadownloads/Global/CHIRPS, and has been validated over East Africa in [17].

14.3.3 Methods

NDVI change maps are created for inter-annual and intra-annual variations for the 2003, 2006, 2009, 2012, 2015, and 2018 MODIS imagery datasets to identify zones of the catchment area that have undergone extensive decline in vegetation coverage. Principal Component Analysis (PCA) is computed on rainfall data to determine if any reduction in vegetation is climate-driven with help of the change maps generated from NDVI data. Google Earth Pro (GEP) imagery from the time of vegetation decrease for each hotspot is obtained to visualise if the decline is anthropogenically-driven. There are large amount of variance in NDVI values among the evaluated years, what can be partly attributed to the low spatial resolution (250×250 m pixels size) of MODIS imagery. However, high temporal resolution of the data permit a valid comparison over time for the NDVI value changes. An NDVI threshold of >0.2 is used to determine whether a pixel is considered vegetation or non-vegetation [16, 45], and the total area of vegetation and non-vegetation for each year then determined by generating binary dataset for each year using ArcGIS environment. This section elaborates on the image-processing procedures used, including; fishnet

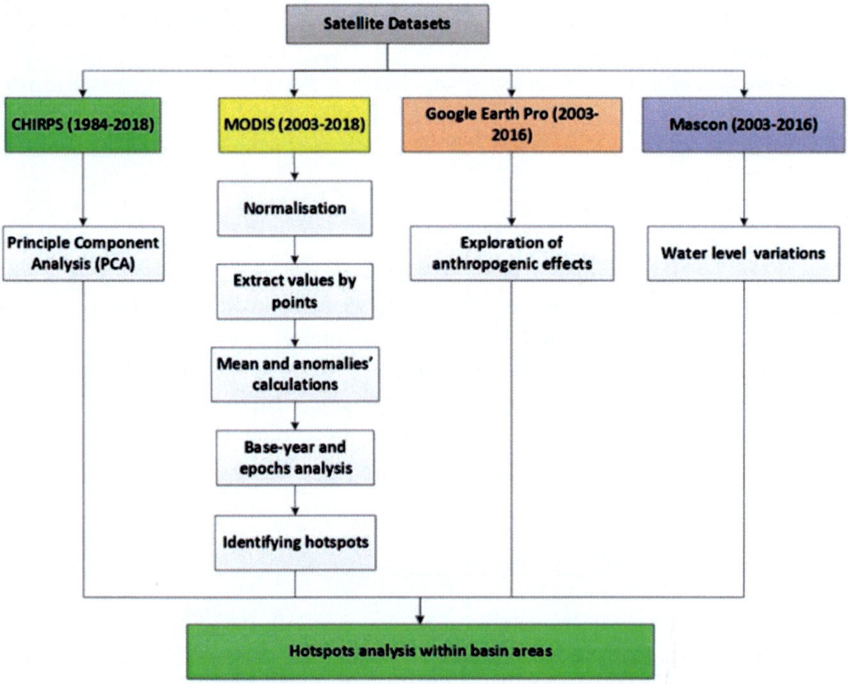

Fig. 14.2 Workflow for the investigation. *Source* Morgan et al. [11]

creation, binary outputs, standardised anomaly, image difference. A structure chart of the methodology is presented in Fig. 14.2.

14.3.3.1 Pre-processing

All 18 of the MODIS images are clipped to the extent of the basin outside of the lake prior to further processing. NDVI values >0.2 are extracted to create binary outputs depicting vegetation and non-vegetation pixels. The same threshold value was used in [16, 45] as a means of representing shrubs and meadows. The NDVI statistics of the images are further normalised to provide the exact NDVI values of all vegetation pixels determined from the binary output, as well as categorising all non-vegetation pixels ≥ 0.2 as a '0' value.

A fishnet of points is then created for each pixel in the LVB area (over 2.6 million; 250 m × 250 m pixels). Within the fishnet dataset, the NDVI values for each pixel is extracted from all images and added to their respective points. The pixel mean NDVI value (i.e., the mean was used for original products for 7 or 10 days interval) for each year (2003, 2006, 2009, 2012, 2015, and 2018) is calculated from the three NDVI values of the original datasets of each year. The overall mean NDVI value of those years is then obtained from these outputs, i.e., the pixel mean for all years is then calculated using the pixel mean values for each year, which in turn allows the overall mean NDVI values to be extracted. Using the pixel mean value of each year, new rasters are derived depicting vegetation and non-vegetation areas using the NDVI threshold of >0.2.

14.3.3.2 Anomaly Calculations

Annual anomalies are computed with respect to the mean change of each pixel. These values are utilised to demonstrate the extent of changes in NDVI of the basin. Two approaches are used to present this; (i) the inter-annual, which highlight short-term 3-year annual trends for NDVI differences, i.e., 2003–2006, 2006–2009 etc., and (ii), the intra-annual maps that present short and long-term trends demonstrating the degree of NDVI difference for each of the years in relation to 2003 (which is set as the base year).

14.3.3.3 Hotspot Significance Maps

The significance of the mean annual changes of NDVI P-value for each pixel is the determining factor in the identification of a "hotspot" signifying reduction in vegetation coverage. Vegetation decreases are indicated for Z-values (i.e., anomally changes) ≤ -2, with the significance of the trends of vegetation changes indicated by P-values < 0.05. P and Z values are calculated for all pixels in the study area where any pixel that fulfils both criteria is interpreted as decreasing at 95% confidence.

14.3.3.4 Principal Component Analysis (PCA)

Principal component analysis (PCA; [28, 44]) is a technique for reducing the dimensionality of datasets, increasing interpretability but at the same time minimizing information loss [8]. It is useful for identifying variance in hydrometeorological parameters [2, 3, 5–7, 18, 22, 25, 51, 52]. In this chapter, it is employed to analyse gridded CHIRPS rainfall variation throughout the catchment area for the 1984–2018 period in order to infer the impacts of climate change on vegetation. As NDVI values are a measure of vegetation greenness, higher NDVI values are dependent on higher water content [19], eutrophication (especially around the lake perimeter, e.g., [31]), the vegetation type (e.g., [41]), density and height, which translated in high NDVI values. Also, in such spatial resolution (250 m) mixed pixels significantly affect the NDVI values on the one hand while on the other hand, the meteorological conditions in the previous days of the selected MODIS images can affect the NDVI values. The patterns determined from the PCA analysis can determine whether the reduction in NDVI values from the 'hotspots' derived from the MODIS imagery processing are climate-driven through the analysis of rainfall variability [5, 6] and comparing with changes in NDVI [41].

14.4 Results and Discussion

14.4.1 Vegetation Analysis Within LVB

The lower mean NDVI values for December 2006 and 2018 (Fig. 14.3a) can be largely attributed to the lower than average rainfalls in Eastern Africa, which were caused by the occurrence of La Ñina events in 2006 and 2016–2017 respectively [24, 50]. Conversely, the higher mean NDVI value for December 2015 (Fig. 14.3a) can be attributed to higher than average rainfalls from the occurrence of an El Ñino event in 2015–2016 [50].

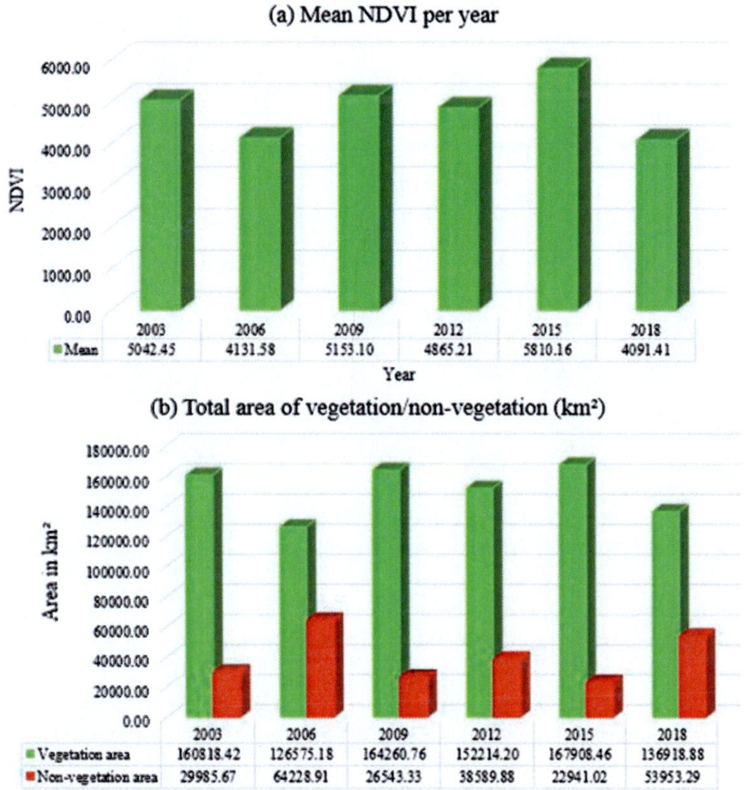

Fig. 14.3 **a** Mean NDVI, and **b**, Area of vegetation/non-vegetation using NDVI >0.2 threshold. *Source* Morgan et al. [11]

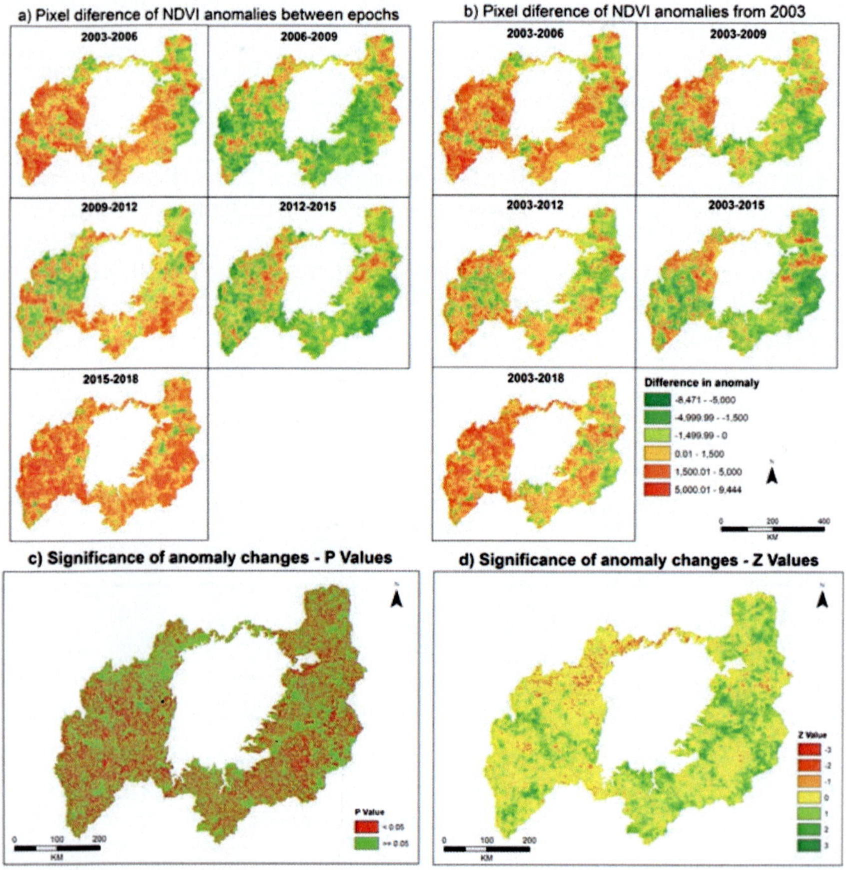

Fig. 14.4 **a** Short-term three-year epoch differences in NDVI, **b** long-term NDVI trends from 2003, **c**, significance of anomaly changes using maps depicting P values, and **d**, significance of anomaly changes using maps depicting Z values. *Source* Morgan et al. [11]

14.4.1.1 Epoch and Base-Year Trend Change Maps

As stated earlier (see Sect. 14.3.3.2), epoch and long-term trends of NDVI changes indicate the potential whereabouts of significant hotspots in the basin. The primary purpose of these maps is to identify areas in the LVB that have undergone drastic transformations in overall vegetation greenness within its respective recorded period.

As the epoch maps highlight the short-term 3-year annual trends, it therefore can highlight the extent in the variation of NDVI values caused by extreme weather events. The 2003–2006 and 2015–2018 maps (Fig. 14.4a) display the magnitude of environmental impact caused by lower than average rainfall from their respective La Ñina events that were highlighted in [24, 50]. Figure 14.4a demonstrates the capacity in which rainfall variations can have on the environment in the short-term. However, vegetation that has yet to be subjected to significant human interference has the capacity to replenish when annual rainfall levels increase. This is displayed in the 2006–2009 period (Fig. 14.4a), which occurred after the La Ñina event of 2006 as well as the 2012–2015 period (Fig. 14.4a), which occurred during the El Ñino event of 2015–2016. The decreases of NDVI values are more widespread and of greater magnitude. In 2003–2006 the impact is generally more profound west of the lake. Significant clusters of reduced NDVI values have been identified in the south-west of the catchment area (Burundi), central to the western side of the lake (Tanzania), north-west of the catchment area (Uganda) and Central Rwanda. East of LVB, there is a large general decrease of NDVI spread across Kenya, which sprawls across into the Tanzanian border and continues nearby to the southern border of the lake.

In the 2015–2018 period, the impact is even more extreme. There are additional areas that have undergone significant NDVI decreases; namely along the northern boundary of the lake as well as more extensively east of the lake across Kenya and Tanzania. However, it must be stated that the extremity of the reduction is harnessed from the contrasting extreme weather events that occurred in 2015 and 2018. As stated in Sect. 14.4.1, higher than average rainfall occurred in the Lake Victoria region due to an El Ñino event in 2015–2016, whereas lower than average rainfall occurred due to a La Ñina event in 2016–2017 [50]. These extreme weather events would be the primary contributor as to why NDVI is higher than normal for 2015, and lower than normal for 2018. This means that when computing the difference in the pixel anomalies inter-annually for 2015–2018, the result would indicate widespread NDVI decrease.

The base-year map (Fig. 14.4b) highlights the long-term trends for NDVI changes for all years from 2003. In a similar vein to the outputs for the epoch maps (Fig. 14.4a), the western side of the basin is substantially more impacted. Within the 15-year time-span; south-western Uganda, north-western Tanzania, central and eastern Rwanda as well as northern Burundi have all been inflicted by this negative trend. East of the lake has also undergone some impact, particularly in an area in south-western Kenya. However, the long-term trend indicates that most of the primary hotspots occur west of the lake, see more details in Sect. 14.4.1.2.

14.4.1.2 Identification of Significant Hotspots

The next step is to determine the whereabouts of statistically significant hotspots that indicate a long-term trend in vegetation decline. The criteria for identifying pixels that would formulate the hotspots are $Z \leq -2$ and $P < 0.05$, indicating significant decrease at 95% confidence. All pixels meeting this criteria are extracted from their respective datasets. All pixels that represent both criteria could potentially be located within clusters of similar pixels. There are such pixels located widespread in the LVB area, however, the purpose is to identify more aggregations, i.e., areas, in which the separation of neighbouring pixels is no greater than 30 km. Due to the large size of the study area, it is understood that using a larger distance threshold could provide a more sustainable output in terms of the number of clusters that will result.

That resultant output when utilising the 30-km separation threshold are 8 significant hotspots (Fig. 14.4c); 5 of which were in Uganda, and 1 each for Kenya, Tanzania and Rwanda respectively. With Uganda containing the most hotspots, those findings correlate with the long-term trends analysed in Sect. 14.4.1.1, in which vegetation located north-west of Lake Victoria was found to have undergone the largest areal decrease in NDVI from 2003–2018. Evidently, parts of the study area that were subjected to short-term vegetation changes were not deemed significant enough in the *P* and *Z* outputs in Fig. 14.4c to be deemed as hotspots.

The results show that there are some correlation between the NDVI hotspots deduced in this investigation and the hotspots discovered in [6] that demonstrated a significant decrease in the surface area of Lake Victoria itself. That study concluded that Winam gulf (Kisii, Kenya), Emin Pasha gulf (Katoro, Tanzania), Mwanza gulf and Birinzi (Kampala and Masaka, Uganda) were the hotspots. Each of these places has expanded their urban environments to various extents since 2003, whilst also showing signs of deforestation and clearing to enable the provision of agriculture and other rural industry. The assumption can be made that this outwards urban expansion has not only reduced all vegetation characteristics but has also necessitated the extraction of nearby freshwater that Lake Victoria provides for these areas.

14.4.2 Rainfall Variability Within LVB

For the 1984–2018 period, a PCA analysis is performed on the rainfall parameters within LVB. Figure 14.5 shows the principal components (PC; time series) and the empirical orthogonal functions (EOF; spatial maps) both of which have to be interpreted together to understand rainfall variability within the LVB. Both PCs and EOFs account for the total variation in the rainfall [18]. The results of the PCA, where PC1 (accounting for 75.2% of the overall variance in rainfall) depicts the dominant seasonal rainfall super-positioned with the annual signal, PC2 (19.1%) shows annual rainfall variation, while PC3 (5.6%) shows extreme rainfall events (i.e., those associated with El Ñino and La Ñina), are consistent, e.g., with the results of [4] who found four modes over the basin for the period 2003–2013. For instance, [4]'s PC1 (representing 63% of total variance of the rainfall) showed a superposition of the annual and seasonal variabilities while PC2 (13%) related to the annual variation and PC3 showed a summation of interannual changes and a linear trend over the basin. The first three EOF modes in Fig. 14.5 are identical to those of [4]. Other similar findings are presented, e.g., in [6, 29].

Identifying if there are climatic drivers in the formation of NDVI hotspots is accomplished through analysing the variation trends derived from the PCA output. In general, overall rainfall decreased throughout the Lake Victoria basin, in some places by as high as 250 mm, such as along the western boundary of the lake and along the south-western boundary of LVB within Burundi. Seasonality caused its most profound rainfall decrease along the southern extremities of LVB. The decrease became gradually less profound progressively north from the southern extent. The

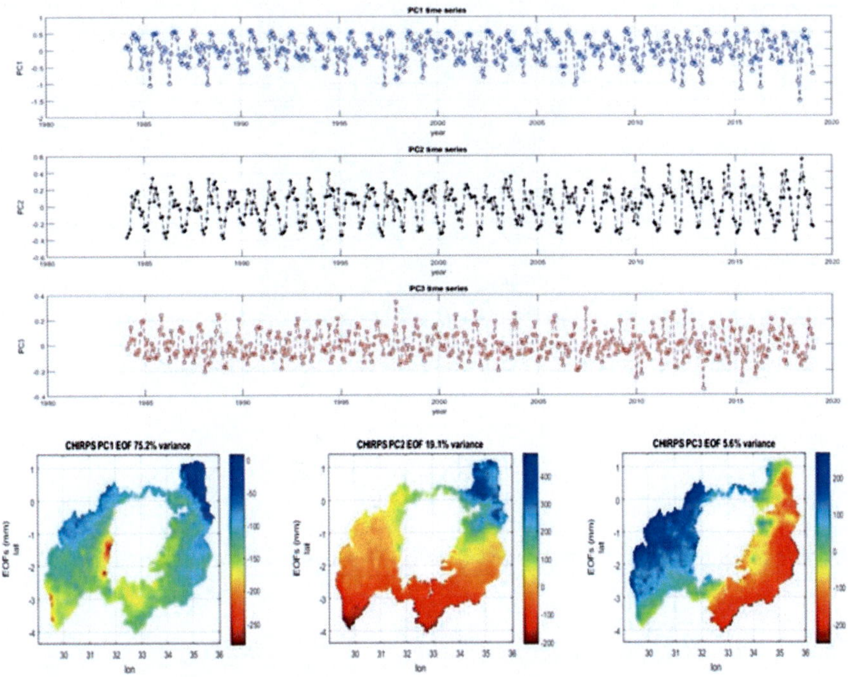

Fig. 14.5 PCA analysis **a** Timeseries of PCA components, and **b**, spatial pattern of components. *Source* Morgan et al. [11]

north-eastern corner overlapping Kenya and Uganda recorded increased rainfall due to seasonal variations. Extreme weather events result in horizontal contrast with increased rainfall recorded over most of the western half of LVB and decreased rainfall over most of the eastern half.

For the NDVI hotspot identified in Kisii (Kenya), the only rainfall variable that could have had any effect in the long-term reduction in NDVI is the La Ñina extreme weather events. However, as that only accounts for 5.6% of rainfall variation, it is safe to assume that the hotspot is the result of a non-climatic driver. For all the Ugandan hotspots, PCA does not provide any meaningful indicator for climate contributing to its formation. Conversely, the total rainfall and seasonality PCs indicate some possibility that they contribute to the formation of the Kigali (Rwanda) and Katoro (Tanzania) hotspots. These PCs account for approximately 94.3% of rainfall variance, which is reason enough to suggest that climatic impact could be meaningful.

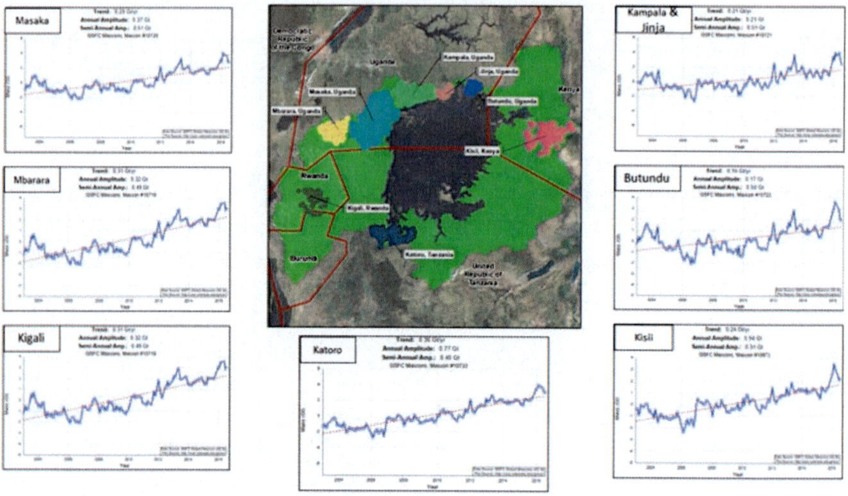

Fig. 14.6 Gravity Recovery and Climate Experiment (GRACE) Mass Concentration mascon analysis of the total water storage changes within the NDVI hotspots. *Source* Morgan et al. [11]

14.4.3 Total Water Storage Changes Within the Hotspots

Annual total water storage (TWS; surface, groundwater, soil moisture and vegetation) changes have been obtained for all Gravity Recovery and Climate Experiment (GRACE) Mass Concentration (mascon) grids where hotspots lie (see Fig. 14.6). The La Ñina event in 2006 presents a decrease in TWS by approximately 2 gigatons for all mascons. In contrast, the El Ñino event in 2015–2016 presented an annual increase of approximately 3–4 gigatons for all mascons.

However, a short-coming in these assessments is their time-span. Analysis of NDVI changes occurred every three years from 2003–2018 while the mascon analysis occurs from January 2003–July 2016. The major detriment for that is that a La Ñina event occurred in 2016–2017, resulting in lower rainfall for Eastern Africa. The effects that this had on TWS changes is not recorded. As December 2018 is the final period for NDVI analysis, the short-term and long-term effects of the La Ñina event is displayed in Sect. 14.4.1.1. With the missing TWS change data, it makes it difficult to determine if rainfall variables are a major driver in the formation of the NDVI hotspots.

14.4.4 Google Earth Pro Imagery

The identified NDVI hotspots from MODIS data is assessed further using the satellite imagery from GEP at different time-scales to provide insight into the extent of the

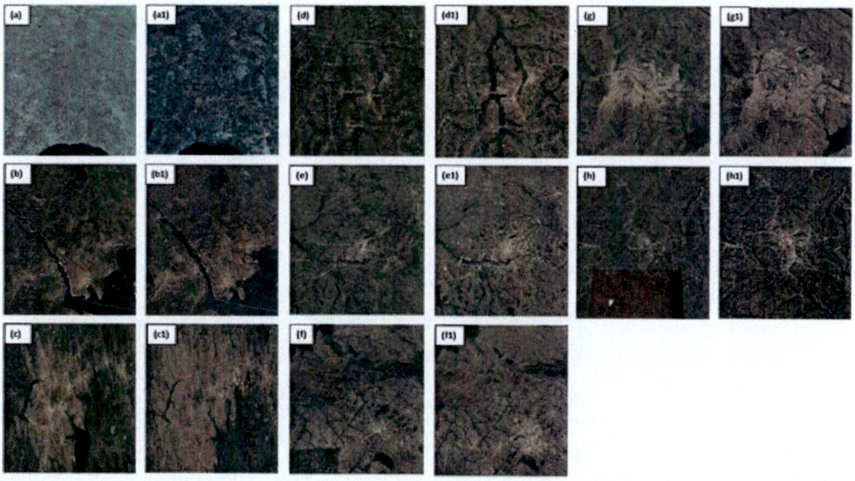

Fig. 14.7 Google Earth Pro imagery with Landsat images as base map **a** Butundu, Uganda 2004 and **a1** Butundu, Uganda 2017; **b** Jinja, Uganda 2003 and **b1** Jinja, Uganda 2016; **c** Kampala, Uganda 2003 and **c1** Kampala, Uganda 2016; **d** Masaka, Uganda 2003 and **d1** Masaka, Uganda 2016; **e** Mbarara, Uganda 2003 and **e1** Mbarara, Uganda 2016; **f** Katoro, Tanzania 2003 and **f1** Katoro, Tanzania 2016; **g** Kigali, Rwanda 2003 and **g1** Kigali, Rwanda 2016; **h** Kisii, Kenya 2003 and **h1** Kisii, Kenya 2018. *Source* Morgan et al. [11]

anthropogenic impacts. From the analysis of Fig. 14.7, it is evident that for the Ugandan hotspots of Jinja, Kampala, Masaka and Mbarara, that outward urban expansion is the primary contributor for the long-term NDVI decline. The same conclusion can be drawn for the Rwandan and Kenyan hotspots also. The Katoro (Tanzania) hotspot can also be seen to have undergone urban expansion, but to a lesser extent. The Ugandan hotspot of Butundu is not necessarily derived from urbanisation, but it was still caused by anthropogenic activities, as it appears that large-scale deforestation and clearing occurred in the region for widespread agricultural practices to begin.

14.5 Concluding Remarks

Following a recent study that indicated reduction of Lake Victoria's surface over the period 1984–2018, this chapter aimed at investigating changes in vegetation cover over the Lake Vectorial Basin (LVB) over the period 2003–2018. To achieve this, the study employed MODIS (Moderate Resolution Imaging Spectroradiometer), Google Earth Pro, CHIRPS (Climate Hazards Group InfraRed Precipitation with station data) precipitation data, Google Earth Pro imagery, Gravity Recovery and Climate Experiment (GRACE)-based Mascon's water storage products, and the statistical method of Principal Component Analysis (PCA). The assumption, here, is that changes in vegetation within LVB is related to the lake's physical dynamics and as such, understanding vegetation changes within the basin and identifying the hotspots where they occur could be essential to the overall management of the lake. The results show that the vegetation within the LVB experienced temporal variations throughout the study period (2003–2018). Specifically, the study found that:

(i) Long-term vegetational changes within LVB over the period 2003–2018 were primarily anthropologically driven, with urbanization expanding at the expense of vegetation as seen from the Google Earth Pro imagery.
(ii) Eight "hotspots" (i.e., areas with significant vegetational changes) in total were identified over LVB: 5 in Uganda, and one each in Kenya (Kisii), Rwanda (Kigali) and Tanzania. Other than the Rwandan and Tanzanian hotspots where climate variability impacts were visible, there is no meaningful evidence presented from the rainfall and Mascon's TWS analysis to suggest that anything other than human processes is causing long-term changes in vegetation characteristics over the other hotspots.
(iii) Out of all the countries within the LVB, it can be said that Uganda has undergone the most profound urbanisation processes since 2003, largely due to the expansion of its major cities such as Kampala, Masaka and Jinja that were identified as hotspots. Small-scale urban expansion also occurred in the Butundu, Mbarara and Katoro cities that do not serve as major urban hubs, but instead service agricultural and industrial practices. The expansion of these regional practices can be attributed to why they have been identified as vegetation hotspots, as clearing of land is required to facilitate these practices.

Understanding the locations of vegetational changes is most profound, as well as the driving forces associated with such changes, in that it provides critical information to major stakeholders regarding future environmental management, policies and planning. Management of Lake Victoria and its basin, therefore, would benefit from such analysis presented in this chapter.

References

1. Agutu NO, Awange JL, Ndehedehe C, Mwaniki M (2020) Consistency of agricultural drought characterization over Upper Greater Horn of Africa (1982–2013): topographical, gauge density, and model forcing influence. Sci Total Env 709:135149
2. Awange JL, Fleming KM, Kuhn M, Featherstone WE, Anjasmara I, Heck B (2011) On the suitability of the 4o x 4o GRACE mascon solutions for remote sensing Australian hydrology. Remote Sens Env 115:864–875. https://doi.org/10.1016/j.rse.2010.11.014
3. Awange JL, Forootan E, Kuhn M, Kusche J, Heck B (2014) Water storage changes and climate variability within the Nile Basin between 2002 and 2011. Adv Water Resour 73:1–25. https://doi.org/10.1016/j.advwatres.2014.06.010
4. Awange JL, Anyah R, Agola N, Forootan E, Omondi P (2013) Potential impacts of climate and environmental change on the stored water of Lake Victoria Basin and economic implications. Water Resour Res 49:8160–8173, IF 3.149. https://doi.org/10.1002/2013WR014350
5. Awange JL, Khandu Forootan E, Schumacher M, Heck B (2016) Exploring hydro-meteorological drought patterns over the Greater Horn of Africa (1979–2014) using remote

sensing and reanalysis products. Adv Water Res 94:45–59. https://doi.org/10.1016/j.advwatres.2016.04.005

6. Awange JL, Saleem A, Sukhadiya RM, Ouma YO, Kexiang H (2019a) Physical dynamics of Lake Victoria over the past 34 years (1984–2018): Is the lake dying? Sci Total Env 658:199–218
7. Awange JL, Hu KX, Khaki M (2019b) The newly merged satellite remotely sensed, gauge and reanalysis-based Multi-Source Weighted-Ensemble Precipitation: Evaluation over Australia and Africa (1981–2016). Science of The Total Environment 670:448–465
8. Awange JL, Palancz B, Völgyesi L (2020) Hybrid imaging and visualization. Employing machine learning with mathematica - Python. Springer Nature International, Berlin
9. Awange JL, Sharifi MA, Ogonda G, Wickert J, Grafarend EW, Omulo MA (2008) The falling Lake Victoria water level: GRACE, TRIMM and CHAMP satellite analysis of the lake basin. Water Res Manag 22(7):775–796
10. Awange JL, Ong'ang'a O (2005) Lake Victoria. Ecology, resources and environment. Springer, Berlin
11. Morgan B, Awange JL, Saleem A, Hu K (2020) Understanding vegetation variability and their "hotspots" within Lake Victoria Basin (LVB: 2003–2018), 122. https://doi.org/10.1016/j.apgeog.2020.102238
12. Baldwin AH, Egnotovich MS, Clarke E (2001) Hydrologic change and vegetation of tidal freshwater marshes: field, greenhouse, and seed-bank experiments. Wetlands 21(4):519–531
13. Chen Y, Song X, Wang S, Huang J, Mansaray LR (2016) Impacts of spatial heterogeneity on crop area mapping in Canada using MODIS data. ISPRS J Photogramm Remote Sens 119:451–461
14. Cheruiyot E, Mito C, Menenti M, Gorte B, Koenders R, Akdim N (2014) Evaluating MERIS-based aquatic vegetation mapping in Lake Victoria. Remote Sens 6(8):7762–7782
15. Detsch F, Otte I, Appelhans T, Hemp A, Nauss T (2016) Seasonal and long-term vegetation dynamics from 1-km GIMMS-based NDVI time series at Mt. Kilimanjaro. Tanzania. Remote Sens Env 178:70–83
16. Duarte L, Teodoro AC, Monteiro AT, Cunha M, Gonçalves H (2018) QPhenoMetrics: an open source software application to assess vegetation phenology metrics. Comput Electr Agric 148:82–94
17. Dinku T, Funk C, Peterson P, Maidment R, Tadesse T, Gadain H, Ceccato P (2018) Validation of the CHIRPS satellite rainfall estimates over eastern Africa. Quart J R Meteorol Soc 144:292–312
18. Dyer TG (1975) The assignment of rainfall stations into homogeneous groups: an application of principal component analysis. Quart J R Meteorol Soc 101(430):1005–1013
19. Foody GM (2003) Geographical weighting as a further refinement to regression modelling: an example focused on the NDVI-rainfall relationship. Remote Sens Env 88(3):283–293
20. Funk C, Peterson P, Landsfeld M, Pedreros D, Verdin J, Shukla S, Husak G, Rowland J, Harrison L, Hoell A, Michaelsen A (2015) The climate hazards infrared precipitation with stations-a new environmental record for monitoring extremes. Sci Data 2:150066. https://doi.org/10.1038/sdata.2015.66
21. Funk C, Verdin A, Michaelsen J, Peterson P, Pedreros D, Husak G (2015b) A global satellite assisted precipitation climatology. Earth Syst Sci Data Discuss 7:1–13. https://doi.org/10.5194/essd-7-275-2015
22. Haroon MA, Rasul G (2009) Principal component analysis of summer rainfall and outgoing long-wave radiation over Pakistan. Pak J Meteorol 5(10)
23. Hill NM, Keddy PA, Wisheu IC (1998) A hydrological model for predicting the effects of dams on the shoreline vegetation of lakes and reservoirs. Env Manag 22(5):723–736
24. Hoell A, Funk C, Zinke J, Harrison L (2017) Modulation of the southern Africa precipitation response to the El Niño Southern Oscillation by the subtropical Indian Ocean dipole. Clim Dyn 48(7–8):2529–2540
25. Hu K, Awange JL, Forootan E, Goncalves RM, Fleming K (2017) Hydrogeological characterisation of groundwater over Brazil using remotely sensed and model products. Sci Total Env 599:372–386. https://doi.org/10.1016/j.scitotenv.2017.04.188

26. Hudon C (1997) Impact of water level fluctuations on St. Lawrence River aquatic vegetation. Can J Fish Aquatic Sci 54(12):2853–2865
27. Jeppesen E, Søndergaard M, Lauridsen TL, Davidson TA, Liu Z, Mazzeo N, Starling F (2012) Biomanipulation as a restoration tool to combat eutrophication: recent advances and future challenges. Advances in ecological research, vol 47. Academic, New York, pp 411–488
28. Jolliffe TI, Cadima J (2016) Principal component analysis: a review and recent developments. Philos Trans A Math Phys Eng Sci 374(2065). https://doi.org/10.1098/rsta.2015.0202
29. Khaki M, Awange J (2019) Improved remotely sensed satellite products for studying Lake Victoria's water storage changes. Sci Total Env 652:915–926
30. Kizza M, Rodhe A, Xu CY, Ntale HK, Halldin S (2009) Temporal rainfall variability in the Lake Victoria Basin in East Africa during the twentieth century. Theor Appl Climatol 98(1–2):119–135
31. Coladello LF, Galo MDLBT, Shimabukuro MH, Ivánová I, Awange J (2020) Macrophytes' abundance changes in eutrophicated tropical reservoirs exemplified by Salto Grande (Brazil): trends and temporal analysis exploiting Landsat remotely sensed data. Appl Geogr 121. https://doi.org/10.1016/j.apgeog.2020.102242
32. Li Z, Kafatos M (2000) Interannual variability of vegetation in the United States and its relation to El Niño/Southern Oscillation. Remote Sens Env 71(3):239–247
33. Love TB, Kumar V, Xie P, Thiaw W (2004) A 20-year daily Africa precipitation climatology using satellite and gauge data, Seattle. In: Proceedings of the 84th AMS Annual Meeting, Conference on Applied Climatology
34. Luthcke SB, Sabaka TJ, Loomis BD, Arendt AA, McCarthy JJ, Camp J (2013) Antarctica, Greenland and Gulf of Alaska land-ice evolution from an iterated GRACE global mascon solution
35. Mati BM, Mutie S, Gadain H, Home P, Mtalo F (2008) Impacts of land-use/cover changes on the hydrology of the transboundary Mara River, Kenya/Tanzania. Lakes & Reserv: Res & Manag 13(2):169–177
36. New T, Xie Z (2008) Impacts of large dams on riparian vegetation: applying global experience to the case of China's Three Gorges Dam. Biodivers Conserv 17(13):3149–3163
37. Nicholson SE, Davenport ML, Malo AR (1990) A comparison of the vegetation response to rainfall in the Sahel and East Africa, using normalized difference vegetation index from NOAA AVHRR. Clim Change 17(2–3):209–241
38. Nilsson, C., & Berggren, K. (2000). Alterations of riparian ecosystems caused by river regulation: dam operations have caused global-scale ecological changes in riparian ecosystems. How to protect river environments and human needs of rivers remains one of the most important questions of our time. BioScience 50(9):783–792
39. Novella NS, Thiaw WM (2013) African rainfall climatology version 2 for famine early warning systems. J Appl Meteorol Climatol 52:588–606. https://doi.org/10.1175/JAMC-D-11-0238.1
40. Okotto-Okotto J, Raburu PO, Obiero KO, Obwoyere GO, Mironga JM, Okotto LG, Raburu EA (2018) Spatio-temporal impacts of Lake Victoria water level recession on the fringing Nyando Wetland. Kenya. Wetlands 38(6):1107–1119
41. Omute P, Corner R, Awange JL (2012) The use of NDVI and its derivatives for monitoring Lake Victoria's water level and drought conditions. Water Res Manag 26(6):1591–1613
42. Park S, Kang D, Yoo C, Im J, Lee MI (2020) Recent ENSO influence on East African drought during rainy seasons through the synergistic use of satellite and reanalysis data. ISPRS J Photogramm Remote Sens 162:17–26
43. Plisnier PD, Serneels S, Lambin EF (2000) Impact of ENSO on East African ecosystems: a multivariate analysis based on climate and remote sensing data. Global Ecol Biogeogr 9(6):481–497
44. Preisendorfer RW (1988) Introduction. In: Mobley C (ed) Principal component analysis in meteorology and oceanography. Elsevier, Amsterdam, pp 1–9
45. Pu R, Gong P, Tian Y, Miao X, Carruthers RI, Anderson GL (2008) Using classification and NDVI differencing methods for monitoring sparse vegetation coverage: a case study of saltcedar in Nevada, USA. Int J Remote Sens 29(14):3987–4011

46. Richard Y, Poccard I (1998) A statistical study of NDVI sensitivity to seasonal and interannual rainfall variations in Southern Africa. Int J Remote Sens 19(15):2907–2920
47. Sabaka TJ, Rowlands DD, Luthcke SB, Boy JP (2010) Improving global mass flux solutions from GRACE through forward modeling and continuous time-correlation. J Res Geophys. https://doi.org/10.1029/2010JB007533
48. Saleem A, Awange JL (2019) Coastline shift analysis in data deficient regions: exploiting the high spatio-temporal resolution Sentinel-2 products. Catena 179:6–19
49. Sand-Jensen K, Riis T, Vestergaard O, Larsen SE (2000) Macrophyte decline in Danish lakes and streams over the past 100 years. J Ecol 88(6):1030–1040
50. Setimela P, Gasura E, Thierfelder C, Zaman-Allah M, Cairns JE, Boddupalli PM (2018) When the going gets tough: performance of stress tolerant maize during the 2015/16 (El Niño) and 2016/17 (La Niña) season in southern Africa. Agric Ecosyst & Env 268:79–89
51. Stathis D, Myronidis D (2009) Principal component analysis of precipitation in Thessaly region (Central Greece). Global NEST J 11(4):467–476
52. Widmann M, Schär C (1997) A principal component and long-term trend analysis of daily precipitation in Switzerland. Int J Climatol 17(12):1333–1356
53. Williams CA, Hanan NP (2011) ENSO and IOD teleconnections for African ecosystems: evidence of destructive interference between climate oscillations. Biogeosciences 8(1):27
54. Zhang Y, Jeppesen E, Liu X, Qin B, Shi K, Zhou Y, Deng J (2017) Global loss of aquatic vegetation in lakes. Earth-Sci Rev 173:259–265
55. Zhao J, Huang S, Huang Q, Wang H, Leng G, Fang W (2020) Time-lagged response of vegetation dynamics to climatic and teleconnection factors. Catena 189:104474

Index

J. Awange, *Lake Victoria Monitored from Space*,
https://doi.org/10.1007/978-3-030-60551-3

N

O

P

R

S

T

U

V